A Romp Tour of Trigonometry

Charles Rambo

*In loving memory of my mom
Lori Riddle-Walker
1960–2017*

Contents

Preface		**vii**
1	**Angles**	**1**
	1.1 Classification of Angles	2
	1.2 Adjacent Angles	5
	1.3 Vertical Angles	8
	1.4 Parallel, Perpendicular, and Transversal Lines	9
	1.4.1 Transversals	10
	1.5 Angles and Triangles	15
	1.6 Exercises	18
2	**Triangles**	**23**
	2.1 Classification of Triangles	23
	2.1.1 Classification by Angles	23
	2.1.2 Classification by Sides	25
	2.2 Congruent Triangles	26
	2.2.1 SSS Postulate	28
	2.2.2 SAS Postulate	29
	2.2.3 ASA Postulate	30
	2.2.4 AAS and HL Theorems	31
	2.3 Similar Triangles	34
	2.4 Right Triangles	38
	2.5 Special Right Triangles	43
	2.6 Exercises	50
3	**Radians, Arc Length, and Rotational Motion**	**59**
	3.1 Directed Angles	59
	3.2 Radian Measure of an Angle	62

3.3	Arc Length	65
3.4	Area of a Sector	67
3.5	Linear and Angular Velocity	69
3.6	Exercises	74

4 Right Triangle Trigonometry — 81
- 4.1 Introduction to Trigonometric Functions 81
  - 4.1.1 Trigonometric Functions and Special Right Triangles 88
- 4.2 Inverse Trigonometric Functions 89
- 4.3 Angles of Elevation and Depression 93
- 4.4 Exercises 98

5 Trigonometry of General Angles — 107
- 5.1 The Six Trigonometric Functions 108
- 5.2 Reference Angles 116
- 5.3 More Evaluation Techniques 121
- 5.4 Finding the Values of Trigonometric Functions . . . 128
- 5.5 Pythagorean Identities 132
- 5.6 Verifying Identities 135
- 5.7 Exercises 138

6 Graphing Trigonometric Functions — 147
- 6.1 Graphing Sine and Cosine 147
- 6.2 Graphing Tangent and Cotangent 155
  - 6.2.1 Graphing Tangent 156
  - 6.2.2 Graphing Cotangent 161
- 6.3 Graphing Secant and Cosecant 165
  - 6.3.1 Graphing Secant 166
  - 6.3.2 Graphing Cosecant 169
- 6.4 Miscellaneous Graphing Problems 171
- 6.5 Exercises 175

7 Using Identities — 183
- 7.1 Sum and Difference Identities 183
  - 7.1.1 Proof of Theorem 7.1 (ii) 188
- 7.2 Other Identities 190
  - 7.2.1 The Cofunction Identities 190
  - 7.2.2 Double Angle Identities 192
  - 7.2.3 Half Angle Identities 196

	7.2.4 Product to Sum and Difference Identities	202
7.3	Verifying Identities	205
7.4	Exercises	208

8 Inverse Trigonometric Functions — 213
- 8.1 Inverses ... 213
  - 8.1.1 Restricting the Domain ... 217
- 8.2 Arc Sine ... 219
- 8.3 Arc Cosine ... 225
- 8.4 Arc Tangent ... 231
- 8.5 Other Inverse Trigonometric Functions ... 238
- 8.6 Graphing Arc Sine, Arc Cosine, and Arc Tangent ... 243
- 8.7 Exercises ... 249

9 Oblique Triangles — 257
- 9.1 Law of Cosines ... 258
  - 9.1.1 Unique Triangle Due to SAS ... 259
  - 9.1.2 Unique Triangle Due to SSS ... 260
- 9.2 Law of Sines in Unambiguous Cases ... 261
  - 9.2.1 Unique Triangle Due to AAS ... 262
  - 9.2.2 Unique Triangle Due to ASA ... 263
  - 9.2.3 Law of Sines to Simplify Calculations ... 264
- 9.3 Law of Sines and SSA ... 267
  - 9.3.1 Given Angle is Obtuse ... 268
  - 9.3.2 Given Angle is Right ... 269
  - 9.3.3 Given Angle is Acute ... 271
  - 9.3.4 Summary ... 276
- 9.4 Proofs of the Laws of Cosines and Sines ... 277
- 9.5 Exercises ... 281

10 Area and Perimeter — 285
- 10.1 Triangles ... 285
- 10.2 Regular Polygons ... 290
- 10.3 Segments of Circles ... 297
- 10.4 Exercises ... 303

11 Vectors — 307
- 11.1 The Basics ... 307
  - 11.1.1 Arithmetic of Vectors ... 310
- 11.2 Bearing ... 318

11.3 Force 321
  11.4 The Dot Product 330
  11.5 Projection 336
  11.6 Work 339
  11.7 Exercises 342

12 Complex Numbers   **353**
  12.1 The Basics 353
    12.1.1 Exponents 362
    12.1.2 The Complex Plane 365
  12.2 Polar Form 367
  12.3 More on Polar Form 369
    12.3.1 Finding Roots 373
  12.4 Exercises 380

13 Polar Coordinates and Equations   **387**
  13.1 Polar Coordinates 387
  13.2 Basic Polar Graphs 394
  13.3 Intermediate Polar Graphs 400
  13.4 Classification of Polar Graphs 407
  13.5 Exercises 411

Appendices   **419**

A Rational Expressions and Equations   **421**

B Radical Expressions   **427**
  B.1 Square Roots 427
  B.2 n-th Roots 432

C Transformations   **435**

D Unit Circle   **443**

E List of Identities   **445**

F Answers   **449**
  F.1 Angles 449
  F.2 Triangles 450
  F.3 Radians, Arc Length, and Rotational Motion 452
  F.4 Right Triangle Trigonometry 454

F.5 Trigonometry of General Angles 457
F.6 Graphing Trigonometric Functions 462
F.7 Using Identities 464
F.8 Inverse Trigonometric Functions 467
F.9 Oblique Triangles 470
F.10 Area and Perimeter 472
F.11 Vectors 474
F.12 Complex Numbers 477
F.13 Polar Coordinates and Equations 480

Glossary   **483**

Preface

Trigonometry is a sore spot for many students. Not only are the topics hard, but STEM majors see them come up again and again in their classes. This book is here to help!

Most trigonometry texts are either extremely expensive, shallow, or old and not well suited for current STEM curricula. My goal is to offer something that fills the hole. This book costs less than a third the price of the average trigonometry textbook. It's modern. And it covers concepts in a detailed and rigorous way.

I have designed this book for STEM majors studying trigonometry. In particular, it is designed so that students going into calculus and physics will have the prerequisite knowledge to do well. The trigonometric identities needed for calculus are emphasized. There is a thorough treatment of polar coordinates. Vectors are addressed and this includes treatment of the dot product. There are several applications of vectors, such as force and work problems, which will be helpful in physics. The material covered is sufficiently robust that it can also be used as a reference.

This book covers trigonometry from a more geometric perspective. The geometric prerequisites aren't just covered–they're emphasized. This is helpful for a variety of reasons.

First, students often forget many of the ideas from geometry. This hinders their performance in trigonometry. Students without much knowledge of geometry frequently lack the vocabulary to describe geometric objects and principles, which is a handicap to building further upon those ideas. Students are also hindered more concretely in the sense that a solid understanding of angles and triangles is essential when working with trigonometric functions.

Second, geometry and trigonometry are highly complementary topics and they provide a much richer experience when studied together. For example, a study of regular polygons is common practice within geometry. This task is simplified by a few area formulas from trigonometry.

Third, a geometric underpinning of trigonometry provides a sounder theoretical framework on which to develop trigonometry. We don't go deeply into geometric postulates, due to space and time constraints, but the text—at very least—takes the reader far enough back into geometry that they would be able trace the ideas used in trigonometry to geometric first principles if they had a solid understanding of the axioms.

Fourth, the introduction of geometry provides students a framework on which to categorize ideas. For example, the triangle congruence postulates and theorems provide a means to determine when a given set of information is sufficient to determine uniqueness. This categorization is very helpful when trigonometry is implemented to solve oblique triangles.

My background in education has been tutoring, not lecturing. Though some may balk at the notion of a tutor producing a trigonometry text instead of a professor, I believe that there are unique advantages that tutors have over professors. In particular, tutors get immediate feedback from students. Students may or may not ask a professor a question during lecture, and even when a student does it is difficult for them to articulate what they do not understand and it is not feasible for a professor to watch a student work out a few problems so the professor can see where the student is struggling. Furthermore, tutors have exposure to a wider range of textbooks. They can see first-hand what material students find most helpful.

Let me highlight a few of the insights gained from tutoring and explain how they influenced this book.

- Students learn more easily when they are first introduced to ideas within a concrete setting as opposed to an abstract one.

Frequently, textbooks are written so abstractly that students do not even look at them for help. For most students, a general understanding of ideas must be built slowly by means of analyses of specific cases. This book utilizes this insight and emphasizes examples as a result. Abstract ideas are covered, but new ideas are always reiterated more concretely.

- Students solve problems based on core ideas and mathematical reasoning. This book solves problems the same way. This is in contrast to math texts' typical approach which heavily relies on formulas. For example, this book only uses one formulation of the Law of Cosines, regardless of whether a side length or an angle measure is found. Most books provide a formulation for sides and another for angles. This makes solving the problems computationally easier, but almost all students forget all but one formulation which completely voids the approach.

- Students need to solve problems to learn mathematics. There is no substitute for problem-solving. More than enough exercises are included to obtain mastery of the material. Furthermore, the exercises range from easy to extremely challenging, so this book will be helpful to trigonometry students at many skill levels.

A few further comments: If you find errors or you think that part of the text is poorly formulated or written, please email me at charles.tutoring@gmail.com. Help from students really improved my previous book *GRE Mathematics Subject Test Solutions: Exams GR1268, GR0568, and GR9768*. Consider contacting me for tutoring if you live in North San Diego County; my website is rambotutoring.com. Please like **Rambo Tutoring** on Facebook.

Thank you so much for checking out my Trigonometry book. I hope you find it informative, useful, and fun.

Charles Rambo
Escondido, California
October 2017

Chapter 1

Angles

Angles are a fundamental aspect of trigonometry. Some knowledge of lines, rays, and line segments is required to obtain a full understanding of angles.

Symbol	Name	Description
$\overleftrightarrow{XY}$	Line	The figure within a plane which contains the points X and Y, extends infinity in both directions, and does not bend. We often refer to lines using lowercase letters as well, e.g. ℓ, m, or n.
$\overrightarrow{XY}$	Ray	The portion of line $\overleftrightarrow{XY}$ which has endpoint X and extends infinitely in the direction of Y.
$\overline{XY}$	Line segment	The portion of line $\overleftrightarrow{XY}$ contained within points X and Y. The length of $\overline{XY}$ is denoted XY.

This chapter assumes basic algebra skills. You will need to solve linear, quadratic, and rational expressions. There will be computations within this chapter, but they can be solved without a calculator.

1.1 Classification of Angles

Definition 1.1 Two rays which meet at a common endpoint form an **angle**. The common endpoint is called the **vertex of the angle**.

There are several ways to denote the angle in the figure:

- ∠B,
- ∠ABC,
- ∠CBA, or
- using a Greek letter, in this case θ.

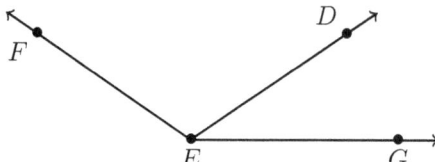

Sometimes using one point to reference an angle is ambiguous. For example, it is not clear where ∠E is in the figure below. Is it ∠DEG, ∠DEF, or ∠FEG? There is no way to know.

When we reference an angle, we are referring to a figure. However, it is helpful to quantify angles based on how much they open up.

Definition 1.2 An **angle measure** is the number which quantifies how much an angle opens up. A protractor is used to obtain angle measures.

Within this chapter, we will quantify angle measures using degree units. In Chapter 3, we will introduce another unit of angle measure, called radians.

To indicate the measure of $\angle A$, we write $m\angle A$. When we use a Greek letter to denote an angle, we will be sloppy and simply use an equals sign to indicate measure, e.g. if angle φ has measure $20°$, we write $\varphi = 20°$.

Definition 1.3 Two angles of the same measure are **congruent**.

We write $\angle X \cong \angle Y$ to indicate that $\angle X$ and $\angle Y$ are congruent.

Definition 1.4

- An **acute angle** has a measure between 0 and $90°$. That is, $\angle W$ is acute if $0 < m\angle W < 90°$.

- A **right angle** has a measure of exactly $90°$. That is, $\angle X$ is right if $m\angle X = 90°$.

- An **obtuse angle** has a measure strictly between $90°$ and $180°$. That is, $\angle Y$ is obtuse if $90° < m\angle Y < 180°$.

- A **straight angle** has a measure of exactly $180°$. That is, $\angle Z$ is straight if $m\angle Z = 180°$.

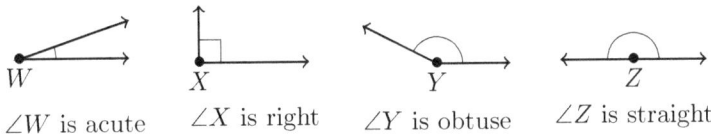

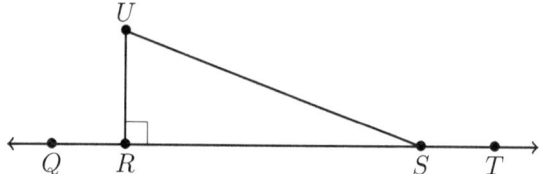

Example 1.1 Classify (a) ∠QRS, (b) ∠QRU, (c) ∠SRU, (d) ∠RUS, (e) ∠USR, and (f) ∠UST as acute, right, or straight.

Solution

(a) ∠QRS is straight.

(b) ∠QRU is right.

(c) ∠SRU is right.

(d) ∠RUS is acute.

(e) ∠USR is acute.

(f) ∠UST is obtuse. ∎

Definition 1.5

- When the sum of two angle measures is 90°, we say the angles are **complementary**.

- When the sum of two angle measures is 180°, we say the angles are **supplementary**.

Sometimes the complement or supplement of an angle is a nonsensical concept. For example, the complement of an angle of measure 100° is nonsensical because 100° is larger than 90°. The supplement of an angle of measure 190° does not make sense either, because 190° is larger than 180°.

Example 1.2 What is the measure of an angle complementary to angle θ, if $\theta = 72°$?

Solution An angle complementary to θ has measure

$$90° - 72° = 18°.$$

Example 1.3 Suppose $\angle T$ and $\angle U$ are supplementary,

$$m\angle T = 4x - 18° \quad \text{and} \quad m\angle U = 57° - x.$$

Find $m\angle T$ and $m\angle U$.

Solution Since $\angle T$ and $\angle U$ are supplementary,

$$m\angle T + m\angle U = 180° \quad \text{implies} \quad 4x - 18° + 57° - x = 180°.$$

From here, a bit of algebra shows $x = 47°$. Hence,

$$m\angle T = 4(47°) - 18° = 170° \quad \text{and} \quad m\angle U = 57° - 47° = 10°.$$

1.2 Adjacent Angles

Definition 1.6 Two angles are **adjacent** when

- they share a vertex,
- they share a common side, and
- they are non-overlapping.

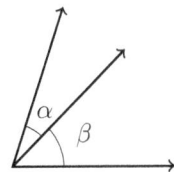

Angles α and β in the figure above are adjacent. Let us take a look at cases where one of the three criteria fails.

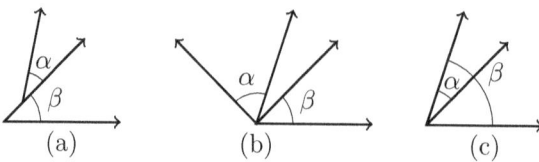

There are no adjacent angles in the three figures. In diagram (a), angles α and β are not adjacent because they do not share a common vertex. In (b), angles α and β are not adjacent because they do not share a common side. And in (c), α and β are not adjacent because angle β overlaps angle α.

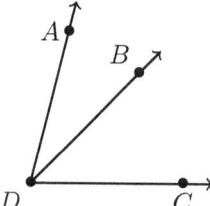

Postulate 1.1 (Angle Addition) *The sum of the measures of adjacent angles is equal to the measure of the angle formed by the non-common rays of the adjacent angles. In the diagram above, this means*
$$m\angle ADC = m\angle ADB + m\angle BDC.$$

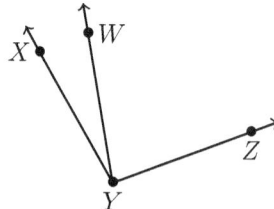

Example 1.4 Suppose $m\angle XYZ = 2t+10°$, $m\angle WYX = 180°-3t$, and $m\angle WYZ = 3t/2 - 2°$. Find the measure of each angle.

Solution Due to the Angle Addition Postulate,

$$m\angle WYX + m\angle WYZ = m\angle XYZ.$$

So,
$$180° - 3t + \frac{3t}{2} - 2° = 2t + 10° \quad \text{implies} \quad t = 48°.$$

Hence,

$$m\angle WYZ = 180° - 3(48°) = 36°, \quad m\angle WYZ = \frac{3(48°)}{2} - 2° = 70°,$$

and
$$m\angle XYZ = 2(48°) + 10° = 106°.$$

■

Definition 1.7 When the non-common sides of two adjacent angles form a straight angle, then the angles are a **linear pair**.

In the figure below, $\angle ABD$ and $\angle CBD$ are a linear pair.

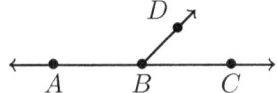

Example 1.5 Suppose $\angle QRS$ and $\angle SRT$ are a linear pair. If $m\angle QRS = 87°$, find $m\angle SRT$.

Solution Since the non-common sides of linear pairs form a straight angle, linear pairs are supplementary. It follows that

$$m\angle SRT = 180° - m\angle QRS = 93°.$$

■

1.3 Vertical Angles

Definition 1.8 Vertical angles are angles on opposite sides of intersecting lines.

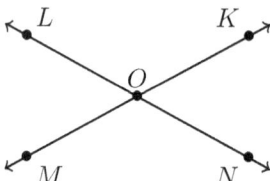

In the diagram, there are two linear pairs. One linear pair is

$$\angle KON \quad \text{and} \quad \angle LOM,$$

and the other is

$$\angle KOL \quad \text{and} \quad \angle NOM.$$

Proposition 1.1 *Vertical angles are congruent.*

Proof We will show that $\angle LOM$ has the same measure as $\angle KON$. Suppose $m\angle LOM = x$. Then $m\angle LOK = 180° - x$, because $\angle LOM$ and $\angle LOK$ are a linear pair. Since $\angle LOK$ and $\angle KON$ are a linear pair as well,

$$m\angle KON = 180° - (180° - x) = x.$$

We conclude

$$\angle LOM \cong \angle KON$$

∎

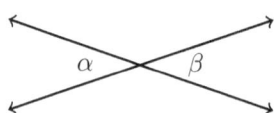

Example 1.6 Suppose $\alpha = 3y - 15°$ and $\beta = 75° - 2y$. Find y.

Solution Because α and β are vertical angles,
$$\alpha = \beta \quad \text{implies} \quad 3y - 15° = 75° - 2y.$$
After a bit of algebra, we find that $y = 18°$. ∎

1.4 Parallel, Perpendicular, and Transversal Lines

Definition 1.9 Geometric objects are **coplanar** when they are contained within the same plane.

In this section, we are interested in coplanar lines.

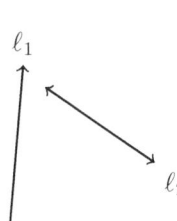

In the figure to the left, lines ℓ_1 and ℓ_2 are coplanar. It is impossible to draw a non-coplanar line on a page, because the page itself is contained within a plane. However, an example of a line non-coplanar to ℓ_1 and ℓ_2 is one that goes out of the page toward you.

Definition 1.10 Parallel lines are coplanar lines which do not intersect.

In two dimensions lines are either parallel or they intersect at some location.

In the diagram to the right, lines m and n are parallel. The arrow within each line indicates that this is the case.

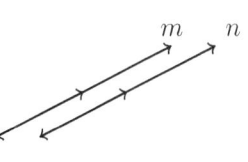

As is visible from the diagram, lines m and n are a fixed distance apart. This is always the case for parallel lines.

We write $m \parallel n$ to indicate that m and n are parallel. Notice that $m \parallel n$ if and only if $n \parallel m$. Furthermore, $m \parallel n$ and $n \parallel p$ implies $m \parallel p$ as long as m and p are unique lines.

9

Definition 1.11 Two lines are **perpendicular** or **orthogonal** if they intersect at a right angle.

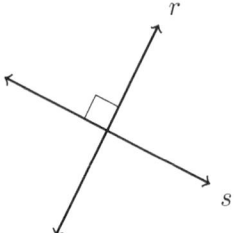

To indicate lines r and s are perpendicular, we use the notation $r \perp s$. Notice $r \perp s$ is true if and only if $s \perp r$ is true.

1.4.1 Transversals

Definition 1.12 A line that intersects two coplanar lines at distinct points is called a **transversal**.

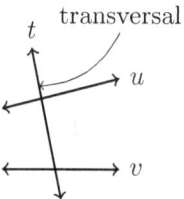

Suppose t is a transversal of u and v, as depicted in the diagram above. The area between u and v is the "interior". The area outside of u and v is the "exterior". Angles on the same side of the transversal are called "same-side angles". Angles on the opposite side the transversal are called "alternate angles". "Corresponding angles" are angles on the same side of the transversal, but one angle is interior and the other is exterior (sliding one line on top of the other would cause the angles to overlap).

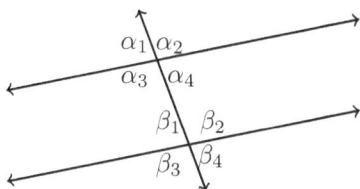

Corresponding angles
α_1 and β_1
α_2 and β_2
α_3 and β_3
α_4 and β_4

Alternate interior angles
α_3 and β_2
α_4 and β_1

Same-side exterior angles
α_1 and β_3
α_2 and β_4

Alternate exterior angles
α_1 and β_4
α_2 and β_3

Same-side interior angles
α_3 and β_1
α_4 and β_2

Postulate 1.2 (Corresponding Angles) *Corresponding angles are congruent, when the transversal intersects two parallel lines.*

Proposition 1.2 *Suppose a transversal intersects two parallel lines. Then each of the following hold:*

(i) Alternate exterior angles are congruent.

(ii) Alternate interior angles are congruent.

(iii) Same-side exterior angles are supplementary.

(iv) Same-side interior angles are supplementary.

Proof We will prove (i) and (iii) and leave the rest as an Exercise. Consider the diagram on the previous page.

(i) We will show $\alpha_1 = \beta_4$. Proving $\alpha_2 = \beta_3$ is nearly identical. We know $\alpha_1 = \beta_1$ because of the Corresponding Angles Postulate. Because β_1 and β_4 are vertical angles, $\beta_1 = \beta_4$. Therefore, $\alpha_1 = \beta_4$.

(iii) We will show $\alpha_1 + \beta_3 = 180°$. Proving $\alpha_2 + \beta_4 = 180°$ is nearly identical. We know $\alpha_1 = \beta_1$ due to the Corresponding Angles

Postulate. Since β_1 and β_3 are a linear pair, $\beta_1 + \beta_3 = 180°$. Substituting α_1 for β_1, we conclude $\alpha_1 + \beta_3 = 180°$. ∎

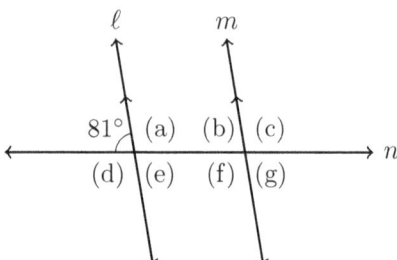

Example 1.7 Suppose ℓ is parallel to m. Find the measures of the labeled angles using the given information.

Solution

(a) The given angle and the angle in position (a) are a linear pair, which implies that they are supplementary. Hence, the angle in position (a) has measure
$$180° - 81° = 99°.$$

(b) The given angle and the angle in position (b) are corresponding angles. Thus, by the Corresponding Angles Postulate, the measure of the angle in position (b) is also $81°$.

(c) The given angle and the angle in position (c) are same-side exterior angles. This implies they are supplementary. Therefore, the measure of the angle in position (c) is
$$180° - 81° = 99°.$$

(d) The given angle and the angle in position (d) form a linear pair. So, they are supplementary. Ergo, the angle in position (d) has measure
$$180° - 81° = 99°.$$

(e) The given angle and the angle in position (e) are vertical angles. As a result, they are congruent. It follows that the measure of the angle located at (e) is 81°.

(f) We have not introduce a name for the relationship between the given angle and the angle in position (f). However, the angle in position (d) and (f) are corresponding angles. So, the Corresponding Angles Postulate tells us that they are congruent. We conclude that the angle in position (f) has measure 99° because that is the measure of the angle in position (d).

(g) The given angle and the angle in position (g) are alternate exterior angles. It follows that they are congruent. We conclude that the angle in position (g) has measure 81°. ∎

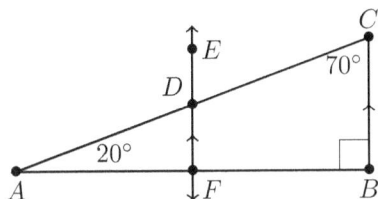

Example 1.8 Suppose $\overline{BC}$ is parallel to $\overleftrightarrow{EF}$. Find the measure of $\angle CDE$.

Solution Think of $\overline{AC}$ as the transversal of $\overleftrightarrow{EF}$ and $\overline{BC}$. Because $\angle BCA$ and $\angle CDE$ are alternate interior angles, it follows that $m\angle CDE = 70°$. ∎

Postulate 1.3 (Converse of Corresponding Angles) *Suppose a transversal intersects two lines such that corresponding angles are congruent. Then the two lines are parallel.*

Proposition 1.3 *Suppose two lines are intersected by a transversal. They are parallel if any of the following is true.*

(i) *Alternate exterior angles are congruent.*

(ii) *Alternate interior angles are congruent.*

(iii) *Same-side exterior angles are supplementary.*

(iv) *Same-side interior angles are supplementary.*

Proof We will prove (i) and leave the rest as an Exercise. Consider the diagram below.

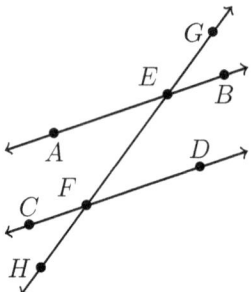

Suppose $\angle BEG \cong \angle CFH$. We know $\angle BEG \cong \angle AEF$ because the angles are vertical. So, $\angle AEF \cong \angle CFH$, which means there is a pair of congruent corresponding angles. Due to the Converse of the Corresponding Angles Postulate, $\overleftrightarrow{AB}$ and $\overleftrightarrow{CD}$ are parallel. The argument to prove $\angle AEG \cong \angle DFH$ implies that $\overleftrightarrow{AB}$ is parallel to $\overleftrightarrow{CD}$ is a nearly identical. ■

Example 1.9 Explain why $\overline{JM}$ is parallel to $\overline{KL}$.

Solution If we consider $\overline{IL}$ as the transversal of $\overline{JM}$ and $\overline{KL}$, then $\angle JML$ and $\angle KLI$ are same-side interior angles. Since $\angle JML$ and $\angle KLI$ are also supplementary, Proposition 1.3 (iv) tells us that $\overline{JM}$ is parallel to $\overline{KL}$. ■

1.5 Angles and Triangles

An angle formed by a vertex of a triangle and its adjacent sides is called an "interior angle." For example, the interior angles of $\triangle ABC$ are

$$\angle A, \quad \angle B, \quad \text{and} \quad \angle C.$$

This section studies the interior angles of a triangle. In particular, it explores relationships between interior angle measures.

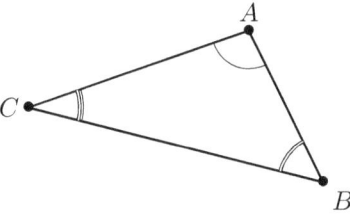

Theorem 1.1 (Triangle Sum) *The sum of the interior angle measures of a triangle is* $180°$. *That is, for any triangle* $\triangle ABC$,

$$m\angle A + m\angle B + m\angle C = 180°.$$

Proof Extend a line through A which is parallel to $\overline{BC}$. Let D and E be points on the line.

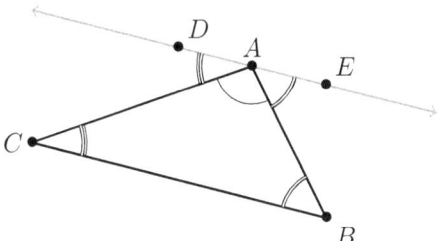

Because $\angle CAD$, $\angle BAC$, and $\angle BAE$ collectively form a straight angle, the Angle Addition Postulate tells us

$$m\angle CAD + m\angle BAC + m\angle BAE = 180°.$$

Due to Proposition 1.2 (ii), we know that $m\angle CAD = m\angle C$ and $m\angle BAE = m\angle B$. So, substitution leads us to conclude

$$m\angle C + m\angle BAC + m\angle B = 180°.$$

Example 1.10 Consider $\triangle QRS$. Suppose $m\angle Q = 94° - 2t$, $m\angle R = 150° - 5t$, and $m\angle S = 3t + 4°$. Find the measure of each angle.

Solution Because of Theorem 1.1,

$$\begin{aligned} m\angle Q + m\angle R + m\angle S &= 180° \\ \Rightarrow \quad 94° - 2t + 150° - 5t + 3t + 4° &= 180° \\ \Rightarrow \quad 248° - 4t &= 180° \\ \Rightarrow \quad t &= 17°. \end{aligned}$$

Hence,

$$m\angle Q = 94° - 2(17°) = 60°, \quad m\angle R = 150° - 5(17°) = 65°,$$

and

$$m\angle S = 3(17°) + 4° = 55°.$$

∎

Theorem 1.2 (Isosceles Triangle) *The angles of a triangle are congruent if and only if the segments opposite have equal length. Furthermore, the measure of an interior angle of a triangle is greater than another if and only if the side opposite the larger angle is longer than the one opposite the smaller.*

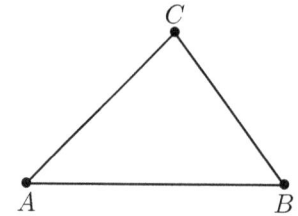

For $\triangle ABC$ Theorem 1.2 translates to

$$\angle A \cong \angle B \quad \text{if and only if} \quad BC = AC,$$

and

$$m\angle A > m\angle B \quad \text{if and only} \quad BC > AC.$$

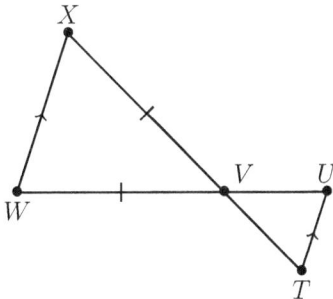

Example 1.11 Suppose $\overline{TU}$ is parallel to $\overline{XW}$ and $\overline{VW}$ has the same length as $\overline{VX}$. If $m\angle U = 76°$, what is the value of $m\angle X$?

Solution Using $\overline{UW}$ as our transversal, $\angle W$ and $\angle U$ are alternate interior angles. It follows that $m\angle W = m\angle U = 76°$. Since $VW = VX$, we know $m\angle X = m\angle W = 76°$. ∎

1.6 Exercises

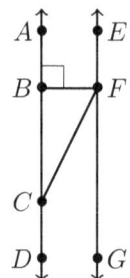

Figure 1

* Exercise 1

Consider Figure 1. Suppose $\overrightarrow{AD}$ is parallel to $\overrightarrow{EG}$. Classify each angle as either acute, right, obtuse, or straight.

(a) ∠ABF
(b) ∠BCF
(c) ∠CFB
(d) ∠EFB
(e) ∠BCD
(f) ∠GFB
(g) ∠CFG
(h) ∠DCF
(i) ∠EFG
(j) ∠CFE

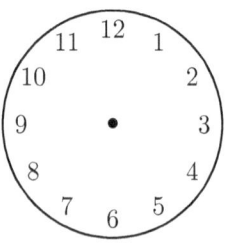

Figure 2

** Exercise 2

At the given times, find the measure of the angle between the minute and hour hand of a clock like the one shown in Figure 2.

(a) 3:00
(b) 6:00
(c) 1:00
(d) 2:00
(e) 10:00
(f) 5:30
(g) 7:45
(h) 11:55

* Exercise 3

Compute the complement and supplement of each angle measure, whenever it makes sense. When the complement or supplement is nonsensical say so.

(a) 20°
(b) 75°
(c) 92°
(d) 80°
(e) 200°
(f) 22.5°

** Exercise 4

Suppose ∠C and ∠D are complementary. Find x using the given information.

(a) $m\angle C = 3x - 17°$
    $m\angle D = 123° - 5x$

(b) $m\angle C = (2x^2 - 12x)°$
    $m\angle D = (7x - 60)°$

(c) $m\angle C = \dfrac{3x^2}{x+20°}$
    $m\angle D = 3x$

** Exercise 5

Suppose $\angle E$ and $\angle F$ are supplementary. Find y assuming the following hold.

(a) $m\angle E = 120° - \dfrac{3y}{4}$
    $m\angle F = \dfrac{9y}{5} - 45°$

(b) $m\angle E = (y^2 - 10y + 150)°$
    $m\angle F = (166 - 15y)°$

(c) $m\angle E = \dfrac{90°(y-60)}{y+10}$
    $m\angle F = (2y)°$

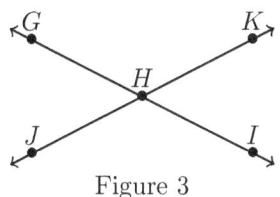

Figure 3

** Exercise 6

Use Figure 3 to answer the following questions.

(a) If $m\angle GHJ = 30°$, what is $m\angle IHK$?

(b) Suppose $m\angle GHJ = 3t/2 - 25°$ and $m\angle IHK = 115° - t$. Find t.

(c) Let $m\angle GHJ = u^2 - 10u$ and $m\angle IHK = 400° - 2u^2$. Find $m\angle GHK$.

(d) What is v, when $m\angle IHK = \left(\dfrac{30v+30}{2v-2}\right)°$ and $\angle IHJ = \left(\dfrac{90v+30}{v-1}\right)°$?

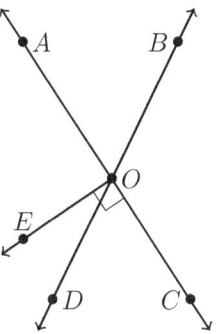

Figure 4

* Exercise 7

Consider Figure 4. Find w using the given information.

(a) $m\angle AOE = 2w$

(b) $m\angle AOB = 2w - 5°$
    $m\angle COD = 43° - w$

(c) $m\angle AOB = (6w^2)°$
    $m\angle BOC = (30w + 36)°$

(d) $m\angle COD = (6w+12)°$
    $m\angle DOE = \left(\dfrac{12w+42}{19-w}\right)°$

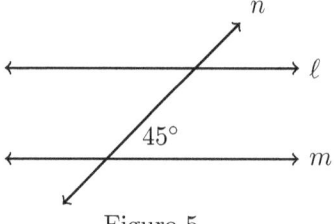

Figure 5

* **Exercise 8**

In Figure 5, suppose ℓ and m are parallel. Find the remaining angles created by the transversal n.

** **Exercise 9**

Consider Figure 1. Suppose $\overrightarrow{AD}$ and $\overrightarrow{EG}$ are parallel. Assume $m\angle BCF = 20°$. Find each of the following.
(a) $m\angle DCF$, (b) $m\angle CFG$, and (c) $m\angle BFC$

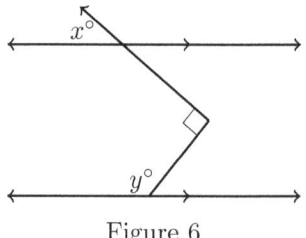

Figure 6

** **Exercise 10**

Use Figure 6 and the given information to find the variable.

(a) If $x° = 40°$, find $y°$.

(b) What is $x°$, when $y° = 140°$?

** **Exercise 11**

Consider Proposition 1.2.
(a) Prove alternate interior angles are congruent.
(b) Prove same-side interior angles are supplementary.

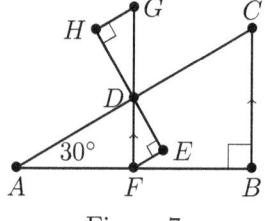

Figure 7

** **Exercise 12**

In Figure 7, find the angle measures listed.

(a) $m\angle C$ (d) $m\angle GDH$
(b) $m\angle EDF$ (e) $m\angle CDG$
(c) $m\angle DFE$ (f) $m\angle DGH$

** Exercise 13

Consider Proposition 1.3.

(a) Prove that congruent alternate interior angles implies the lines are parallel.

(b) Prove that supplementary same-side exterior angles implies the lines are parallel.

(c) Prove that supplementary same-side interior angles implies the lines are parallel.

** Exercise 14

Consider $\triangle IJK$.

(a) Find $m\angle K$ when $m\angle I = 15°$ and $m\angle J = 52°$.

(b) Find $m\angle I$ when $m\angle J = 22°$ and $\angle K$ is right.

(c) Suppose $m\angle I = m\angle J = m\angle K$. What is the measure of each angle?

(d) Find x, if $m\angle I = 3x - 16°$, $m\angle J = 76° - x$, and $m\angle K = 10x$.

(e) Let $m\angle I = (y^2 + 10y + 20)°$, $m\angle J = (2y^2 + 100)°$, and $m\angle K = (50 - 3y)°$. What is the value(s) of y?

(f) Find the possible value(s) of z, when
$$m\angle I = \frac{120°}{z+1},$$
$$m\angle J = \frac{180°}{z+1},$$
and $m\angle K = 15°(z+4)$.

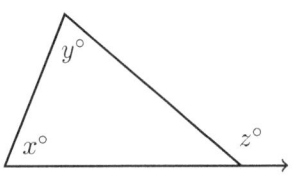

Figure 8

** Exercise 15

In Figure 8, find the remaining variable.

(a) $x = 50$ and $y = 34$

(b) $x = 35$ and $z = 100$

(c) $y = 24$ and $z = 120$

** Exercise 16

Suppose x, y, and z are as shown in figure 8. Prove
$$x° + y° = z°.$$

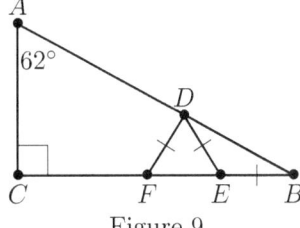

Figure 9

** Exercise 17

In Figure 9, $BE = DE = DF$. Find the measure of $\angle CFD$.

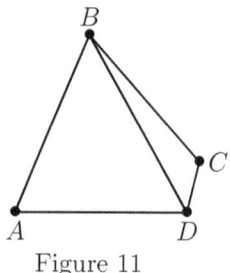

Figure 11

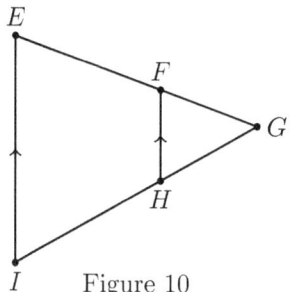

Figure 10

** Exercise 18

Consider Figure 10. Assume $\overline{FH}$ is parallel to $\overline{EI}$.

(a) Let $FG = GH$ and $m\angle HFE = 130°$. Find $m\angle I$.

(b) Say that $m\angle E = 70°$ and $EG = IG$. What is $m\angle GHF$?

(c) Suppose $m\angle E = 45°$ and $m\angle FHI = 135°$. Suppose $EG = 2t$ and $GI = 18 - t$. Find t.

** Exercise 19

Use Figure 11 to complete the problems below. Note the diagram is not necessarily drawn to scale.

(a) Rank the interior angle measures of $\triangle ABD$ from least to greatest. Assume $AB = 2$, $AD = 4$ and $BD = 3$.

(b) Rank the interior angle measures of $\triangle BCD$ from least to greatest. Suppose $BD = 7$, $BC = 5$, and $CD = 3$.

(c) Let

$$m\angle A = 50°,$$
$$m\angle ABD = 70°,$$
$$m\angle CBD = 11°,$$
$$\text{and} \quad m\angle BDC = 45°.$$

Rank the sides from least length to greatest.

Chapter 2

Triangles

In this chapter, we will study triangles. In particular, we provide two ways of classifying triangles and we analyze key properties of triangles. Readers will need a complete understanding of the properties of angles discussed in Chapter 1 as well as a proficient understanding of algebra. Ratios and radicals are utilized frequently, so unfamiliar readers are advised to consult Appendices A and B. Calculators are not necessary for this chapter.

2.1 Classification of Triangles

We will classify triangles two ways: By angles and by sides.

2.1.1 Classification by Angles

Definition 2.1

- An **acute triangle** has three acute interior angles.
- The interior angles of an **equiangular triangle** are congruent.
- When one of the interior angles of a triangle is right, it is a **right triangle**.
- An **obtuse triangle** has exactly one obtuse interior angle.

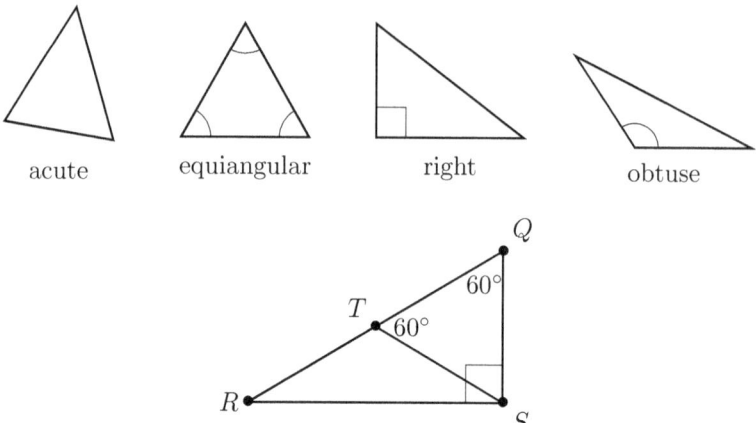

acute   equiangular   right   obtuse

Example 2.1 Classify (a) $\triangle QRS$, (b) $\triangle QST$, and (c) $\triangle RST$ according to their angle measures.

Solution

(a) Since $\angle QSR$ is right, $\triangle QRS$ is a right triangle.

(b) Because the sum of the measures of the interior angles of a triangle is 180°,

$$\begin{aligned} m\angle QST + m\angle Q + m\angle QTS &= 180° \\ \Rightarrow \quad m\angle QST + 60° + 60° &= 180° \\ \Rightarrow \quad m\angle QST &= 60° \end{aligned}$$

Hence, all of the angles of $\triangle QST$ are congruent, so it must be an equiangular triangle.

(c) Since $\angle RTS$ and $\angle QTS$ are a linear pair, they are supplementary. It follows that

$$m\angle RTS = 180° - m\angle QTS \quad \text{implies} \quad m\angle RTS = 120°.$$

Because $\angle RTS$ is an obtuse angle, $\triangle RST$ is an obtuse triangle. ∎

2.1.2 Classification by Sides

Definition 2.2

- Each side of a **scalene triangle** has a distinct length, i.e. no two sides have equal lengths.
- An **isosceles triangle** has two sides of the same length.
- All three sides of an **equilateral triangle** have the same length.

We have defined isosceles triangles so that all equilateral triangles are isosceles. Some books say isosceles triangles have *exactly* two sides of equal length. Our definition makes is easier to prove triangles are isosceles; simply prove two sides have the same length. The alternative definition would also require one to prove the third side length is not equal to the others.

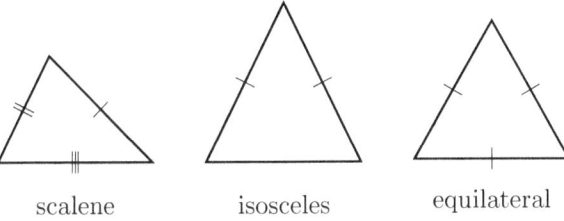

scalene   isosceles   equilateral

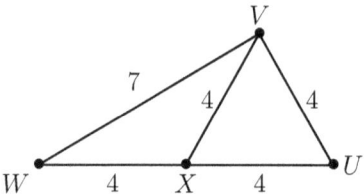

Example 2.2 Classify (a) $\triangle UVX$, (b) $\triangle VWX$, and $\triangle UVW$ by means of their sides.

(a) Because all of the side lengths of $\triangle UVX$ are equal, it must be an equilateral triangle.

(b) Two of the side lengths of $\triangle VWX$ are the same. So, $\triangle VWX$ is an isosceles triangle.

(c) Since $UW = 4 + 4 = 8$, no two sides of $\triangle UVW$ have equal length. Hence, it is a scalene triangle. ■

2.2 Congruent Triangles

Definition 2.3 Two triangles are **congruent**, if there is a correspondence between the vertices of the two triangles such that corresponding angles are congruent and corresponding sides have the same length.

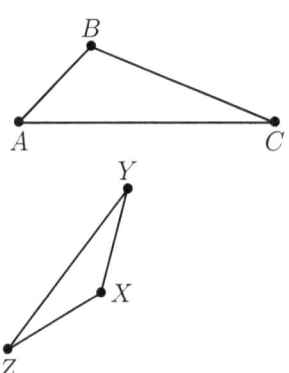

As usual for congruence statements, the symbol $\cong$ is used to denote congruent triangles. So,

$$\triangle ABC \cong \triangle XYZ$$

if and only if

$$\begin{array}{lll} \angle A \cong \angle X & BC = YZ \\ \angle B \cong \angle Y & AC = XZ \\ \angle C \cong \angle Z & AB = XY. \end{array}$$

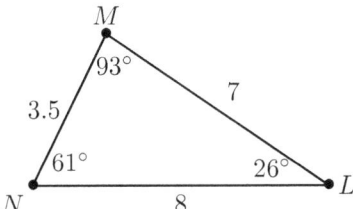

Example 2.3 Suppose $\triangle LMN \cong \triangle OPQ$. Find all the side lengths and angle measures of $\triangle OPQ$.

Solution Because $\triangle LMN \cong \triangle OPQ$,

$$L \leftrightarrow O, \quad M \leftrightarrow P, \quad \text{and} \quad N \leftrightarrow Q.$$

Hence,

$$m\angle O = m\angle L = 26°, \quad m\angle P = m\angle M = 93°, \quad m\angle Q = m\angle N = 61°,$$
$$OP = LM = 7, \quad OQ = LN = 8, \quad \text{and} \quad PQ = MN = 3.5.$$

∎

The next few subsections examine sufficient criteria to prove congruence between two triangles. This information is intrinsically valuable and has been studied for centuries. However, we are particularly intersected in congruent triangles because of Chapter 9.

In Chapter 9, we will find the lengths and angle measures of triangles using various givens. Our study of triangle congruence allows us determine when enough information is provided to determine uniqueness

2.2.1 SSS Postulate

Postulate 2.1 (Side-Side-Side Congruence Postulate, SSS)
Suppose there is a correspondence between the vertices of two triangles such that corresponding sides have equal length. Then the triangles are congruent.

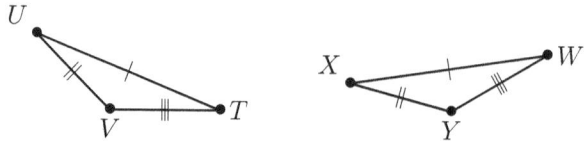

Using the SSS triangle congruence postulate, we conclude

$$\triangle TUV \cong \triangle WXY.$$

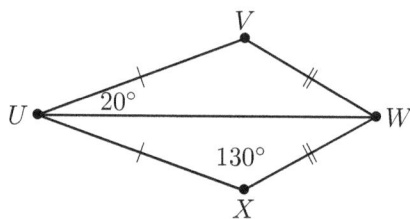

Example 2.4 Write a congruence statement. Use it to determine the unknown angle measures in each triangle.

Solution Because $UV = UX$, $WV = WX$, and $UW = UW$, SSS tells us
$$\triangle UVW \cong \triangle UXW.$$
This implies

$$m\angle V = m\angle X = 130° \quad \text{and} \quad m\angle WUX = m\angle WUV = 20°.$$

Because the sum of the interior angle measures of a triangle is 180°,

$$m\angle VUW + m\angle V + m\angle UWV = 180° \quad \text{implies} \quad m\angle UWV = 30°.$$

Congruence of the two triangles leads us to conclude $m\angle UWX = 30°$ as well. ∎

2.2.2 SAS Postulate

Postulate 2.2 (Side-Angle-Side Congruence, SAS) *Suppose there is a correspondence between the vertices of two triangles such that there are two pairs of equal length sides, and the angles between the sides in each triangle are congruent. Then the two triangles are congruent.*

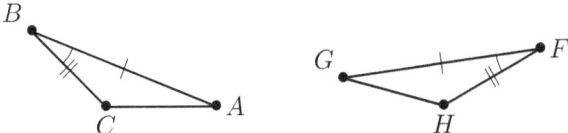

Using the SAS triangle congruence postulate, we conclude

$$\triangle ABC \cong \triangle GFH.$$

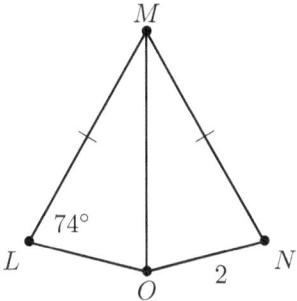

Example 2.5 Suppose $\overline{MO}$ is an angle bisector of $\angle LMN$, $m\angle LMN = 60°$, and $LM = NM$. Make a congruence statement between the two triangles. Find the length LO, and all the interior angles.

Solution We will prove $\triangle LMO \cong \triangle NMO$ using the SAS congruence postulate. We are given $LM = NM$, it is clear $MO = MO$, and $\angle LMO \cong \angle NMO$ due to the definition of a bisector. Hence, $\triangle LMO \cong \triangle NMO$ due to SAS.

Now we can use the congruence correspondence and the other given information to find the needed lengths and angles. We know $LO = NO = 2$. Because $\overline{MO}$ is an angle bisector, $m\angle LMO$ and $m\angle NMO$ are both $60°/2 = 30°$. To find $m\angle LOM$, we will use the fact that the sum of the interior angle measures of a triangle is $180°$. So,

$$m\angle L + m\angle LMO + m\angle LOM = 180° \quad \text{implies} \quad m\angle LOM = 76°$$

It follows that $m\angle NOM = 76°$ as well, and

$$m\angle LON = 2(76°) = 152°.$$

■

2.2.3 ASA Postulate

Postulate 2.3 (Angle-Side-Angle Congruence, ASA) *Suppose there is a correspondence between the vertices of two triangles such that there are two pairs of congruent angles, and the lengths of the sides between the two angles in each triangle are equal. Then the two triangles are congruent.*

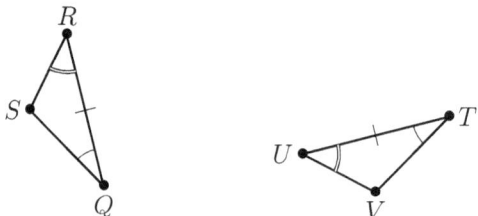

Using the ASA triangle congruence postulate, we conclude

$$\triangle QRS \cong \triangle TUV.$$

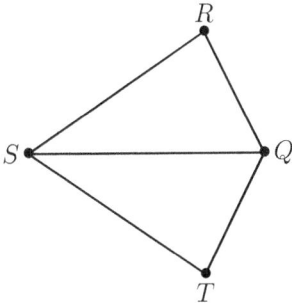

Example 2.6 Suppose $\overline{QS}$ is an angle bisector of $\angle RQT$ and $\angle RST$. If $RS = x + 7$ and $TS = 12 - 4x$, what is the value of x.

Solution Since $\overline{QS}$ is an angle bisector of $\angle RQT$, we know $\angle RQS \cong \angle TQS$. The same reasoning shows $\angle QSR \cong \angle QST$. We have $QS = QS$, because a segment is always equal to itself. Due to ASA, we conclude $\triangle QRS \cong \triangle QTS$.

Because corresponding sides of congruent triangles have equal length, $RS = TS$. It follows that $x + 7 = 12 - 4x$. Then a bit of algebra shows $x = 1$. ∎

2.2.4 AAS and HL Theorems

Theorem 2.1 (Angle-Angle-Side, AAS) *Suppose there is a correspondence between the vertices of two triangles such that there are two pairs of corresponding congruent angles and a pair of corresponding sides of equal length which are not contained between the angles. Then the two triangles are congruent.*

Proof This follows immediately from ASA. The sum of the measures of the interior angles of a triangle is 180°, so two pairs of corresponding angles being congruent implies the last pair of corresponding angles are congruent.

∎

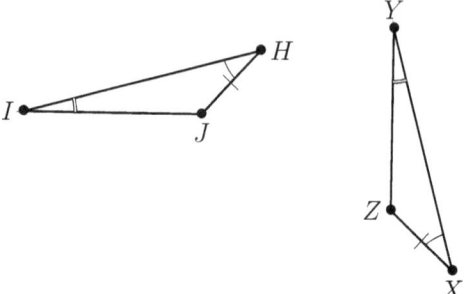

Using AAS, we conclude

$$\triangle HIJ \cong \triangle XYZ.$$

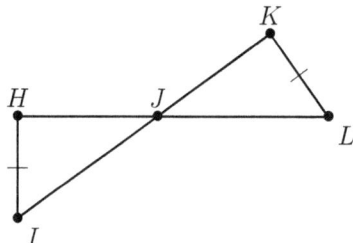

Example 2.7 Consider $\triangle HIJ$ and $\triangle JKL$. Suppose $\angle I \cong \angle K$, and $HI = LK$. Prove J is the midpoint of $\overline{HL}$.

Solution We know $\angle HJI \cong \angle LJK$ because they are vertical angles. Since we are given $\angle I \cong \angle K$ and $HI = LK$, AAS tells us $\triangle HIJ \cong \triangle LKJ$.

Because corresponding sides of congruent triangles have equal lengths, $HJ = LJ$. Ergo, J is the midpoint of $\overline{HL}$. ∎

Definition 2.4 Consider a right triangle.

- The side opposite the right angle is the **hypotenuse**.
- The other two sides of a right triangle are its **legs**.

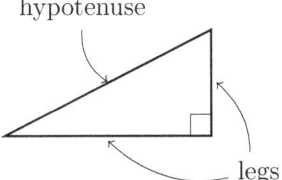

Because of the Isosceles Triangle Theorem (Theorem 1.2), we know that the hypotenuse is always the longest side of a right triangle.

Theorem 2.2 (Hypotenuse-Leg, HL) *Suppose two right triangles have hypotenuses of equal length and one pair of legs of equal length. Then the triangles are congruent.*

Proof Suppose $\triangle ABC$ and $\triangle DEF$ are right with $\angle C$ and $\angle F$ being the right angles of their respective triangles. Further, suppose $AB = DE$ and $AC = DF$, i.e. suppose the hypotenuses and a pair of legs have equal lengths.

Consider line $\overleftrightarrow{BC}$. Let G be the point on $\overleftrightarrow{BC}$ outside of $\triangle ABC$ such that $EF = CG$.

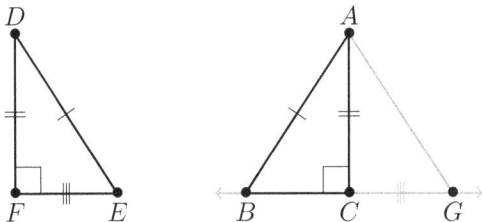

By assumption $AC = DF$. We know $\angle ACG \cong \angle F$, because they are both right. Furthermore, we constructed G so that $CG = EF$. It follows that $\triangle AGC \cong \triangle DEF$ due to SAS.

Since $AB = DE$ and $DE = AG$, we have $AB = AG$. It follows that $\angle ABC \cong \angle AGC$ due to the the Isosceles Triangle Theorem (Theorem 1.2). We know that $\angle ACB \cong \angle ACG$, because both angles are right. Clearly, $AC = AC$. Then AAS congruence theorem leads us to conclude $\triangle ABC \cong \triangle AGC$. Thus,

$$\triangle ABC \cong \triangle AGC \quad \text{and} \quad \triangle AGC \cong \triangle DEF$$

implies
$$\triangle ABC \cong \triangle DEF.$$

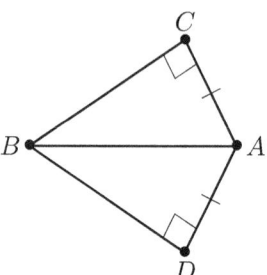

Example 2.8 Suppose that $AC = AD$. Let $\angle C$ and $\angle D$ be right. Prove that $\overline{AB}$ is an angle bisector of $\angle CBD$.

Solution Since $AC = AD$ and $AB = AB$, HL tells us that $\triangle ABC \cong \triangle ABD$. It follows that $\angle CBA \cong \angle DBA$. Hence, $\overline{AB}$ is an angle bisector of $\angle CBD$.

2.3 Similar Triangles

Definition 2.5 Two triangles are **similar**, if there is a correspondence between the vertices such that corresponding angles are congruent and the ratios of corresponding side lengths are equal.

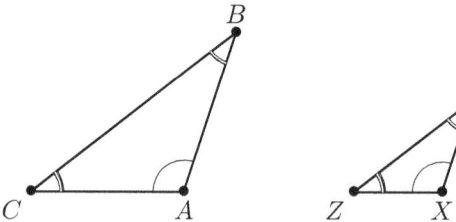

The symbol $\sim$ denotes similarity. So, $\triangle ABC \sim \triangle XYZ$ if and only if
$$\angle A \cong \angle X, \quad \angle B \cong \angle Y, \quad \angle C \cong \angle Z,$$
and
$$\frac{AB}{XY} = \frac{AC}{XZ} = \frac{BC}{YX}.$$
Much like congruence statements, the order in which the vertices are listed shows the correspondence.

Intuitively, when two triangles are similar, it means that they are a scaled image of each other. One triangle can be converted into the other by scaling the sides (scaling leaves angles unchanged).

Example 2.9 Suppose $\triangle LMN \sim \triangle OPQ$. If $LM = 5$, $OP = 10$, and $MN = 15$, find PQ.

Solution Since $\triangle LMN$ is similar to $\triangle OPQ$, we know the ratios of corresponding sides are equal. The similarity statement tells us that LM corresponds to OP and MN corresponds to PQ. So,
$$\frac{LM}{OP} = \frac{MN}{PQ} \quad \text{implies} \quad \frac{5}{10} = \frac{15}{PQ}.$$
Then some algebra shows that $PQ = 30$. ■

Postulate 2.4 (Angle-Angle Similarity, AA) *Suppose there is a correspondence between two triangles such that two pairs of corresponding angles are congruent. Then the two triangles are similar.*

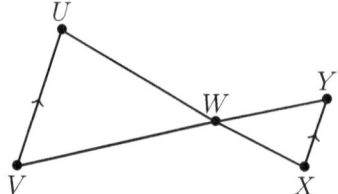

Example 2.10 Assume $\overline{UV}$ is parallel to $\overline{XY}$. Let $WU = 12$, $WX = 9$, and $WV = 16$. Find WY.

Solution Because $\overline{UV}$ is parallel to $\overline{XY}$, alternate interior angles are congruent. It follows that $\angle U \cong \angle X$ and $\angle V \cong \angle Y$.

Hence by the AA similarity postulate,

$$\triangle UVW \sim \triangle XYW$$

It follows that

$$\frac{WY}{WV} = \frac{WX}{WU} \quad \text{implies} \quad \frac{WY}{16} = \frac{9}{12}.$$

After a bit of algebra, we conclude $WY = 12$. ∎

Theorem 2.3 (Side-Side-Side Similarity, SSS) *Suppose there is a correspondence between the vertices of two triangles such that the ratios of corresponding side lengths are equal. Then the triangles are similar.*

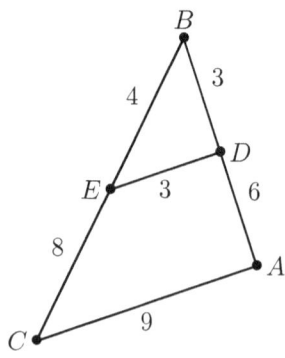

Example 2.11 Prove that $\overline{AC}$ is parallel to $\overline{DE}$.

Solution We know
$$AB = 6 + 3 = 9 \quad \text{and} \quad BC = 4 + 8 = 12.$$
It follows that
$$\frac{BD}{AB} = \frac{BE}{BC} = \frac{DE}{AC} = \frac{1}{3}.$$
Hence, $\triangle ABC \sim \triangle DBE$ due to the SSS similarity theorem. It follows that
$$\angle A \cong \angle BDE \quad \text{and} \quad \angle C \cong \angle BED$$
due to the definition of similarity.

By the Converse of Corresponding Angles Postulate, we know $\overline{AC}$ is parallel to $\overline{DE}$. ∎

Theorem 2.4 (Side-Angle-Side Similarity, SAS) *Suppose there is a corresponds between the vertices of two triangles such that the ratios of two pairs of corresponding side lengths are equal, and the angles between the sides in each triangle are congruent. Then the two triangles are similar.*

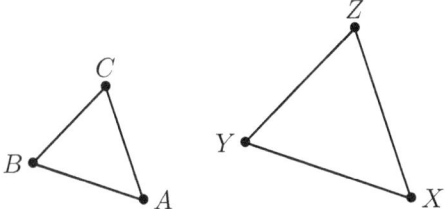

For the triangles above, the SSS similarity postulate says
$$\frac{BC}{YZ} = \frac{AC}{XZ} \quad \text{and} \quad \angle C \cong \angle Z,$$
allows us to conclude
$$\triangle ABC \sim \triangle XYZ.$$

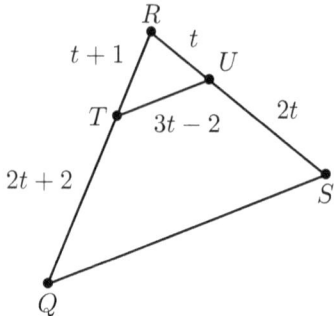

Example 2.12 Find QS in terms of t.

Solution Is clear

$$QR = t + 1 + 2t + 2 = 3t + 3 \quad \text{and} \quad RS = t + 2t = 3t.$$

Because

$$\frac{QR}{TR} = \frac{3t+3}{t+1} = 3 \quad \text{and} \quad \frac{RS}{RU} = \frac{3t}{t} = 3,$$

the ratio of two pairs of corresponding sides is fixed. Furthermore, the angles between the sides in each triangle are congruent due to the fact that $\angle R \cong \angle R$. Hence, the SAS similarity theorem tells us

$$\triangle QRS \sim \triangle TRU.$$

Thus,

$$\frac{QS}{TU} = 3 \quad \text{implies} \quad \frac{QS}{3t-2} = 3.$$

Solving for QS leads us to $QS = 9t - 6$. ∎

2.4 Right Triangles

Consider a right triangle and the altitude perpendicular to its hypotenuse. The altitude creates two similar right triangles, due to the AA similarity postulate. It also follows that the larger right triangle is similar to both of the smaller ones, because of the AA similarity postulate.

Let us see why this is the case using $\triangle IJK$.

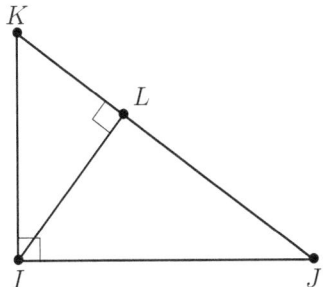

Altitude $\overline{IL}$ creates $\triangle LJI$ and $\triangle LIK$. Because the two acute angles of a right triangle are complementary, a bit of thought leads us to conclude

$$\angle K \cong \angle JIL \quad \text{and} \quad \angle J \cong \angle LIK.$$

Then an application of the AA similarity postulate gives

$$\triangle LJI \sim \triangle LIK \sim \triangle IJK.$$

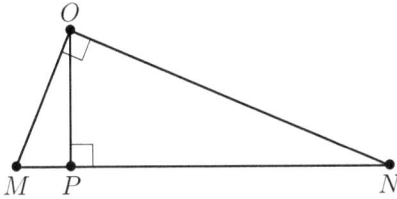

Example 2.13 Consider $\triangle MNO$. Suppose $MN = 169$, $OP = 60$, and $MO = 65$. Find the remaining side lengths of the triangles.

Solution We know $\triangle MNO \sim \triangle MOP$. It follows that

$$\frac{MN}{MO} = \frac{NO}{OP} \quad \text{implies} \quad \frac{169}{65} = \frac{NO}{60}.$$

From here, a bit of easy algebra shows $NO = 156$. Furthermore,

$$\frac{MO}{MN} = \frac{MP}{MO} \quad \text{implies} \quad \frac{65}{169} = \frac{MP}{65}.$$

Then some algebra shows $MP = 25$.

It is clear
$$NP = MN - MP$$
$$= 169 - 25$$
$$= 144.$$

∎

Similar triangles are so powerful that even the Pythagorean Theorem follows from them. Before we introduce the theorem and proof, let us introduce a new convention.

> Within a triangle, a lower case letter denotes the length of the side opposite the vertex of the corresponding uppercase letter.

For example, in $\triangle XYZ$,
$$x = YZ, \quad y = XZ, \quad \text{and} \quad z = XY.$$

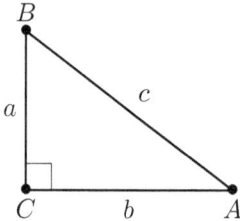

Theorem 2.5 (Pythagorean) *Consider $\triangle ABC$. Suppose $\angle C$ is the right angle. Then*
$$a^2 + b^2 = c^2.$$

Proof Let us draw the altitude from vertex C to $\overline{BA}$. Let D be the point where it intersects $\overline{BA}$. For convenience, say $h = CD$.

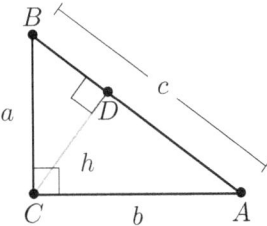

The idea behind this proof is to use the fact that

area of $\triangle CBD$ + area of $\triangle ACD$ = area of $\triangle ABC$.

The area of $\triangle ABC$ is
$$\frac{ch}{2}.$$

To find the area of $\triangle CBD$, we will find BD. Because $\triangle ABC \sim \triangle CBD$,
$$\frac{BD}{BC} = \frac{BC}{BA} \quad \text{implies} \quad \frac{BD}{a} = \frac{a}{c}.$$
It follows that $BD = a^2/c$. So, the area of $\triangle CBD$ is
$$\frac{a^2 h}{2c}.$$

To find the area of $\triangle ACD$, we will find AD. Since $\triangle ABC \sim \triangle ACD$,
$$\frac{AD}{AC} = \frac{AC}{AB} \quad \text{implies} \quad \frac{AD}{b} = \frac{b}{c}.$$
It follows that $AD = b^2/c$. So, the area of $\triangle ACD$ is
$$\frac{b^2 h}{2c}.$$

Using our area identity from before,
$$\frac{a^2 h}{2c} + \frac{b^2 h}{2c} = \frac{ch}{2}.$$
Then multiplying by $2c/h$ yields
$$a^2 + b^2 = c^2.$$

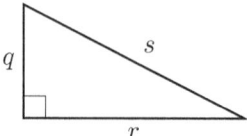

Example 2.14 Consider $\triangle QRS$ above. If $q = t+1$, $r = t+8$, and $s = t+9$, what is the value of t?

Solution Using the Pythagorean Theorem,

$$q^2 + r^2 = s^2 \quad \text{implies} \quad (t+1)^2 + (t+8)^2 = (t+9)^2.$$

Hence,

$$\begin{aligned} t^2 + 2t + 1 + t^2 + 16t + 64 &= t^2 + 18t + 81 \\ \Rightarrow \quad 2t^2 + 18t + 65 &= t^2 + 18t + 81 \\ \Rightarrow \quad t^2 &= 16 \\ \Rightarrow \quad t &= \pm 4. \end{aligned}$$

It is impossible for t to equal -4, because that would imply q equals $-4+1 = -3$. We conclude that $t = 4$ is the only solution. ∎

Proposition 2.1 (Converse of Pythagorean Theorem) *Suppose three sides of a triangle have lengths a, b, and c, where $a \leq b \leq c$. Then the triangle is right if*

$$a^2 + b^2 = c^2.$$

Proof It is easy to construct a right triangle of side lengths a, b, and c. Due to the SSS congruence postulate, this triangle is congruent to any triangle of side lengths a, b, and c. This implies that all triangle of this form are right, because corresponding angles of congruent triangles are congruent.

∎

Example 2.15 Determine whether a triangle of side lengths (a) 3, 4, and 5, and (b) a triangle of side lengths 5, 7, and 11 are right.

Solution

(a) Because
$$3^2 + 4^2 = 25 = 5^2,$$
the corresponding triangle must be right.

(b) Since
$$5^2 + 7^2 = 74 \neq 121 = 11^2,$$
there is no right triangle with the given side lengths.

∎

2.5 Special Right Triangles

Proposition 2.2 (45° – 45° – 90° Special Right Triangle) *A triangle is a 45° – 45° – 90° special right triangle if and only if the relationship between the sides and angles in the diagram holds for some $t > 0$.*

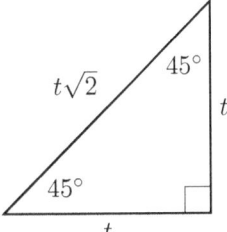

There are several ways to conclude that a right triangle is a 45° – 45° – 90° special right triangle. It suffices to know that the right triangle has ...

- ... an angle of measure 45°.
- ... two legs of equal length.
- ... a hypotenuse of length $\sqrt{2}$ times the length of a leg.

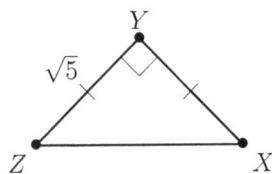

Example 2.16 Suppose $YZ = YZ$. Find the remaining side lengths.

Solution This is a $45° - 45° - 90°$ special right triangle because it is a right triangle with two legs of equal length. Furthermore, we see that the relationship between the sides described in Proposition 2.2 holds for $t = \sqrt{5}$. This implies $z = \sqrt{5}$ and
$$y = \sqrt{5} \cdot \sqrt{2} = \sqrt{10}.$$

∎

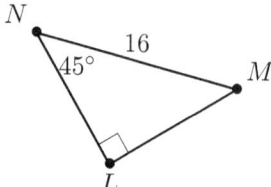

Example 2.17 What are the remaining side lengths?

Solution Since this is a $45° - 45° - 90°$ special right triangle, $\ell = 16 = t\sqrt{2}$. This implies

$$t = \frac{16}{\sqrt{2}}$$
$$= \frac{16}{\sqrt{2}} \cdot \frac{\sqrt{2}}{\sqrt{2}}$$
$$= \frac{16\sqrt{2}}{2}$$
$$= 8\sqrt{2}.$$

We conclude $m = 8\sqrt{2}$ and $n = 8\sqrt{2}$. ∎

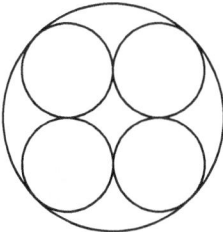

Example 2.18 Suppose the radius of the large circle is 3, and the shaded circles are radii of equal length. Find the total shaded area.

Solution Suppose the radius of each of the small circles is r. Let us zoom in on the top left circle.

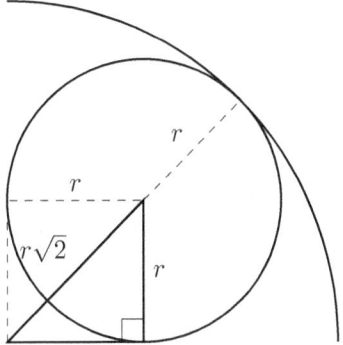

Since the two legs of the right triangle drawn are both r, it is a $45° - 45° - 90°$ special right triangle and its hypotenuse is $r\sqrt{2}$. It follows that $r + r\sqrt{2} = 3$. So,

$$\begin{aligned} r &= \frac{3}{1+\sqrt{2}} \\ &= \frac{3}{1+\sqrt{2}} \cdot \frac{1-\sqrt{2}}{1-\sqrt{2}} \\ &= \frac{3(1-\sqrt{2})}{1-2} \\ &= 3\sqrt{2} - 3. \end{aligned}$$

It follows that the area of one shaded circle is

$$\pi(3\sqrt{2} - 3)^2 = \pi(18 - 18\sqrt{2} + 9) = 9(3 - 2\sqrt{2})\pi.$$

Since the total area is four times the area of one shaded circle, we conclude that the final answer is

$$36(3 - 2\sqrt{2})\pi.$$

∎

Proposition 2.3 (30° − 60° − 90° Special Right Triangle) *A triangle is a 30° − 60° − 90° special right triangle if and only if the relationship between the sides and angles in the diagram hold for some $t > 0$.*

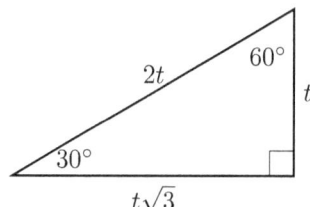

As was the case for 45° − 45° − 90° special right triangles, there are several ways to conclude that a right triangle is a 30° − 60° − 90° special right triangle. It suffices to know that the right triangle ...

- ... has an interior angle of measure either 30° or 60°.
- ... is such that the ratio of the larger to smaller leg is $\sqrt{3}$.
- ... is such that the ratio of a leg to the hypotenuse is either $\sqrt{3}/2$ or $1/2$.

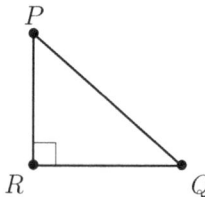

Example 2.19 The right triangle in the figure above is not to scale. Use the given information to find the missing sides and angles.

(a) $p = 5$ and $m\angle P = 30°$
(b) $p = 2\sqrt{3}$ and $q = 6$
(c) $q = 7$ and $r = 14$

Solution

(a) Since $m\angle P = 30°$, $m\angle R = 90°$ and
$$m\angle P + m\angle Q + m\angle R = 180°,$$
a bit of algebra shows $m\angle Q = 60°$. As such, we have a $30° - 60° - 90°$ special right triangle.

It follows that $t = p = 5$, which implies
$$q = 5\sqrt{3} \quad \text{and} \quad r = 2(5) = 10.$$

(b) Since $p = 2\sqrt{3}$ and $q = 6$,
$$r^2 = q^2 + p^2$$
$$= (6)^2 + \left(2\sqrt{3}\right)^2$$
$$= 36 + 4(3)$$
$$= 48.$$

This implies
$$r = \sqrt{48}$$
$$= \sqrt{16 \cdot 3}$$
$$= \sqrt{16} \cdot \sqrt{3}$$
$$= 4\sqrt{3}.$$

Due to the relationship between the sides, Proposition 2.3 leads us to conclude $m\angle P = 30°$ and $m\angle Q = 60°$.

(c) Since $p = 7$ and $r = 14$,
$$r^2 = q^2 + p^2 \quad \text{implies} \quad 14^2 = q^2 + 7^2.$$

A bit of algebra shows

$$\begin{aligned}q &= \sqrt{147} \\ &= \sqrt{49 \cdot 3} \\ &= \sqrt{49} \cdot \sqrt{3} \\ &= 7\sqrt{3}.\end{aligned}$$

Proposition 2.3 tells us we have a $30° - 60° - 90°$ special right triangle, where $m\angle P = 30°$ and $m\angle Q = 60°$.

∎

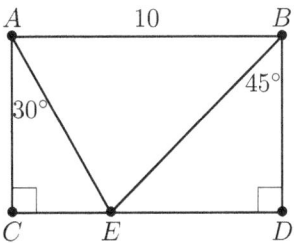

Example 2.20 Find AC in rectangle $ABCD$.

Solution For convenience, let $y = AC$. We know
$$CD = CE + DE \quad \text{and} \quad CD = 10.$$

Using special right triangles,
$$CE = \frac{y}{\sqrt{3}} = \frac{y\sqrt{3}}{3} \quad \text{and} \quad DE = BD = y.$$

It follows that
$$y + \frac{y\sqrt{3}}{3} = 10$$
$$\Rightarrow \frac{3y + y\sqrt{3}}{3} = 10$$
$$\Rightarrow \frac{y(3 + \sqrt{3})}{3} = 10$$
$$\Rightarrow y = \frac{30}{3 + \sqrt{3}}$$
$$= \frac{30(3 - \sqrt{3})}{9 - 3}$$
$$= 5(3 - \sqrt{3}).$$

Hence,
$$AC = 5(3 - \sqrt{3}). \qquad \blacksquare$$

2.6 Exercises

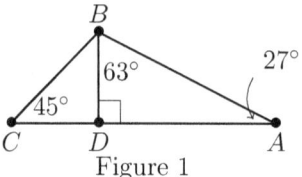

Figure 1

* Exercise 1

Consider Figure 1. Classify (a) $\triangle ABC$, (b) $\triangle ABD$, and (c) $\triangle BCD$ according to their interior angles.

** Exercise 3

Suppose $\triangle IJK$ is equiangular. Find x when ...

(a) ... x is the measure of each interior angle.

(b) ... $m\angle I = (78 - 2x)°$.

(c) ... $m\angle J = (24x^2 - 108x)°$.

(d) ... $m\angle K = \left(\dfrac{10x - 5}{x - 1}\right)°$.

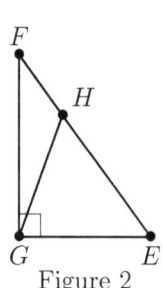

Figure 2

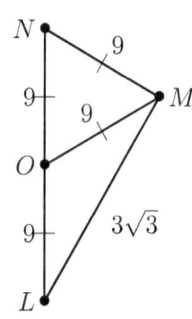

Figure 3

* Exercise 2

Use Figure 2 and classify (a) $\triangle EFG$, (b) $\triangle EGH$, and (c) $\triangle FGH$ according to their interior angles.

* Exercise 4

Consider Figure 3. Classify (a) $\triangle LMN$, (b) $\triangle LMO$, and (c) $\triangle OMN$ according to their side lengths.

* **Exercise 5**

Explain why a triangle is equiangular if and only if it is equilateral.

* **Exercise 6**

Why is it impossible for a triangle to have more than one obtuse interior angle?

* **Exercise 7**

Suppose $\triangle QRS \cong \triangle TUV$.

(a) Which pairs of angles must be congruent?

(b) Which pairs line segments must have equal length?

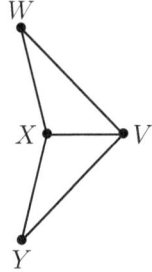

Figure 4

** **Exercise 8**

Write a congruence statement based on Figure 4 and the given information. Justify your congruence statement with a theorem or postulate.

(a) $\overline{VX}$ is an angle bisector of $\angle WVY$ and $\angle WXY$.

(b) $VW = VY$ and $WX = YX$.

(c) $\overline{VX}$ is an angle bisector of $\angle WVY$ and $VW = VY$.

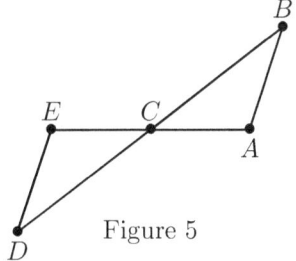

Figure 5

** **Exercise 9**

Find congruence statement using Figure 5 and the given information. Justify your congruence statement with a theorem or postulate.

(a) $\overline{AB}$ is parallel to $\overline{DE}$ and $AC = EC$.

(b) $AC = EC$ and $BC = DC$.

(c) $\angle B \cong \angle D$ and C is the midpoint of $\overline{AE}$.

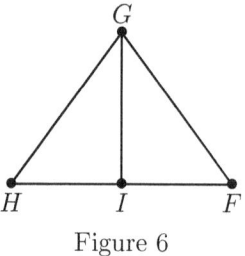

Figure 6

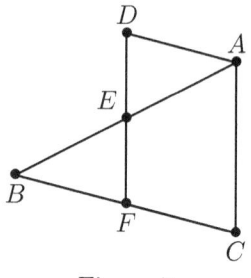

Figure 7

** Exercise 10

Write a congruence statement based on Figure 6 and the given information. Justify your congruence statement with a theorem or postulate.

(a) ∠FIG ≅ ∠HIG and FG = HG.

(b) ∠FIG ≅ ∠HIG and FI = HI.

(c) FI = HI and FG = HG.

(d) ∠FIG ≅ ∠HIG and ∠F ≅ ∠H.

*** Exercise 11

Side-side-angle, or SSA, does not determine congruence between two triangles. Draw two non-congruent triangles with two pairs of equal side lengths and a pair of congruent angles, not included within the sides.

** Exercise 12

Consider Figure 7. Write a similarity statement using the given information. Justify your similarity statement with its appropriate theorem or postulate.

(a) $\overline{AC}$ is parallel to $\overline{DF}$

(b) $\overline{AD}$ is parallel to $\overline{BC}$

(c) $\dfrac{DE}{FE} = \dfrac{AE}{BE}$

(d) $\dfrac{AB}{BE} = \dfrac{BC}{BF} = \dfrac{AC}{EF}$

** Exercise 13

Use Figure 7 to answer the following.

(a) Suppose $BE = 6$, $EF = 4$, $BF = 8$, $AE = 5$, $DE = 10/3$. Find AD.

(b) Let $AE = 1$, $CF = 1.2$, $BE = 2$, $BF = 2.4$, and $AC = 3$. What is EF?

(c) Assume $\overline{AD}$ is parallel to $\overline{BC}$. If $AD = 95$, $BF = 100$, and $BE = 120$, then what is the value of AE?

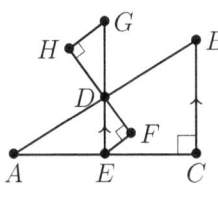

Figure 8

** Exercise 14

Consider Figure 8. Suppose $AB = 2\sqrt{34}$, $AC = 10$, $BC = 6$, $m\angle A = 31°$, and $\overline{BC}$ is parallel to $\overline{EG}$.

(a) Find all the sides and angles of $\triangle DEF$ if $DE = 3$.

(b) Find all the sides and angles of $\triangle DGH$ if $HD = 3.2$.

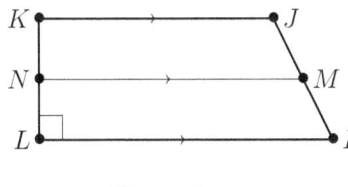

Figure 9

*** Exercise 15

In Figure 9, suppose $IL = 5$, $JK = 4$, and $KL = 2$. Further, suppose $MN = w$ and $NL = h$. Assume $\overline{KJ}$ is parallel to $\overline{MN}$ and $\overline{IL}$.

(a) Write h as a function of w.

(b) The area of a trapezoid is

$$\frac{h(b_1 + b_2)}{2},$$

where b_1 and b_2 are the bases of the trapezoid and h is the height. Write the area of trapezoid $IMNL$ as a function of w.

(c) If $JI = \sqrt{5}$, write IM as a function of w.

Hint: Use similar triangle techniques to establish ratios.

** Exercise 16

A woman stands x meters away from a 2.5-meter lamp. If the woman is 1.6 meters tall, write the length of her shadow in terms of x.

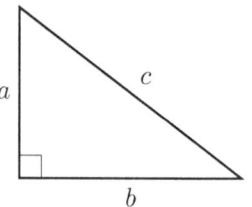

Figure 11

*** Exercise 17

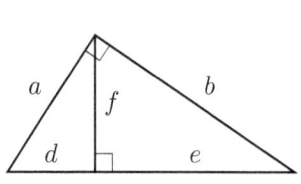

Figure 10

Let the variables be as shown in Figure 10. Find the remaining side lengths using the given information.

(a) $a = 6$ and $b = 8$

(b) $c = 17$ and $d = 64/17$

(c) $e = 12$ and $f = 5$

(d) $c = 17$ and $f = 120/17$

(e) $a = 5$ and $d = 25/13$

** Exercise 18

Consider Figure 11. Find the missing side.

(a) $a = 6$ and $b = 8$

(b) $b = 15$ and $c = 17$

(c) $a = 25$ and $c = 65$

(d) $a = 6$ and $b = 9$

(e) $b = 13$ and $c = 17$

(f) $a = 7$ and $c = 15$

** Exercise 19

Use Figure 11 and the given information to find t.

(a) $a = 8$, $b = t$, and $c = t + 2$

(b) $a = t + 6$, $b = 2t + 6$, and $c = 4t + 3$

(c) $a = t$, $b = t + 1$, and $c = 3t - 4$

** Exercise 20

Determine whether the given side lengths could be the sides of a right triangle.

(a) 10, 24, and 26

(b) 39, 52, and 65

(c) 32, 60, and 62

** Exercise 21

A **Pythagorean Triple** is an ordered triplet of positive integers (a, b, c) such that

$$a^2 + b^2 = c^2.$$

For example, $(3, 4, 5)$ is a Pythagorean Triple. Prove that if (a, b, c) is a Pythagorean Triple, then so is (ka, kb, kc) for all positive integers k.

Use the given information to find the remaining side lengths.

(a) $\ell = 5$

(b) $m = \sqrt{6}$

(c) $n = 7\sqrt{2}$

(d) $n = \sqrt{3}$

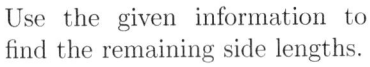

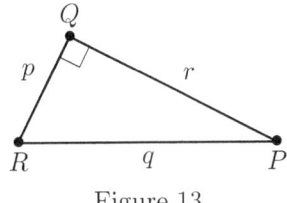

Figure 13

Figure 12

** Exercise 22

Consider Figure 12. Suppose $m\angle L = 45°$ and $m\angle M = 45°$.

** Exercise 23

In Figure 13, suppose $m\angle P = 30°$ and $m\angle R = 60°$. One length of $\triangle PQR$ is given. What are the remaining side lengths?

(a) $p = 4$

(b) $r = 2\sqrt{3}$

(c) $q = 16$

(d) $p = \sqrt{6}$

(e) $r = 7\sqrt{5}$

(f) $q = \dfrac{\sqrt{3}}{6}$

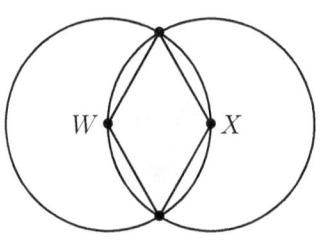

Figure 14

Figure 15

*** Exercise 24

Consider Figure 14. Suppose circle W and circle X both have a radius of 2. Find the area of the shaded region.

*** Exercise 25

A circle of radius 2 is inscribed within an equilateral triangle.

(a) Find the ratio of the radius of the circle to the altitude of the equilateral triangle.

(b) What is the area of the triangle?

Hint: The radius of an inscribed circle is perpendicular to the boundary of the equilateral triangle at the points of tangency.

*** Exercise 26

In Figure 15, suppose the radius of the large circle is 10, and the four gray circles have the same radius. Find the total area of the gray circles. Hint: The radius of a circle is perpendicular any of the circle's tangent lines at the point of tangency.

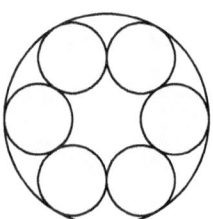

Figure 16

*** Exercise 27

Consider Figure 16. Suppose the radius of the large circle is 6, and the radii of the six gray circles are equal. Calculate the ratio of the gray area to the white area. Hint: The radius of a circle is perpendicular to any of the circle's tangent lines at the point of tangency.

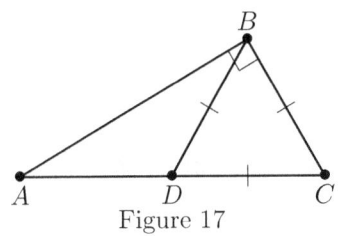

Figure 17

*** **Exercise 28**

In Figure 17, suppose $\triangle BCD$ is equilateral and $AB = 6$. Find the area of $\triangle ABD$.

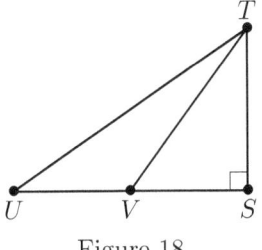

Figure 18

*** **Exercise 29**

Consider Figure 18. Suppose $m\angle U = 45°$ and $m\angle SVT = 60°$.

(a) What is UV when $ST = 7\sqrt{3}$?

(b) Find UV if $VS = 5$.

(c) Assume $UT = 15\sqrt{2}$. Then TV is equal to what?

(d) Find ST if $UV = 8$.

*** **Exercise 30**

In Figure 18, let $m\angle U = 30°$ and $m\angle STV = 45°$.

(a) Find UV given $VS = 10$.

(b) Suppose $UV = 2$. What is VT?

(c) If $UV = 10$, then TU equals what?

(d) Find UV given the area of $\triangle TUV$ is $100(\sqrt{3}+1)$.

57

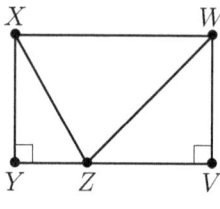

Figure 19

*** Exercise 31

Consider Figure 19. Suppose $m\angle VWZ = 60°$ and $m\angle YXZ = 45°$

(a) If $XY = 5$, then VY is equal to what?

(b) Say $WX = 20$. Find XZ.

(c) Suppose the area of rectangle $VWXY$ is $1 + \sqrt{3}$. What is WZ?

(d) Assume the area of $\triangle VWZ$ is $18\sqrt{3}$. Find the area of $\triangle XYZ$.

*** Exercise 32

Use Figure 19 to solve the following problems.

(a) Suppose $XY = 18$, $m\angle VWZ = 45°$, and $m\angle YXZ = 30°$. What is the value of VY?

(b) Let $WX = 72$, $m\angle YXZ = 30°$, and $m\angle VWZ = 45°$. Find XY.

(c) Assume $WX = 100$, $m\angle WXZ = 45°$, and $m\angle VZW = 30°$. Find XZ.

(d) Suppose the area of $\triangle XYZ$ is $8\sqrt{3}$. If $m\angle YXZ = 30°$ and $m\angle XWZ = 45°$, then the area of $\triangle WXZ$ is equal to what?

Chapter 3

Radians, Arc Length, and Rotational Motion

This chapter will further elaborate on our understanding of angles. We will provide a richer definition of an angle. Then we will introduce radian measures and their applications. Knowledge of Chapters 1 and 2 is assumed. This chapter contains a modest amount of calculator problems.

3.1 Directed Angles

In Chapter 1, we reviewed the concept of an angle taught in geometry class. However, this definition is not sufficiently robust for our purposes. For example, the angles students study in geometry have measure no greater than 180°, and they do not have an orientation. Let us introduce a richer formulation of an angle.

Definition 3.1 Consider rays $\overrightarrow{OX}$ and $\overrightarrow{OY}$. A rotation of $\overrightarrow{OX}$ about O which terminates at $\overrightarrow{OY}$ is the **directed angle** $\angle XOY$. Ray $\overrightarrow{OX}$ is the **initial side** and $\overrightarrow{OY}$ is the **terminal side** of $\angle XOY$.

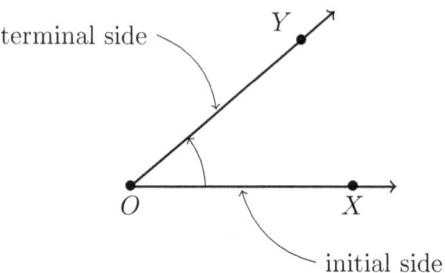

Because any angle from geometry can be a directed angle, we will not introduce special notation for directed angles. Simply assume an angle is directed when it is necessary given the context.

Like angles from geometry, directed angles have measures. The measure of the directed angle $\angle X$ is denoted $m\angle X$. However, unlike angles from geometry, it makes sense for a directed angle to be signed. Let us introduce a convention.

> A directed angle has *positive* measure if its initial side rotates *counterclockwise* to its terminal side.

> A directed angle has *negative* measure if its initial side rotates *clockwise* to its terminal side.

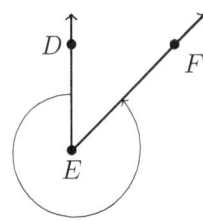

In the diagram to the left, $\angle DEF$ has positive measure. This is because its initial side $\overrightarrow{ED}$ rotates counterclockwise to its terminal side $\overrightarrow{EF}$.

In contrast, $\angle GHI$ has negative measure because its initial side $\overrightarrow{HG}$ rotates clockwise to its terminal side $\overrightarrow{HI}$.

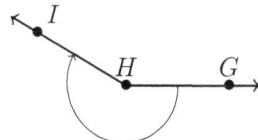

Angles of measure greater than 180° or less than −180° are a natural concepts within the context of directed angles. Indeed, careful measurements would reveal

$$m\angle DEF = 315° \quad \text{and} \quad m\angle GHI = -240°.$$

The initial side of a directed angle can rotate more than one revolution before it reaches its terminal side. This allows a directed angle to have measure greater than 360° or less than −360°. However, it also introduces some ambiguity as to which angle we are referring. Context should alleviate this ambiguity, and when it does not assume the angle rotates less than one revolution.

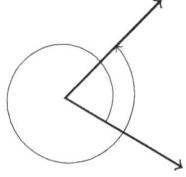

Example 3.1 Write the corresponding degree measure for each rotation.

(a) half a clockwise revolution

(b) two and a third counterclockwise revolutions

(c) seven and five-eighths clockwise revolutions

Solution

(a) Suppose α is the degree measure of half a clockwise revolution. Then
$$\frac{\alpha}{-1/2} = \frac{360°}{1} \quad \text{implies} \quad \alpha = -180°.$$

(b) Let β be the degree measure of two and a third counterclockwise revolutions. We know $2\frac{1}{3} = 7/3$, so
$$\frac{\beta}{7/3} = \frac{360°}{1} \quad \text{implies} \quad \beta = 840°.$$

(c) Assume γ is degree measure of seven and five-eights clockwise revolutions. Since $-7\frac{5}{8} = -61/8$,
$$\frac{\gamma}{-61/8} = \frac{360°}{1} \quad \text{implies} \quad \gamma = -2745°.$$

∎

Note that the sign of the revolution is determined by whether it rotates counterclockwise or clockwise.

3.2 Radian Measure of an Angle

Definition 3.2 An **arc** is a portion of a circle.

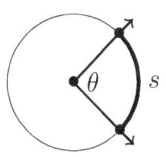
We say arc s "subtends" central angle θ or central angle θ is "subtended" by arc s.

Notice the natural correspondence between central angle measures and arc lengths on any particular circle: Match each arc length with the measure of a central angle it subtends. Indeed, even when an angle rotates more than one revolution, the correspondence holds; just count the length of the overlapped portion of the arc as many times as necessary. When an angle rotates clockwise, consider the length of the arc to be negative.

Definition 3.3 Consider an arbitrary angle α. Place its vertex at the center of a circle of radius 1. Suppose α is subtended by some arc, which we will call s. The **radian measure** of α is the length of the arc s.

Usually, we either leave radian measures unit-less or denote them by the unit "rad".

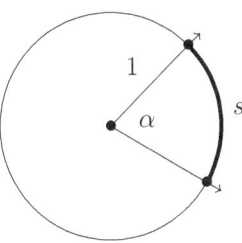

Let us consider some radian measures. We know that the circumference of a circle is $2\pi r$. This implies the radian measure of an angle of degree measure 360° is $2\pi(1) = 2\pi$. It is also clear that 0° corresponds to 0 radians. In general, we can use proportions to develop a relationship between an angle measure and a radian measure. Suppose an angle has degree measure α and radian measure s. Then

$$\frac{\alpha}{s} = \frac{360°}{2\pi} \quad \text{implies} \quad \frac{\alpha}{s} = \frac{180°}{\pi}.$$

Proposition 3.1

(i) *To convert from degrees to radians, multiply the angle measure by*
$$\frac{\pi}{180°}.$$

(ii) *To convert from radians to degrees, multiply the radian measure by*
$$\frac{180°}{\pi}.$$

Example 3.2 Convert into radian measures.
(a) 30°, (b) 45°, and (c) 60°.

Solution

(a) We will utilize Proposition 3.1 (i) to convert the degree measures to radians. We have
$$30° \cdot \left(\frac{\pi}{180°}\right) = \frac{\pi}{6}.$$

(b) Because of Proposition 3.1 (i),
$$45° \cdot \left(\frac{\pi}{180°}\right) = \frac{\pi}{4}.$$

(c) Proposition 3.1 (i) allows us to conclude
$$60° \cdot \left(\frac{\pi}{180°}\right) = \frac{\pi}{3}.$$

■

Example 3.3 Convert into degree measures.
(a) π, (b) $2\pi/3$, and (c) 3.

Solution

(a) We will use Proposition 3.1 (ii) to convert the radian measures to degrees. We have

$$\pi \cdot \left(\frac{180°}{\pi}\right) = \frac{180°}{1} = 180°.$$

(b) Proposition 3.1 (ii) tells us

$$\frac{2\pi}{3} \cdot \left(\frac{180°}{\pi}\right) = 120°.$$

(c) We utilize Proposition 3.1 (ii) to conclude

$$3 \cdot \left(\frac{180°}{\pi}\right) = \frac{540°}{\pi} \approx 171.887°.$$

■

3.3 Arc Length

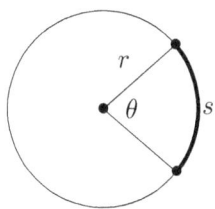

Proposition 3.2 *Consider a circle of radius r. Let θ be the radian measure of a central angle. Then the central angle is subtended by an arc of length*

$$s = r\theta$$

for $0 \leq \theta \leq 2\pi$.

Proof We will prove this proposition with ratios. The arc length of the entire circle is $2\pi r$ and the central angle corresponding to the entire circle has radian measure 2π. Hence,

$$\frac{s}{\theta} = \frac{2\pi r}{2\pi} \quad \text{implies} \quad s = r\theta.$$

∎

Example 3.4 Suppose the central angle of a circle has measure $40°$. If the radius of the circle is 36, what is the length of the corresponding arc?

Solution To use Proposition 3.2 we need to convert $40°$ to radians:

$$40° \cdot \left(\frac{\pi}{180°}\right) = \frac{2\pi}{9}.$$

It follows that the arc length is

$$s = 36\left(\frac{2\pi}{9}\right) = 8\pi.$$

∎

Example 3.5 Suppose the length of an arc is 20π and the radius of the circle is 15.

(a) Find the radian measure of the central angle subtended by the arc.

(b) What is the degree measure of the central angle subtended by the arc?

Solution

(a) Let θ be the radian measure of the central angle. Since the arc has length $s = 20\pi$ and the radius is 15,

$$20\pi = 15\theta \quad \text{implies} \quad \theta = \frac{4\pi}{3}.$$

(b) This is an application of Proposition 3.1 (ii):

$$\frac{4\pi}{3} \cdot \left(\frac{180°}{\pi}\right) = 240°.$$

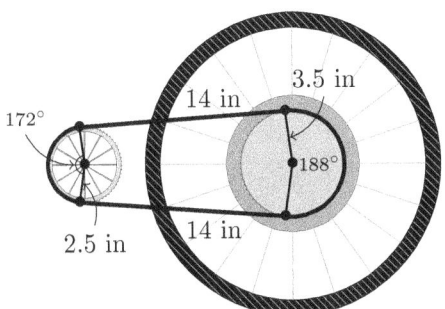

Example 3.6 The chain of a bicycle travels along the front and rear sprockets. To the nearest inch, how long is the chain?

Solution The first step is to find the amount of chain touching each sprocket. Let us convert the given degree measures into radians. Since the question is looking for an approximate answer, we round where needed:

$$172° \cdot \left(\frac{\pi}{180°}\right) \approx 3.0 \text{ rad} \quad \text{and} \quad 188° \cdot \left(\frac{\pi}{180°}\right) \approx 3.3 \text{ rad}.$$

It follows that that the lengths of the chain touching each sprocket are
$$2.5(3.0) = 7.5 \text{ in} \quad \text{and} \quad 3.5(3.3) \approx 11.6 \text{ in},$$
respectively.

Hence, the total length of the chain is about
$$7.5 + 14 + 11.6 + 14 \approx 47 \text{ in}$$
or about 3 feet 11 inches. ∎

3.4 Area of a Sector

Definition 3.4 A **sector of a circle** is the region of a circle bound between two radii and their intercepted arc.

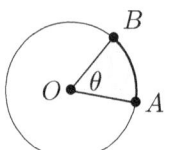

In the diagram to the left, sector ABO is bounded by the radii $\overline{AO}$ and $\overline{BO}$ as well as the arc AB. Arc AB subtends central angle θ. Our next proposition allows us to find the area of sectors like sector ABO.

Proposition 3.3 *Consider a circle of radius r, and let θ be the radian measure of the central angle. Suppose $0 \leq \theta \leq 2\pi$. Then the area of the sector bounded by angle θ and the arc which subtends θ is*

$$A = \frac{r^2 \theta}{2}.$$

Proof The area of the entire circle is πr^2, which corresponds to a radian measure of 2π. Using ratios, we have

$$\frac{A}{\theta} = \frac{\pi r^2}{2\pi} \quad \text{implies} \quad A = \frac{r^2 \theta}{2}.$$

■

Example 3.7 Consider a circle of diameter 10, and suppose central angle φ has measure $25°$. Find the area of the sector bounded by φ and the arc which subtends φ.

Solution We will use Proposition 3.3. Let us convert $25°$ to radians:

$$25° \left(\frac{\pi}{180°}\right) = \frac{5\pi}{36}.$$

The radius of the circle is $10/2 = 5$. So, the area of the sector bounded by φ and the arc which subtends φ is

$$\frac{5^2(5\pi/36)}{2} = \frac{125\pi/36}{2/1}$$
$$= \frac{125\pi}{36} \cdot \frac{1}{2}$$
$$= \frac{125\pi}{72}.$$

Example 3.8 Suppose the area of a sector is 144π, and the corresponding central angle has measure $30°$. What is the length of the circle's radius?

Solution The first step is to convert $30°$ to radians:
$$30° \cdot \left(\frac{\pi}{180°}\right) = \frac{\pi}{6}.$$

Hence, Proposition 3.3 tells us
$$\frac{r^2(\pi/6)}{2} = 144\pi \quad \text{implies} \quad \frac{\pi r^2}{12} = 144\pi.$$

A bit of algebra shows
$$r^2 = 1728,$$
which implies
$$r = \sqrt{1728}$$
$$= \sqrt{576 \cdot 3}$$
$$= \sqrt{576} \cdot \sqrt{3}$$
$$= 24\sqrt{3}.$$

■

3.5 Linear and Angular Velocity

Recall linear velocity from algebra class. When a point moves uniformly, its linear velocity is
$$v = \frac{d}{t},$$
where d is the distance the point travels in t units of time.

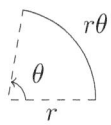

69

When the point moves along an arc of radius r, Proposition 3.2 tells us that the distance it travels is $d = r\theta$, where θ is the radian measure that the point rotates in t units of time. As a result, the linear velocity is
$$v = \frac{r\theta}{t}.$$

This formulation of velocity is suitable in some contexts. However, notice that an object which rotates equally as fast will not have the same linear velocity if its radius is different.

Let us elaborate on this further. Consider points A and B which lie on $\overrightarrow{OB}$. Further, suppose $\overrightarrow{OB}$ rotate about O.

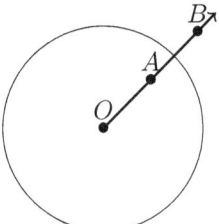

After a bit of thought, it is clear that the linear velocity of point A is less than that of B, because $AO < BO$. However, in some sense, both points are moving at the same rate. More specifically, since both points lie on $\overrightarrow{OB}$, the change in their angle measures over any time interval t is the same. Our next definition provides a vocabulary for this formulation of velocity.

Definition 3.5 When an object rotates about a point, the object's **angular velocity** is the rate of change of θ with respect to time, where θ is the central angle subtended by the initial and final position of the object.

This definition is too broad for our purposes. Our next proposition makes angular velocity more tractable.

Proposition 3.4 *Suppose an object rotates at a uniform rate of θ radians per t units of time. Then the angular velocity of the object is*
$$\omega = \frac{\theta}{t}.$$

Example 3.9 A point rotates $\pi/7$ radians in two minutes. Find its angular velocity.

Solution The angular velocity of the point is

$$\omega = \frac{\pi/7}{2} = \frac{\pi}{14} \text{ rad/min}.$$

■

Example 3.10 An object rotates at an angular velocity of $\pi/3$ radians per second. How many complete revolutions does the object travel in 100 seconds?

Solution We know

$$\omega = \frac{\theta}{t} \quad \text{implies} \quad \theta = t\omega.$$

It follows that the object rotates

$$\theta = 100 \left(\frac{\pi}{3}\right) = \frac{100\pi}{3} \text{ rad}$$

in 100 seconds.

All that is left is to convert to revolutions. There are 2π radians in one revolution. Therefore, the object travels

$$\frac{100\pi/3}{2\pi} = \frac{50}{3} = 16\frac{2}{3} \text{ rev}$$

in 100 seconds. We conclude that the object completes 16 revolutions in 100 seconds.

■

Example 3.11 Tim and Natalia play on a merry-go-round which has a radius of three meters. Tim sits half-way between the center of the merry-go-round and the outside edge, and Natalia sits at the edge. The merry-go-round makes a complete revolution every two seconds.

(a) Find Tim and Natalia's rotational velocities measured in radians per second.

(b) Compute their linear velocities.

71

Solution

(a) The rotational velocity is not affected by the position of each person on the merry-go-round. So, we only need to find one value. They rotate one complete revolution every two seconds, which means they both rotate $\theta = 2\pi$ radians every $t = 2$ seconds. Hence,

$$\omega = \frac{2\pi}{2} = \pi \text{ rad/sec.}$$

(b) Tim and Natalia's linear velocities are different because Tim is closer to the center of the merry-go-round than Natalia. The distance Tim travels per revolution is $2\pi(1.5) = 3\pi$, since he sits midway between the center and the outside edge. Hence, Tim's linear velocity is

$$v = \frac{3\pi}{2} \text{ m/sec.}$$

The distance Natalia travels per revolution is $2\pi(3) = 6\pi$, because she sits at the outside edge of the merry-go-round. Thus, Natalia's linear velocity is

$$v = \frac{6\pi}{2} = 3\pi \text{ m/sec.}$$

∎

Example 3.12 Consider the bicycle in Example 6. If the larger sprocket is rotating 100 revolutions per minute, what is the rotational velocity of the smaller sprocket?

Solution Since the two sprockets share the same belt, they have the same linear velocity. As a result, our strategy is to find the linear velocity of the larger sprocket and use it to find the rotational velocity of the smaller one.

The larger sprocket travels $2\pi(3.5) = 7\pi$ inches each revolution. Since it travels 100 revolutions in a minute, its linear velocity is

$$\frac{100(7\pi)}{1} = 700\pi \text{ in/min.}$$

The smaller sprocket travels 2.5θ inches per minute, where θ is the radian measure of the angle between the starting and final position of a point on the cycle. Therefore, the sprocket's linear velocity is $2.5\theta/1 = 2.5\theta$ inches per minute.

We are ready to find the smaller sprocket's angular velocity. Since the linear velocity of the two sprockets are equal, we have

$$2.5\theta = 700\pi \quad \text{implies} \quad \theta = 280\pi \text{ rad}.$$

Hence, the rotational velocity of the smaller sprocket is

$$\omega = \frac{280\pi}{1} = 280\pi \text{ rad/min}.$$

Since our question was formulated in terms of revolutions, we will convert our result into revolutions per minute. There are 2π radians in one revolution. Therefore, the rotational velocity is

$$\omega = \frac{280\pi \text{ rad}}{\text{min}} \cdot \frac{\text{rev}}{2\pi \text{ rad}} = 140 \text{ rev/min}.$$

■

3.6 Exercises

** Exercise 1

Find the degree measure of each directed angle.

(a) 3 clockwise revolutions

(b) 1/4 counterclockwise revolutions

(c) 3/4 clockwise revolutions

(d) 15/16 counterclockwise revolutions

(e) π clockwise revolutions

(f) 17/6 counterclockwise revolutions

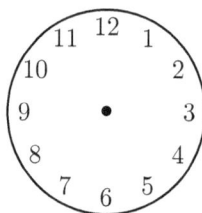

Figure 1

** Exercise 2

In Figure 1, the initial side of the directed angle θ is the minute hand, and the terminal side is the hour hand. If $0 \leq \theta < 360°$, find the measure of θ at the given times.

(a) 3:00

(b) 9:00

(c) 6:00

(d) 11:30

** Exercise 3

Consider Figure 1. Suppose the initial side of the directed angle φ is the minute hand, and the terminal side is the hour hand. Find the measure of φ at the given times. Suppose $-180° < \varphi \leq 180°$.

(a) 1:00

(b) 8:00

(c) 6:00

(d) 7:30

* Exercise 4

Convert to degrees.

(a) $\dfrac{\pi}{3}$

(b) $\dfrac{\pi}{4}$

(c) π

(d) $\dfrac{\pi}{6}$

(e) 2π

(f) $\dfrac{\pi}{2}$

* Exercise 5

Convert to radians. Write in terms of π.

(a) 30°

(b) 180°

(c) 60°

(d) 45°

(e) 360°

(f) 90°

* **Exercise 6**

Convert to degrees.

(a) $-\dfrac{6\pi}{5}$
(b) $\dfrac{17\pi}{10}$
(c) $-\dfrac{2\pi}{3}$
(d) $\dfrac{6\pi}{5}$
(e) $-\dfrac{9\pi}{5}$
(f) $\dfrac{7\pi}{10}$

* **Exercise 7**

Convert to radians. Write in terms of π.

(a) $337.5°$
(b) $-190°$
(c) $195°$
(d) $-285°$
(e) $150°$
(f) $-22.5°$

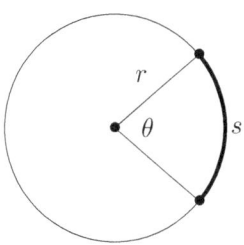

Figure 2

** **Exercise 8**

In Figure 2, find the missing variable using the given information. Write the measure of the central angle in degrees whenever θ is not given.

(a) $r = 10$ and $\theta = \dfrac{\pi}{4}$

(b) $r = 22$ and $\theta = 15°$

(c) $r = 12$ and $s = \dfrac{17\pi}{4}$

(d) $s = 15\pi$ and $\theta = 10°$

** **Exercise 9**

Janet is measuring the distance she travels by bicycle each day. She attached a device to her bicycle wheel which counts the number of revolutions her wheel makes. Janet's bicycle wheel has a diameter 74 cm. Find the distance Janet traveled supposing her device has the following reading at the end of the day. Round your answer to the nearest kilometer.

(a) 3441
(b) 2150
(c) 3392
(d) 1250

*** Exercise 10

Consider Figure 1. Suppose the minute hand is 6 cm long and the hour hand is 5 cm long.

(a) What is the distance the tip of the minute hand travels each minute?

(b) Find the distance the tip of the hour and travels in an hour.

(c) Over 15 minutes, how far does the tip of the minute hand travel?

(d) What is the distance the tip of the hour hand travels in 15 minutes?

(e) How far does the minute hand travel in 2 hours?

(f) Find the distance the tip of the hour hand travels in 45 minutes.

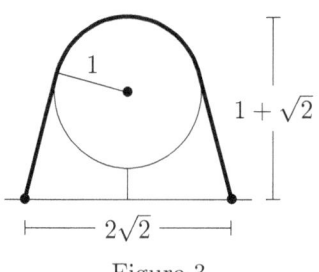

Figure 3

*** Exercise 11

Consider Figure 3. Calculate the length of the thick band. Hint: The radius of a circle is perpendicular to any tangent line at the point of tangency.

** Exercise 12

In Figure 2, suppose A is the area of the sector bounded by two radii and arc s. Of the variables A, r, and θ, two will be given. Find the third. Write the central angle in degrees whenever θ is not given.

(a) $r = 10$ and $\theta = \dfrac{\pi}{8}$

(b) $r = 15$ and $\theta = 315°$

(c) $r = 14$ and $A = 28\pi$

(d) $A = \dfrac{55\pi}{12}$ and $\theta = 66°$

(e) $A = \dfrac{121\pi}{3}$ and $\theta = 80°$

(f) $A = \dfrac{15\pi}{8}$ and $r = \dfrac{5\sqrt{3}}{2}$

** Exercise 13

Consider Figure 1. Suppose A is the area of the sector bounded by two radii and arc s. Prove that
$$A = \frac{rs}{2}.$$

** Exercise 14

Let A, r, and s represent the same variables as in Exercise 13.

Two of the three variables will be given, use the formula in Exercise 13 to find the third.

(a) $r = 2$ and $s = \dfrac{8\pi}{3}$

(b) $A = 18\pi$ and $r = 6$

(c) $A = \dfrac{49\pi}{2}$ and $s = 7\pi$

** Exercise 15

A small pizza has a diameter of 6" and a giant one has a diameter of 24". The small and giant pizzas are cut into 4 and 16 slices, respectively. The parents of twins want to give their kids an equal amount of pizza. If one of the twins is given a giant slice, how many small slices should the parents give to the other child? Assume the small and giant pizzas are equally thick.

the dashed line is equal to the area on the right.

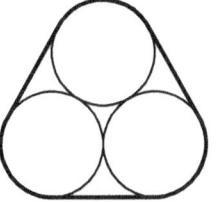

Figure 5

*** Exercise 17

In Figure 5, suppose each of the three small circles has a radius of length 10.

(a) Find the length of the thick outer band in the diagram.

(b) Calculate the gray area.

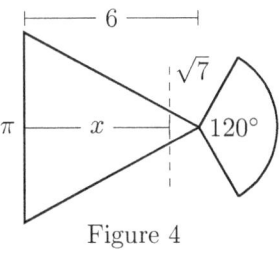

Figure 4

*** Exercise 16

Consider Figure 4. Find x so that the area on the left side of

** Exercise 18

Suppose ω is rotational velocity, θ is the angle of rotation, and t is the time the rotation takes. Two of the three variables will be given. Find the third.

(a) $t = 10$ sec and $\theta = \dfrac{5\pi}{6}$ rad

(b) $t = 4$ min and $\theta = 50°$

(c) $t = 50$ days and $\theta = 215$ rev

(d) $\theta = \dfrac{2\pi}{3}$ rad and $\omega = \dfrac{17}{6}$ rev/sec

(e) $\theta = 200°$ and $\omega = 7$ rev/min

(f) $t = \dfrac{1}{2}$ sec and $\omega = 2\pi$ rad/sec

** Exercise 19

Suppose an object orbits a point in a circular path of radius r. Let v be the linear velocity, and ω be the angular velocity of the object. Suppose ω is measured in radians per unit of time. Prove

$$v = r\omega.$$

** Exercise 20

Let r, v, and ω be as they are defined in Exercise 19. Two of the three variables will be given. Use the formula in Exercise 19 to find the third.

(a) $r = 10$ ft and $\omega = \dfrac{\pi}{4}$ rad/min

(b) $v = 300000$ km/sec and $\omega = 4\pi$ rad/sec

(c) $r = 10$ cm and $v = 100$ cm/hour

(d) $r = \pi$ in and $\omega = 20$ rad/year

** Exercise 21

Use the formula in Exercise 19 to find an easier way to solve Example 12.

** Exercise 22

Consider Figure 1. Suppose the minute hand is six centimeters long and the hour hand is five centimeters long. Compute ...

(a) ... the rotational and linear velocity of the minute hand.

(b) ... the rotational and linear velocity of the hour hand.

Write the rotational velocity in radians per minute and the linear velocity in centimeters per minute.

** Exercise 23

Kate and Dave play on a merry-go-round which has a twelve meter diameter. Kate sits two meters from the center and Dave sits on the outside edge. The merry-go-round makes a complete revolution every fifteen seconds. Calculate ...

(a) ... Kate's rotational velocity in radians per second.

(b) ... Dave's rotational velocity in radians per second.

(c) ... Kate's linear velocity in meters per second.

(d) ... Dave's linear velocity in meters per second.

Planet	Distance to sun (10^6 km)	Earth days/rev
Mercury	58	88
Venus	108	225
Earth	150	365
Mars	228	687
Jupiter	778	4330
Saturn	1470	10751
Uranus	2871	20686
Neptune	4497	60156

* Exercise 24

Consider the table above. Use a calculator to compute the rotational and linear velocity of each planet. Assume the orbits are perfect circles.

** Exercise 25

A belt attached to a pulley moves at a rate of $5\pi/4$ meters per minute. The belt rotates the pulley two revolutions per minute. Compute the diameter of the pulley.

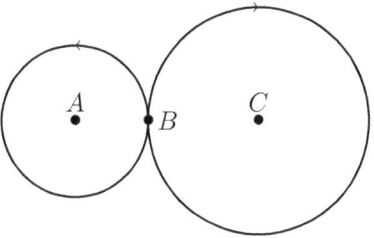

Figure 6

** Exercise 26

In Figure 6, circle A and circle C are tangent at B. Suppose circle A rotates counterclockwise at a rate of three revolutions per minute, and circle C rotate clockwise at a rate of two revolutions per minute. Calculate

$$\frac{AB}{BC}.$$

*** Exercise 27

Consider Figure 7. Suppose the rear wheel of the bicycle has diameter 24", the pedals have a radius 9", the rear sprocket has a radius of $2\frac{1}{2}$" and the sprocket attached to the pedals has a radius of 6". The bicycle travels 100'. Use a calculate to find the following.

(a) How many times did the rear wheel rotate?

(b) How many times did the pedals rotate?

Figure 7

Chapter 4

Right Triangle Trigonometry

In this chapter, we will study the trigonometry of right triangles. For pedagogic reasons, a definition of the trigonometric functions will be delayed until Chapter 5. We assume knowledge of Chapters 1, 2, and 3. Readers unfamiliar with square roots may find Appendix B helpful. Because trigonometric functions require either a calculator or a table to be evaluated at most angle measures, it is assumed that the reader has access to a scientific calculator and knowledge of its basic functionality.

4.1 Introduction to Trigonometric Functions

The three trigonometric functions that we will study in this chapter are sine, cosine, and tangent. Their values at $\angle A$ are denoted by

$$\sin A, \quad \cos A, \quad \text{and} \quad \tan A,$$

respectively.

For the time being, we will rely on our calculators, instead of a definition, for the correspondence between angle measures and the outputs of each trigonometric function.

Example 4.1 Use a calculate to find each of the following. Round to three decimal places.

(a) sin 72°

(b) cos 22°

(c) tan 27°

(d) 3 sin 10°

Solution Make sure your calculate is set to degree mode when you evaluate.

(a) sin 72° ≈ 0.951

(b) cos 22° ≈ 0.927

(c) tan 27° ≈ 0.510

(d) 3 sin 10° ≈ 0.521

■

Our next theorem, Theorem 4.1, is very powerful because it relates the values obtained from evaluation of trigonometric functions to ratios of a right triangle's side lengths. As a result of Theorem 4.1, when given a side length and an angle measure of a right triangle, we can find the other two side lengths using only a bit of algebra.

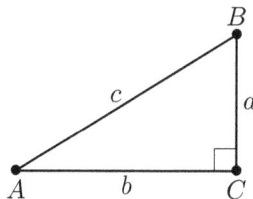

Theorem 4.1 *For any $\triangle ABC$ where $\angle C$ is right, the following trigonometric ratios hold.*

$$\sin A = \frac{a}{c} \quad and \quad \sin B = \frac{b}{c}$$

$$\cos A = \frac{b}{c} \quad\quad\quad\quad \cos B = \frac{a}{c}$$

$$\tan A = \frac{a}{b} \quad\quad\quad\quad \tan B = \frac{b}{a}.$$

Many students remember these relationships by means of a mnemonic device such as the following.

SohCahToa

Sine is **o**pposite over **h**ypotenuse.
Cosine is **a**djacent over **h**ypotenuse.
Tangent is **o**pposite over **a**djacent.

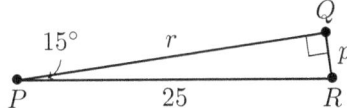

Example 4.2 Find the lengths of the remaining sides of the triangle.

Solution Let us start by finding p. We know the length of the hypotenuse and we want to find the length of the leg opposite $\angle P$. Theorem 4.1 tells us sine relates these sides. In particular, it says

$$\sin 15° = \frac{p}{25} \quad \text{implies} \quad p = 25 \sin 15° \approx 6.470.$$

Let us find r. We know the hypotenuse and we want to find the length of the leg adjacent to $\angle P$. According to Theorem 4.1, cosine relates these sides. Specifically,

$$\cos 15° = \frac{r}{25} \quad \text{implies} \quad r = 25 \cos 15° \approx 24.148.$$

∎

Notice that we could have used tangent to find r, but that would require the value of p which we had previously found. This is a mathematically legitimate strategy. However, using p to find r would result in a less accurate value of r, due to rounding error. We recommend using the given information as much as possible.

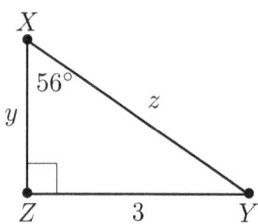

Example 4.3 Find y and z in the diagram.

Solution Let us find y. We know the length of the leg opposite $\angle X$, and we want to find the length of the leg adjacent $\angle X$. The trigonometric function which relates these sides is tangent. We have
$$\tan 56° = \frac{3}{y} \quad \text{implies} \quad y = \frac{3}{\tan 56°} \approx 2.024.$$

All that is left is to find z. We know the length of the side opposite $\angle X$, and we want to find the hypotenuse. The trigonometric function which relates these sides is sine. In particular,
$$\sin 56° = \frac{3}{z} \quad \text{implies} \quad z = \frac{3}{\sin 56°} \approx 3.619.$$ ∎

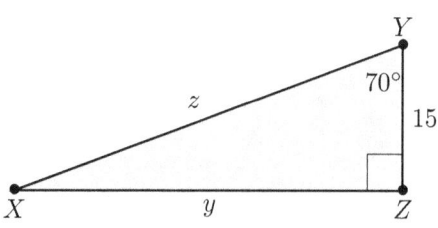

Example 4.4 Find (a) the area and (b) the perimeter of $\triangle XYZ$.

Solution

84

(a) Recall that the area of a triangle is

$$\frac{1}{2}bh,$$

where b is the base and h is the height. If we treat x as the length of the height, then y is the length of the base. As result, our task is to find y. Notice

$$\tan 70° = \frac{y}{15} \quad \text{implies} \quad y = 15\tan 70°.$$

It follows that the area of $\triangle XYZ$ is

$$\frac{1}{2}(15\tan 70°)(15) = \frac{225\tan 70°}{2} \approx 309.091.$$

(b) We already know $y = 15\tan 70°$, so all that we need is z. We have

$$\cos 70° = \frac{15}{z} \quad \text{implies} \quad z = \frac{15}{\cos 70°}.$$

Thus, the perimeter of $\triangle XYZ$ is

$$x + y + z = 15 + 15\tan 70° + \frac{15}{\cos 70°} \approx 100.069.$$

∎

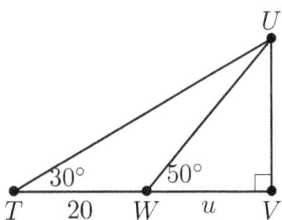

Example 4.5 What is the value of u?

Solution The idea is to use Theorem 4.1 on $\triangle UVW$ and $\triangle TUV$. The theorem allows us to write UV in terms of u two ways, which lets us establish an equation. Then we can solve for u.

In $\triangle UVW$, the length UV is opposite $\angle UWV$, and $VW = u$ is adjacent $\angle UWV$. The trigonometric function which relates these sides is tangent. Therefore,

$$\tan 50° = \frac{UV}{u} \quad \text{implies} \quad UV = u \tan 50°.$$

Consider $\triangle TUV$. We know UV is opposite $\angle T$, and $TV = u + 20$ is adjacent $\angle T$. Once again, tangent relates these sides. It follows that

$$\tan 30° = \frac{UV}{u+2} \quad \text{implies} \quad UV = (u+20) \tan 30°.$$

Since $UV = UV$, we have

$$\begin{aligned}
&& u \tan 50° &= (u+20) \tan 30° \\
\Rightarrow && u \tan 50° &= u \tan 30° + 20 \tan 30° \\
\Rightarrow && u \tan 50° - u \tan 30° &= 20 \tan 30° \\
\Rightarrow && u(\tan 50° - \tan 30°) &= 20 \tan 30° \\
\Rightarrow && u &= \frac{20 \tan 30°}{\tan 50° - \tan 30°} \\
&& &\approx 18.794.
\end{aligned}$$

∎

Because there is a correspondence between radian measures and degree measures, evaluation of trigonometric functions at radian measures makes sense. The value of a trigonometric function at a radian measure is simply equal to the trigonometric function evaluated at the corresponding degree measure.

As a practical matter, evaluation of trigonometric functions at radian measures is simply a matter of adjusting your calculator to the correct mode.

Example 4.6 Evaluate.

(a) $\sin \dfrac{\pi}{5}$

(b) $12 \cos \dfrac{3\pi}{7}$

(c) $5 \tan \dfrac{\pi}{11}$

(d) $\sin 1$

Solution

(a) $\sin\dfrac{\pi}{5} \approx 0.588$

(b) $12\cos\dfrac{3\pi}{7} \approx 2.670$

(c) $5\tan\dfrac{\pi}{11} \approx 1.468$

(d) $\sin 1 \approx 0.841$

■

In Example 6 (d), there was no π in the expression. However, we know that we are evaluating a radian measure because there is no degree symbol.

In general, we advise readers to be mindful of the units. Using the incorrect setting changes the output. For example,

$$\tan 15° \approx 0.268 \quad \text{and} \quad \tan 15 \approx -0.856.$$

The first result is tangent evaluated at the degree measure 15°, and the second is tangent evaluated at the radian measure 15. The latter corresponds to a degree measure of about 859.437°.

4.1.1 Trigonometric Functions and Special Right Triangles

We can use special right triangles to find the exact values of our three trigonometric functions evaluated at 30°, 45°, and 60°.

Proposition 4.1

θ	30°	45°	60°
$\sin\theta$	$\dfrac{1}{2}$	$\dfrac{\sqrt{2}}{2}$	$\dfrac{\sqrt{3}}{2}$
$\cos\theta$	$\dfrac{\sqrt{3}}{2}$	$\dfrac{\sqrt{2}}{2}$	$\dfrac{1}{2}$
$\tan\theta$	$\dfrac{\sqrt{3}}{3}$	1	$\sqrt{3}$

Proof We will prove the first column, and leave the rest as exercises. Consider the 30°-60°-90° special right triangle.

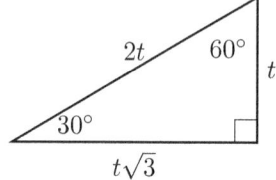

Then

$$\sin 30° = \frac{t}{2t}, \quad \cos 30° = \frac{t\sqrt{3}}{2t}, \quad \text{and} \quad \tan 30° = \frac{t}{t\sqrt{3}}$$

$$= \frac{1}{2} \qquad\qquad = \frac{\sqrt{3}}{2} \qquad\qquad = \frac{1}{\sqrt{3}} \cdot \frac{\sqrt{3}}{\sqrt{3}}$$

$$\qquad\qquad\qquad\qquad\qquad\qquad\qquad\qquad\qquad = \frac{\sqrt{3}}{3}.$$

∎

This leads us to make a corresponding radian version of Proposition 4.1.

Proposition 4.2

θ	$\dfrac{\pi}{6}$	$\dfrac{\pi}{4}$	$\dfrac{\pi}{3}$
$\sin \theta$	$\dfrac{1}{2}$	$\dfrac{\sqrt{2}}{2}$	$\dfrac{\sqrt{3}}{2}$
$\cos \theta$	$\dfrac{\sqrt{3}}{2}$	$\dfrac{\sqrt{2}}{2}$	$\dfrac{1}{2}$
$\tan \theta$	$\dfrac{\sqrt{3}}{3}$	1	$\sqrt{3}$

Example 4.7 Evaluate.

(a) $\tan 45°$

(b) $\sin \dfrac{\pi}{3}$

(c) $\cos 60°$

(d) $\tan \dfrac{\pi}{6}$

Solution This is an application of Proposition 4.1 and Proposition 4.2.

(a) $\tan 45° = 1$

(b) $\sin \dfrac{\pi}{3} = \dfrac{\sqrt{3}}{2}$

(c) $\cos 60° = \dfrac{1}{2}$

(d) $\tan \dfrac{\pi}{6} = \dfrac{\sqrt{3}}{3}$

∎

4.2 Inverse Trigonometric Functions

This section is an introduce to inverse trigonometric functions. In particular, we introduce arc sine, arc cosine, and arc tangent. To denote these functions evaluated at x, we write

$$\arcsin x, \quad \arccos x, \quad \text{and} \quad \arctan x,$$

respectively.

We will introduce a partial definition of arc sine, arc cosine, and arc tangent here. A more robust definition is contained in Chapter 8.

Definition 4.1 Suppose that $0 \leq \theta \leq 90°$.

- The **arc sine** function $f(x) = \arcsin x$ is defined by the relationship
$$\arcsin x = \theta \quad \text{if} \quad \sin \theta = x$$
for $-1 \leq x \leq 1$.

- The **arc cosine** function $f(x) = \arccos x$ is defined by the relationship
$$\arccos x = \theta \quad \text{if} \quad \cos \theta = x$$
for $-1 \leq x \leq 1$.

- The **arc tangent** function $f(x) = \arctan x$ is defined by the relationship
$$\arctan x = \theta \quad \text{if} \quad \tan \theta = x$$
for $x \geq 0$.

There is alternative notation for these functions. In particular,
$$\sin^{-1} x = \arcsin x, \quad \cos^{-1} x = \arccos x, \quad \text{and} \quad \tan^{-1} x = \arctan x.$$

We prefer the "arc" notation, because it is less confusing. The -1 exponent in the alternative notation is sometimes incorrectly interpreted as the reciprocal of the trigonometric function. For example, sometimes students confuse $\sin^{-1} x$ and $1/\sin x$.

Example 4.8 Find the degree measure of each value.
(a) $\arcsin \dfrac{1}{2}$ (b) $\arccos 5$ (c) $\arctan 2$ (d) $\cos^{-1} \dfrac{2}{3}$

Solution

(a) $\arcsin \dfrac{1}{2} = 30°$

(b) $\arccos 5$ is undefined because 5 is not in the interval $[-1, 1]$.

(c) $\arctan 2 \approx 63.435°$

(d) $\cos^{-1} \dfrac{2}{3} \approx 48.190°$

■

Example 4.9 What is the radian measure of each value?
(a) $\sin^{-1} 7$ (b) $\arccos \dfrac{2}{5}$ (c) $\tan^{-1} \sqrt{3}$ (d) $\arcsin \dfrac{6}{11}$

Solution

(a) $\sin^{-1} 7$ is undefined because 7 is no in the interval $[-1, 1]$.

(b) $\arccos \dfrac{2}{5} \approx 1.159$ rad

(c) $\tan^{-1} \sqrt{3} \approx 1.047$ rad

(d) $\arcsin \dfrac{6}{11} \approx 0.577$ rad

Example 4.10 Suppose $\sin x = 0.7$, $\cos y = 8/11$, and $\tan z = 3$. Find the degree measure of (a) x, (b) y, and (c) z. Assume the angle measures are acute.

Solution We will utilize Definition 4.1 for (a), (b), and (c).

(a) Because $\sin x = 0.7$, we have $x = \arcsin 0.7 \approx 44.427°$.

(b) Due to the fact that $\cos y = 8/11$, the definition of arc cosine tells us $y = \arccos(8/11) \approx 43.342°$.

(c) Since $\tan z = 3$, we conclude $z = \arctan 3 \approx 71.565°$.

■

Problems like those in Example 10 could require radian measures as well. However, the only change in procedure is to convert your calculator to radian mode.

Example 4.11 Let $\cos A = 5/7$. Find the radian measure of $\angle A$.

Solution We make sure to change our calculator to radian mode. Then

$$\cos A = \frac{5}{7} \quad \text{implies} \quad m\angle A = \arccos \frac{5}{7} \approx 0.775 \text{ rad}.$$

■

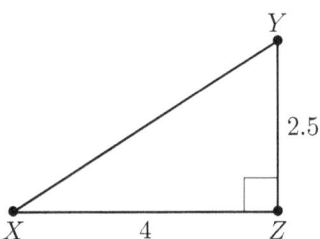

Example 4.12 Find the degree measures of $\angle X$ and $\angle Y$.

Solution Theorem 4.1 and Definition 4.1 tell us that

$$\tan X = \frac{2.5}{4} \quad \text{implies} \quad m\angle X = \arctan \frac{2.5}{4} \approx 32.005°$$

and

$$\tan Y = \frac{4}{2.5} \quad \text{implies} \quad m\angle Y = \arctan \frac{4}{2.5} \approx 57.995°.$$

∎

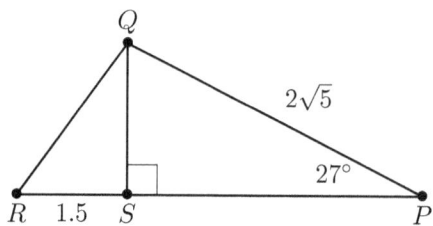

Example 4.13 Find the degree measure of $\angle R$.

Solution Once we find QS, we can use arc tangent to find $m\angle R$. We know
$$QS = 2\sqrt{5}\sin 27° \approx 2.030.$$
It follows that
$$\tan R = \frac{2.030}{1.5} \quad \text{implies} \quad m\angle R \approx \arctan \frac{2.030}{1.5} \approx 53.543°.$$

∎

4.3 Angles of Elevation and Depression

Definition 4.2

- An **angle of depression** is formed by a horizontal ray and another ray below it.
- An **angle of elevation** is formed by a horizontal ray and another ray above it.

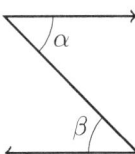

In the diagram on page 93, α is an angle of depression and β is an angle of elevation.

Example 4.14 A man is standing 500 feet away from a tall building. If the angle of elevation to the top of the build from his perspective is $63°$, how tall is the building?

Solution The first step is to draw a diagram. We assume that the building is perpendicular to the horizontal; this assumption is standard practice when solving this type of problem. We neglect the height of the man, because the question did not provide it.

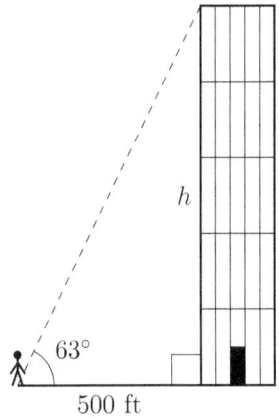

Let h be the height of the building. Then

$$\tan 63° = \frac{h}{500} \quad \text{implies} \quad h = 500 \tan 63° \approx 981.305.$$

Hence, the building is about 981 feet tall. ∎

Example 4.15 A canoe is tethered to the floor of a dock by a rope of length 1.5 meters. The canoe is 1 meter below the floor of the dock. Calculate the angle of depression of the rope.

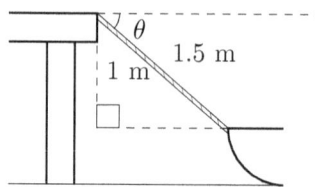

Solution Proposition 1.2 tells us that alternate interior angles are congruent. So, the angle opposite the 1-meter side in the triangle above has measure θ. Hence,

$$\sin\theta = \frac{1}{1.5} \quad \text{implies} \quad \theta = \arcsin\frac{1}{1.5} \approx 41.810°.$$

Thus, the angle of depression is about 41.810°. ∎

Example 4.16 Tom, Jim, and Gwen are hiking in a steep canyon 195 meters deep. Tom is just starting, Jim has been hiking for awhile but has not reached the bottom, and Gwen is already at the bottom. Tom looks down at an angle of depression of 45° to see Jim. Gwen looks up to see Jim at an angle of elevation of 58°. What is Jim's vertical distance from Tom, if Tom and Gwen are a horizontal distance of 150 meters apart and Jim is horizontally between Gwen and Tom?

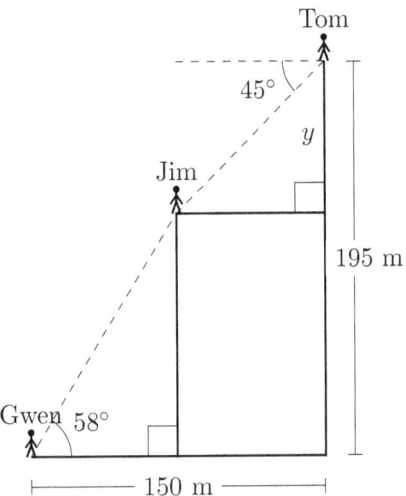

Solution To find the vertical distance between Jim and Tom y, we will set up two right triangles which contain y. Then we can use trigonometry and algebra to find y.

The first right triangle we consider is the one which has Jim and Tom each standing at a vertex. Using the fact that alternate interior angles are congruent and the $45° - 45° - 90°$ special right triangle, we know the horizontal distance between Tom and Jim is also y.

The second triangle we consider is the one which has Gwen and Jim each standing at a vertex. It is clear from our previous work that the horizontal distance between Jim and Gwen is $150 - y$ and the vertical distance between Jim and Gwen is $195 - y$. This gives us the following triangle.

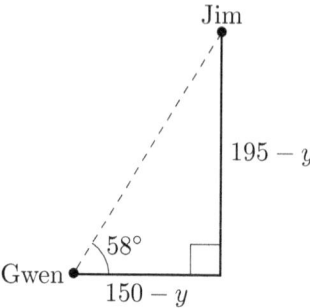

All that is left is a computation:

$$\begin{aligned}
\tan 58° &= \frac{195-y}{150-y} \\
\Rightarrow \quad (150-y)\tan 58° &= 195-y \\
\Rightarrow \quad 150\tan 58° - y\tan 58° &= 195-y \\
\Rightarrow \quad y - y\tan 58° &= 195 - 150\tan 58° \\
\Rightarrow \quad y(1-\tan 58°) &= 195 - 150\tan 58° \\
\Rightarrow \quad y &= \frac{195 - 150\tan 58°}{1-\tan 58°} \\
&\approx 75.042.
\end{aligned}$$

We conclude that the vertical distance between Jim and Tom is about 75 meters. ■

4.4 Exercises

* Exercise 1

Evaluate. Round to three decimal places.

(a) $\tan 47°$ (d) $5\cos 42°$
(b) $\sin 17°$ (e) $6\tan 1°$
(c) $\sin 83°$ (f) $4\cos 15°$

* Exercise 2

Evaluate. Round to three decimal places.

(a) $\sin \dfrac{8\pi}{9}$ (d) $\tan 1.5$
(b) $\tan \dfrac{\pi}{7}$ (e) $2\sin \dfrac{7\pi}{9}$
(c) $4\cos \dfrac{3\pi}{10}$ (f) $5\cos 0.2$

* Exercise 3

Consider Figure 1. Suppose $a = 3$, $b = 4$, and $c = 5$. Find the following.

(a) $\cos B$ (d) $\sin A$
(b) $\tan A$ (e) $\tan B$
(c) $\sin B$ (f) $\cos A$

* Exercise 4

Repeat Exercise 3, but assume $a = 2$, $b = 3$, and $c = \sqrt{13}$.

* Exercise 5

In Figure 1, say $m\angle A = 34°$. Use a calculator and the given side length to find the lengths of the remaining two sides.

(a) $a = 7$ (c) $c = 20$
(b) $b = 11$ (d) $a = \sqrt{7}$

* Exercise 6

Consider Figure 1. Suppose $m\angle B = 50°$. Use a calculator and the given side length to find the lengths of the remaining two sides.

(a) $a = 18$ (c) $c = 102$
(b) $b = 5$ (d) $b = \dfrac{7}{5}$

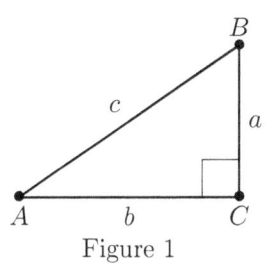

Figure 1

** Exercise 7

Use Figure 1 and the given information to find (i) the area and (ii) the perimeter of of $\triangle ABC$.

(a) Suppose $m\angle A = 25°$ and $c = 19$.

(b) Assume $m\angle B = 40°$ and $a = 25$.

(c) Say $m\angle A = 35°$ and $a = 10$.

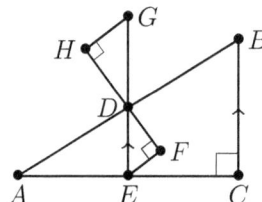

Figure 2

** Exercise 8

In Figure 2, assume $m\angle A = 40°$, $DE = 2$, $DF = DH$, and $\overline{BC}$ is parallel to $\overline{EG}$. Find each of the following.

(a) DF
(b) EF
(c) GH
(d) DG

** Exercise 9

Consider Figure 2. Let $m\angle B = 70°$, $GH = 5$, $DF = \frac{3}{4}DH$, and $\overline{BC}$ is parallel to $\overline{EG}$. What are each of the following?

(a) DG
(b) DH
(c) EF
(d) DE

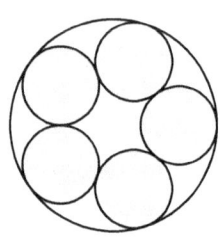

Figure 3

*** Exercise 10

In Figure 3, suppose the radius of the large circle is 12, and the five gray circles have the same radius. Find the total area of the gray circles. Hint: Tangent lines of circles are perpendicular to radii at points of tangency.

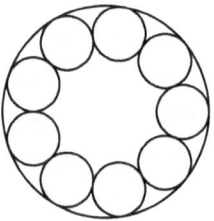

Figure 4

*** Exercise 11

Consider Figure 4. Say the radius of the large circle is 10, and the radii of the nine gray circles are equal. Calculate the ratio of the gray area to the white area. Hint: Tangent lines of circles are perpendicular to radii at points of tangency.

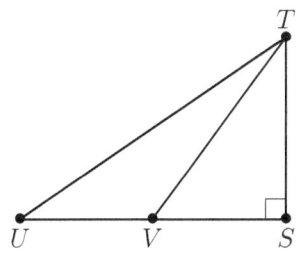

Figure 5

* Exercise 12

Find the exact value.

(a) $\sin 60°$
(b) $\cos 45°$
(c) $\tan 30°$
(d) $\cos 30°$
(e) $\sin 30°$
(f) $\tan 60°$

* Exercise 13

What is the exact value?

(a) $\sin \dfrac{\pi}{6}$
(b) $\tan \dfrac{\pi}{4}$
(c) $\cos \dfrac{\pi}{6}$
(d) $\tan \dfrac{\pi}{3}$
(e) $\cos \dfrac{\pi}{4}$
(f) $\sin \dfrac{\pi}{4}$

** Exercise 14

Prove the second and third column of Proposition 4.1.

*** Exercise 15

In Figure 5, let $m\angle U = 40°$ and $m\angle SVT = 63°$.

(a) What is UV if $SV = 5$?

(b) Say $ST = 12$. Find UV.

(c) Assume $TU = 21$. What is TV?

(d) Find ST supposing $UV = 8$.

*** Exercise 16

Use Figure 5 and suppose $m\angle U = 35°$ and $m\angle STV = 75°$.

(a) Suppose $UV = 12$. Then TU equals what?

(b) Find UV if $SV = 12$.

(c) Say $UV = 2$. What is TV?

(d) Assume the area of $\triangle TUV$ is 300. What is SV?

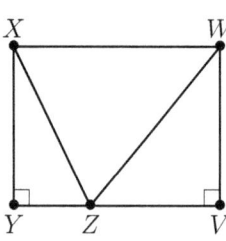

Figure 6

*** Exercise 17

Consider Figure 6. Suppose $m\angle VWZ = 40°$ and $m\angle YXZ = 27°$

(a) Assume $WX = 15$. Then XZ is equal to what?

(b) Find VY when $XY = 10$.

(c) Let the area of $\triangle VWZ$ be 33.984. What is the area of $\triangle XYZ$?

(d) Suppose the area of rectangle $VWXY$ is 113.962. Calculate the length WZ.

*** Exercise 18

Use Figure 6 to solve the following problems.

(a) Say $WX = 50$, $m\angle WXZ = 61°$, and $m\angle VZW = 25°$. Find XZ.

(b) Assume $XY = 18$, $m\angle VWZ = 42°$, and $m\angle YXZ = 29°$. What is the value of VY?

(c) Let $WX = 80$, $m\angle YXZ = 28°$, and $m\angle VWZ = 41°$. Then the length XY is what?

(d) Suppose the area of $\triangle XYZ$ is 4.807. If $m\angle YXZ = 31°$ and $m\angle XWZ = 44°$, calculate the area of $\triangle WXZ$.

* **Exercise 21**

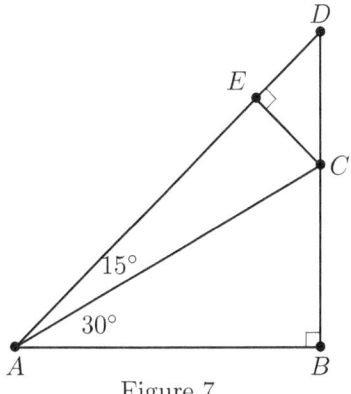

Figure 7

What is the radian measure? Some expressions are undefined.

(a) $\sin^{-1} \dfrac{\sqrt{5}}{3}$ (d) $\arcsin \dfrac{3}{\sqrt{3}}$

(b) $\arccos 15$ (e) $\cos^{-1} 0.7$

(c) $\tan^{-1} \dfrac{1}{2}$ (f) $\arctan 15$

*** **Exercise 19**

With the aid of Figure 7, find the exact values of the following.

(a) $\sin 15°$

(b) $\cos 15°$

(c) $\tan 15°$

Hint: Notice that $m\angle BAD = 45°$ and $m\angle D = 45°$ and use special right triangles.

* **Exercise 20**

Find the degree measure. Some expressions are undefined.

(a) $\arcsin 0.2$ (d) $\sin^{-1} 4$

(b) $\cos^{-1} \sqrt{\dfrac{1}{7}}$ (e) $\arccos 0.25$

(c) $\arctan 5$ (f) $\tan^{-1} \dfrac{1}{3}$

* **Exercise 22**

Find the *degree* measure of α. Assume that α is an acute angle.

(a) $\sin \alpha = \dfrac{3}{7}$

(b) $\cos \alpha = 0.2$

(c) $\tan \alpha = \dfrac{17}{13}$

(d) $\sin \alpha = 0.5$

(e) $\cos \alpha = \dfrac{\sqrt{2}}{2}$

(f) $\tan \alpha = \sqrt{3}$

* Exercise 23

Compute the *radian* measure of β. Assume that β is an acute angle.

(a) $\sin\beta = \dfrac{\sqrt{2}}{5}$

(b) $\cos\beta = \dfrac{5}{3\sqrt{7}}$

(c) $\tan\beta = 0.78$

(d) $\sin\beta = \dfrac{\sqrt{2}}{2}$

(e) $\cos\beta = \dfrac{\sqrt{3}}{2}$

(f) $\tan\beta = \dfrac{\sqrt{3}}{3}$

* Exercise 24

Use Figure 1 and the given information to find the degree measure of $\angle A$.

(a) $b = 10$ and $c = 11$

(b) $a = 3$ and $b = 5$

(c) $a = 1$ and $c = 7$

(d) $b = 5$ and $c = 9$

* Exercise 25

Consider Figure 1. Find the radian measure of $\angle B$, using the given information.

(a) $a = 2$ and $c = 5$

(b) $a = 7$ and $b = 15$

(c) $b = 5$ and $c = 10$

(d) $a = 2$ and $b = 3$

* Exercise 26

Use Propositions 4.1 and 4.2 to fill in the table with the appropriate (a) degree measures and (b) radian measures.

x	$\dfrac{1}{2}$	$\dfrac{\sqrt{2}}{2}$	$\dfrac{\sqrt{3}}{2}$
$\arcsin x$			
$\arccos x$			

* Exercise 27

Use Propositions 4.1 and 4.2 to find the exact solution to each equation. Write your answer (a) using degrees and (b) using radians.

- $\tan\alpha = 1$,
- $\tan\beta = \sqrt{3}$, and
- $\tan\gamma = \dfrac{\sqrt{3}}{3}$.

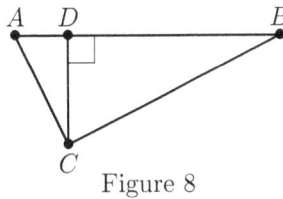

Figure 8

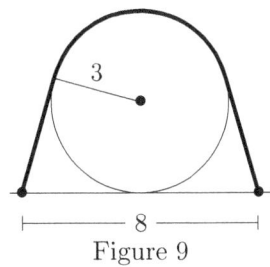

Figure 9

*** **Exercise 28**

Use the given information to find the degree measure of the angle in Figure 8.

(a) Suppose $m\angle B = 27°$, $BD = 4$, and $AC = \sqrt{5}$. Find $m\angle A$.

(b) Let $m\angle BCD = 55°$, $BC = 6\sqrt{10}$, and $AD = 3$. What is $m\angle ACD$?

(c) If $AB = 10$, $m\angle A = 61°$, what is $m\angle B$?

(d) Assume $m\angle A = 25°$, $BC = 4.5$, and $AC = 2\sqrt{5}$. Find $m\angle B$.

(e) Say $m\angle B = 45°$, $AB = 7$, and $AC = 5$. What is $m\angle A$, if its measure is greater than $45°$?

*** **Exercise 29**

Consider Figure 9. Calculate the length of the thick band. Hint: Tangent lines of circles are perpendicular to radii at points of tangency.

** **Exercise 30**

An observer sees a helicopter that is a horizontal distance of 1 mile from her location. The angle of elevation from the observer's perspective to the helicopter is $10°$.

(a) What is the distance between the helicopter and the ground?

(b) Find the distance between the observer and the helicopter.

** Exercise 31

A yogi bends at the waist and has a straight back and neck. Suppose his back is bent to an angle of depression of 60°, and the yogi is 168 cm tall. The yogi's back and head comprise 50% of his total height.

(a) What is the distance between the tip of the yogi's head and his legs?

(b) How far is the tip of the yogi's head from the ground?

** Exercise 32

An airplane ascends from the ground at an angle of elevation of 25°. Suppose there are 200 feet of runway left when the plane lifts off.

(a) Calculate the distance the plane has traveled since lift off when it is at the end of the runway.

(b) How far is the plane from the ground when it is at the end of the runway?

** Exercise 33

A daredevil is constructing a ramp for a stunt. She designs the ramp to have a horizontal length of 5 feet and a vertical height of 3 feet.

(a) Find the slant length of the ramp.

(b) What is the angle of elevation of the ramp?

** Exercise 34

A young boy is inspecting a slide. The vertical ladder to the top of the slide is 4 feet. The boy estimates that the angle of depression from the top of the ladder to the bottom of the slide is 35°.

(a) Calculate the length of the slide.

(b) How far is the bottom of the ladder from the bottom of the slide?

(c) The boy observes that it takes about 0.729 seconds for a colleague to slide down. What was the colleague's average speed?

** Exercise 35

A straight 10-mile road goes up a hill. The change in elevation between the top and bottom of the hill is 0.5 miles. Calculate the angle of elevation of the road.

** Exercise 36

A tire swing hangs from a rope of length 10 feet. The rope is tied to a branch 12 feet from the ground. A girl pulls the tire back so that it is five feet from the ground. What is the angle of depression from the branch to the tire? Assume the rope is taut.

*** Exercise 37

An observer standing on a 500-meter building looks down at another building. The angle of depression from the observer to the top of the other building is $20°$ and the angle of depression to the bottom of the building is $50°$.

(a) Calculate the height of the other building.

(b) Find the distance between the buildings.

*** Exercise 38

Heather and Sabrina are 300 feet apart and are on opposite sides of a tall tree. The angle of elevation of Heather to the top of the tree is $20°$ and the angle of elevation of Sabrina to the tree is $30°$. How tall is the tree? Assume both women are 5.5 feet tall.

*** Exercise 39

The angle of elevation between Vaibhav and a mountain is $35°$. He walks 200 meters closer to the mountain and his angle of elevation is $45°$. Calculate the height of the mountain.

Chapter 5

Trigonometry of General Angles

This chapter defines the trigonometric functions we studied in Chapter 4. These definitions allow us to evaluate our trigonometric functions at angles of arbitrary measure, albeit only exactly at a limited number of values. We will introduce three more trigonometric functions as well: secant, cosecant, and cotangent. After our definitions are provided, the rest of the chapter will study properties of trigonometric functions. Inverse trigonometric functions will be discussed in Chapter 8. Students will not need a calculator.

5.1 The Six Trigonometric Functions

The xy-coordinate system develops a correspondence between points and ordered pairs (x, y).

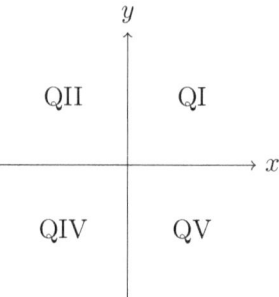

- Quadrant I (QI) is the set of points in the xy-plane such that $x > 0$ and $y > 0$.

- Quadrant II (QII) is the set of points in the xy-plane such that $x < 0$ and $y > 0$.

- Quadrant III (QIII) is the set of points in the xy-plane such that $x < 0$ and $y < 0$.

- Quadrant IV (QIV) is the set of points in the xy-plane such that $x > 0$ and $y < 0$.

Points on either of the two axes are not in any quadrant. For example, the point $(5, 0)$ does not belong to a quadrant; it is on the x-axis. Similarly, the point $(0, -2)$ is not in a quadrant; it is on the y-axis.

Example 5.1 State the quadrant or axis in which each of the points lies.

(a) $T(-4, 2)$

(b) $U(1, 3)$

(c) $V(\sqrt{8}, -1)$

(d) $W\left(\dfrac{\sqrt{2}}{2}, 0\right)$

(e) $X(-2, -3)$

(f) $Y(0, -\pi)$

(g) $Z\left(-\dfrac{11}{7}, \dfrac{3}{2}\right)$

Solution Let us graph these points, so we can see their location. We will graph the points carefully. However, knowing whether each coordinate is positive, negative, or zero is enough to find the quadrant or axis in which the point lies.

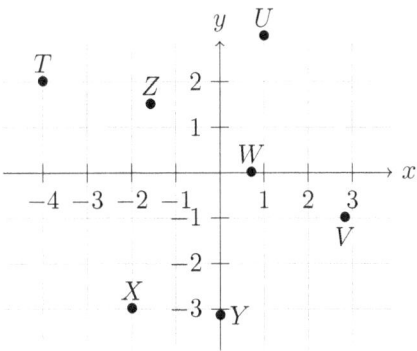

From here, the conclusions follow easily:

(a) T is in quadrant II.

(b) U is in quadrant I.

(c) V is in quadrant IV.

(d) W is on the x-axis

(e) X is in quadrant III.

(f) Y is on the y-axis.

(g) Z is in quadrant II.

∎

Definition 5.1 The **unit circle** is the set of points (x, y) of distance 1 from the origin. In other words, the unit circle is the set of points (x, y) such that
$$x^2 + y^2 = 1.$$

Our goal is to create a correspondence between directed angles and the unit circle. To achieve this, we will establish a convention for the location of directed angles' initial sides. This allows the position of the terminal side to be completely determined by the measure of the angle.

Definition 5.2 A **standard position angle** is an angle whose initial side lies on the positive x-axis.

Let us make the following convention:

> Assume all angles are in standard position when they are used within the context of trigonometric functions unless there is a reason to suppose otherwise.

Because of this convention, we can be sloppy with our language. We often refer to angles simply by their measures, e.g. we say things like "angle 30°". We often say an angle "lies" in a particular quadrant; when we do this, it is understood that we are referring to the terminal side of a standard position angle.

We are ready to formulate our correspondence between standard position angles and points on the unit circle: Let the angle θ correspond to the point (x, y) where the terminal side of θ intersects the unit circle.

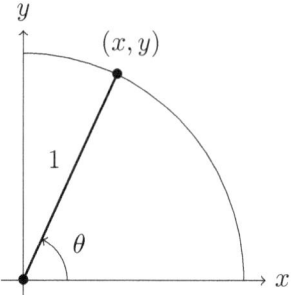

Example 5.2 Find the points on the unit circle corresponding to (a) $\theta = 90°$, (b) $\theta = 180°$, and (c) $\theta = 30°$.

Solution

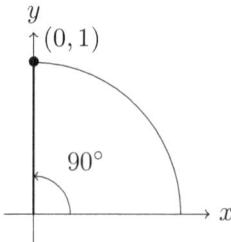

(a) As can be seen from the diagram above, the point corresponding to $\theta = 90°$ is $(0, 1)$.

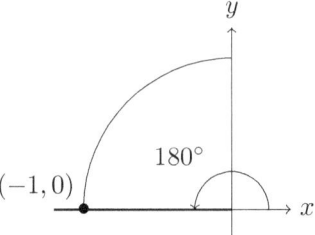

(b) The diagram illustrates that $\theta = 180°$ corresponds to $(-1, 0)$.

(c) We need the $30° - 60° - 90°$ special right triangle.

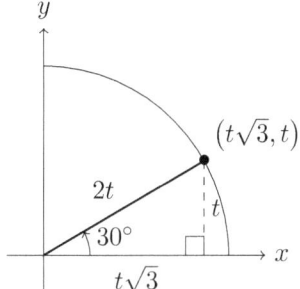

Since the radius of the unit circle is 1, we have

$$2t = 1 \quad \text{implies} \quad t = \frac{1}{2}.$$

Hence, the point corresponding to $\theta = 30°$ is $\left(\sqrt{3}/2, 1/2\right)$.

Using the $45°$–$45°$–$90°$ and $30°$–$60°$–$90°$ special right triangle we can fill in some key points in the first quadrant of the unit circle. Then symmetry gives us points in the other three quadrants.

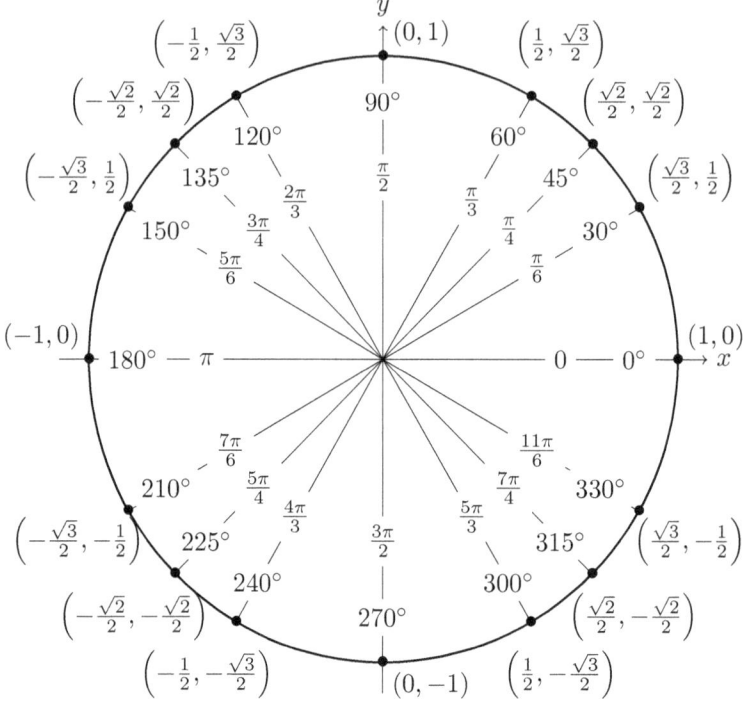

Definition 5.3 Suppose θ is an angle in standard position and (x, y) is the corresponding point on the unit circle. Define

$$\sin\theta = y, \quad \cos\theta = x, \quad \text{and} \quad \tan\theta = \frac{y}{x}.$$

A useful relationship that follows immediately from the above is that
$$\tan\theta = \frac{\sin\theta}{\cos\theta}.$$

Example 5.3 Find $\sin\theta$, $\cos\theta$, and $\tan\theta$ for (a) $\theta = 0$, (b) $\theta = \pi/2$, and (c) $\theta = 3\pi/2$ using the unit circle. When tangent is undefined say so.

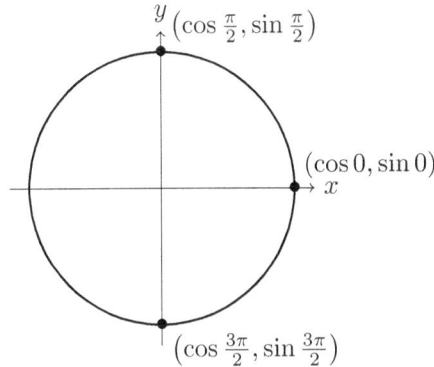

Solution

(a) We know $\theta = 0$ corresponds to the point $(1, 0)$. Therefore,

$$\sin 0 = 0, \quad \cos 0 = 1 \quad \text{and} \quad \tan 0 = \frac{0}{1} = 0.$$

(b) Since $\theta = \pi/2$ corresponds to the point $(0, 1)$,

$$\sin \frac{\pi}{2} = 1 \quad \text{and} \quad \cos \frac{\pi}{2} = 0.$$

Tangent is undefined at $\pi/2$, because $\tan \theta = y/x$ and $x = 0$.

(c) Because $\theta = 3\pi/2$ corresponds to the point $(0, -1)$, we have

$$\sin \frac{3\pi}{2} = -1 \quad \text{and} \quad \cos \frac{3\pi}{2} = 0.$$

Tangent is undefined at $3\pi/2$, because $\tan \theta = y/x$ and $x = 0$.

∎

We can utilize the unit circle on page 112 to evaluate trigonometric functions at more sophisticated angle measures.

Example 5.4 Find the exact value of each of the following.
(a) $\sin 120°$, (b) $\cos \dfrac{5\pi}{4}$, and (c) $\tan 330°$.

Solution

(a) We see that the point corresponding to 120° is $(-1/2, \sqrt{3}/2)$. Sine is the y-coordinate of this point. Hence,
$$\sin 120° = \frac{\sqrt{3}}{2}.$$

(b) The point corresponding to $5\pi/4$ is $(-\sqrt{2}/2, -\sqrt{2}/2)$. Cosine is the x-coordinate of this point. Thus,
$$\cos \frac{5\pi}{4} = -\frac{\sqrt{2}}{2}.$$

(c) The point corresponding to 330° is $(\sqrt{3}/2, -1/2)$. Tangent is the ratio of the y- and x-coordinates. It follows that
$$\tan 330° = \frac{-1/2}{\sqrt{3}/2} = -\frac{\sqrt{3}}{3}.$$

■

Example 5.5 Suppose
$$\cos \theta = -\frac{\sqrt{3}}{2}.$$
Find all θ that satisfy this equation for $0 \le \theta < 2\pi$.

Solution If $\cos \theta = -\sqrt{3}/2$, then the points corresponding to θ on the unit circle must satisfy $x = -\sqrt{3}/2$. Via inspection of the unit circle, we see that if the x-coordinate of the point is $-\sqrt{3}/2$ and $0 \le \theta < 2\pi$, then
$$\theta = \frac{5\pi}{6} \quad \text{or} \quad \theta = \frac{7\pi}{6}.$$

■

Definition 5.4 Suppose θ is an angle in standard position whose terminal side intersects the unit circle at (x, y). Define
$$\sec \theta = \frac{1}{x}, \quad \csc \theta = \frac{1}{y}, \quad \text{and} \quad \cot \theta = \frac{x}{y}.$$

114

Definition 5.5 An **identity** is a statement of equality between mathematical expressions, which holds for all values of the variables contained within the domains of each expression.

Identities are the subject of Chapter 7. However, we will introduce our first few in this chapter.

Proposition 5.1 (Reciprocal Identities) *For θ a degree or radian measure each expression holds whenever it is defined.*

- $\sec\theta = \dfrac{1}{\cos\theta}$
- $\csc\theta = \dfrac{1}{\sin\theta}$
- $\cot\theta = \dfrac{\cos\theta}{\sin\theta}$
- $\cot\theta = \dfrac{1}{\tan\theta} = \dfrac{\cos\theta}{\sin\theta}$

This proposition follows from simple substitutions, so we omit a formal proof.

Example 5.6 Evaluate each of the follow.
(a) $\sec\dfrac{7\pi}{4}$, (b) $\csc 60°$, and (c) $\cot\dfrac{5\pi}{6}$.

Solution

(a)
$$\sec\frac{7\pi}{4} = \frac{1}{\cos(7\pi/4)}$$
$$= \frac{1}{\sqrt{2}/2}$$
$$= \frac{2}{\sqrt{2}} \cdot \frac{\sqrt{2}}{\sqrt{2}}$$
$$= \sqrt{2}.$$

(b)
$$\csc 150° = \frac{1}{\sin 150°}$$
$$= \frac{1}{1/2}$$
$$= 2.$$

(c)
$$\cot\frac{5\pi}{6} = \frac{\cos(5\pi/6)}{\sin(5\pi/6)}$$
$$= \frac{-\sqrt{3}/2}{1/2}$$
$$= -\sqrt{3}.$$

∎

Example 5.7 Assume $0 \leq \varphi < 360°$.
$$\cot \varphi = -\sqrt{3}.$$
Solve for φ.

Solution We know that cotangent of φ is the ratio of the x- and y-coordinates of the points corresponding to φ on the unit circle. To help us identify the appropriate points, note that
$$-\sqrt{3} = -\frac{\sqrt{3}/2}{1/2}.$$
Then, via inspection of the unit circle, we see the ratio of the x- and y-coordinates of either $\left(-\sqrt{3}/2, 1/2\right)$ or $\left(\sqrt{3}/2, -1/2\right)$ produces a quotient of $-\sqrt{3}$. Angles $150°$ and $330°$ correspond to the points $\left(-\sqrt{3}/2, 1/2\right)$ and $\left(\sqrt{3}/2, -1/2\right)$, respectively. Hence,
$$\varphi = 150° \quad \text{or} \quad \varphi = 330°.$$

∎

5.2 Reference Angles

Committing the entire unit circle to memory is a challenge for many students. In this section, we will develop techniques to dramatically reduce the necessary amount of memorization.

Let us begin with a discussion of the sign of sine, cosine, and tangent. Cosine and sine correspond to the x- and y-coordinates on the unit circle, respectively, so their signs depend on the quadrant in which the terminal side of the angle lies.

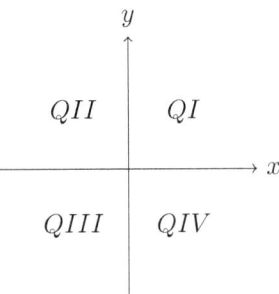

The following table lists the quadrants in which sine, cosine, and tangent are positive and negative.

Quadrant	Positive	Negative
I	sine, cosine, and tangent	none
II	sine	cosine and tangent
III	tangent	sine and cosine
IV	cosine	sine and tangent

Many students remember this using a mnemonic device. For example,

A Smart **T**rig **C**lass

All trigonometric functions are positive in quadrant I.
Sine is positive in quadrant II.
Tangent is positive in quadrant III.
Cosine is positive in quadrant IV.

Example 5.8 Determine in which quadrant θ lies via the signs of the given trigonometric functions.

(a) $\cos\theta > 0$ and $\sin\theta < 0$

(b) $\tan\theta > 0$ and $\csc\theta < 0$

(c) $\cot\theta < 0$ and $\sec\theta > 0$

Solution

(a) Since $\cos\theta > 0$, we know θ lies in quadrant I or IV. Because $\sin\theta < 0$, we conclude θ lies in quadrant III or IV. The only quadrant held in common is quadrant IV. Hence, θ lies in quadrant IV.

(b) Since $\tan\theta > 0$, it follows that θ lies in quadrant I or III. Due to the fact that $\csc\theta < 0$ implies $\sin\theta < 0$, it must be the case that θ lies in quadrant III or IV. By the process of elimination, θ lies in quadrant III.

(c) If $\cot\theta < 0$, then $\tan\theta < 0$, which means that θ lies in quadrant II or IV. Because $\sec\theta > 0$ is the same as saying $\cos\theta > 0$, we know θ lies in quadrant I or IV. Thus, θ must lie in quadrant IV. ∎

Definition 5.6 Consider the directed angle θ in standard position. Suppose the terminal side of θ lies within a quadrant. The **reference angle** θ_R is the acute angle formed by the terminal side of θ and either the positive or negative x-axis.

Note that the acuteness of θ_R determines whether the positive or negative x-axis forms a side of θ_R. Only the closer of the two makes an acute angle with the terminal side of θ.

We can find simple formulas for θ_R, if we suppose $0 < \theta < 2\pi$. When the terminal side of θ is in quadrant I, the closest x-axis is the positive one. So, the measure of the reference angle is

$$\theta_R = \theta.$$

The other three quadrants are a bit more complicated.

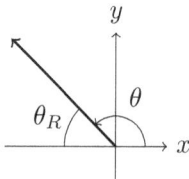

When the terminal side of θ is in quadrant II, θ_R is the angle formed by the negative x-axis and terminal side of θ. Its measure is

$$\theta_R = \pi - \theta.$$

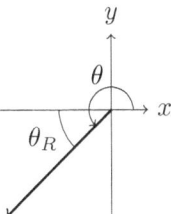

When the terminal side of θ is in quadrant III, θ_R is the angle formed by the negative x-axis and the terminal side of θ. Its measure is
$$\theta_R = \theta - \pi.$$

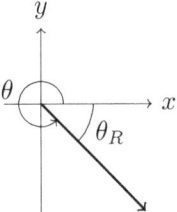

When the terminal side of θ is in quadrant IV, θ_R is the angle formed by the positive x-axis and the terminal side of θ. Its measure is
$$\theta_R = 2\pi - \theta.$$

Proposition 5.2 summarizes the above.

Proposition 5.2 *For an angle θ of radian measure between 0 and 2π or of degree measure between 0 and 360°, the following table provides the measure of the reference angle.*

θ_R of θ	QI	QII	QIII	QIV
Radian measure	θ	$\pi - \theta$	$\theta - \pi$	$2\pi - \theta$
Degree measure	θ	$180° - \theta$	$\theta - 180°$	$360° - \theta$

The six trigonometric functions evaluated at θ and θ_R have the same magnitude. So, to evaluate a trigonometric function at an angle measure, evaluate it at its reference angle and change the sign as needed.

Example 5.9 Evaluate

(a) $\tan \dfrac{7\pi}{4}$

(b) $\sin 120°$

(c) $\csc 210°$

(d) $\cos \dfrac{5\pi}{3}$

Solution

(a) Since
$$\frac{3\pi}{2} < \frac{7\pi}{4} < 2\pi,$$
$7\pi/4$ lies in quadrant IV. This implies that tangent is negative. Furthermore, the reference angle is
$$\theta_R = 2\pi - \frac{7\pi}{4} = \frac{\pi}{4}.$$
Since $\tan(\pi/4) = 1$, we have
$$\tan \frac{7\pi}{4} = -\tan \frac{\pi}{4} = -1.$$

(b) Due to the fact that
$$90° < 120° < 180°,$$
$120°$ lies in quadrant II. Sine is positive in quadrant II. Furthermore, the reference angle is
$$\theta_R = 180° - 120° = 60°.$$
We know $\sin 60° = \sqrt{3}/2$. Therefore,
$$\sin 120° = \sin 60° = \frac{\sqrt{3}}{2}.$$

(c) The angle 210° lies in quadrant III, because
$$180° < 210° < 270°.$$
It follows that sine is negative, which means its reciprocal cosecant is also negative. Furthermore, the reference angle is
$$\theta_R = 210° - 180° = 30°.$$
We have $\sin 30° = 1/2$. So,
$$\csc 210° = -\csc 30°$$
$$= -\frac{1}{\sin 30°}$$
$$= -\frac{1}{1/2}$$
$$= -2.$$

(d) The inequality
$$\frac{3\pi}{2} < \frac{5\pi}{3} < 2\pi,$$
tells us that $5\pi/3$ lies in quadrant IV. As a result, we expect the output to be positive, because cosine is positive in quadrant IV. Furthermore, the reference angle is
$$\theta_R = 2\pi - \frac{5\pi}{3} = \frac{\pi}{3}.$$
Because $\cos(\pi/3) = 1/2$,
$$\cos \frac{5\pi}{3} = \cos \frac{\pi}{3} = \frac{1}{2}.$$

∎

5.3 More Evaluation Techniques

In this section, we will learn to evaluate trigonometric functions for $\theta \geq 360°$ and $\theta < 0$.

Definition 5.7 The function f is **periodic** with period $p > 0$ if p is the smallest number such that
$$f(x + p) = f(x)$$

for all x in the domain of f.

Proposition 5.3 *All six trigonometric functions are periodic. Their periods are given in the table below.*

	Degree period	Radian period
$\cos x$	$360°$	2π
$\sin x$	$360°$	2π
$\tan x$	$180°$	π
$\sec x$	$360°$	2π
$\csc x$	$360°$	2π
$\cot x$	$180°$	π

Using Proposition 5.3, it is not difficult to see that the output of a trigonometric function is not affected by adding or subtracting integer multiples of its period to the input. We can use this fact to convert the input into a value between 0 and $360°$, and then utilize the techniques discussed previously to evaluate.

Example 5.10 Evaluate.

(a) $\cot(-1710°)$

(b) $\cos 5\pi$

(c) $\tan\left(-\dfrac{4\pi}{3}\right)$

(d) $\csc 585°$

Solution

(a) Cotangent has period $180°$ when evaluated using degrees, and $-1710 \div 180 = -9\frac{1}{2}$. So, we will add $180°(10) = 1800°$ to the input:
$$\cot(-1710°) = \cot(-1710° + 1800°) = \cot 90° = 0.$$

(b) Cosine has period 2π when evaluated using radians, and $5\pi \div (2\pi) = 2\frac{1}{2}$. As a result, we will subtract $2\pi(2) = 4\pi$ from the input:
$$\cos 5\pi = \cos(5\pi - 4\pi) = \cos \pi = -1.$$

(c) Tangent has period π when evaluated using radians, and $-\frac{4\pi}{3} \div \pi = -1\frac{1}{3}$. To make the number inside positive, we will add $\pi(2) = 2\pi$ to the input:
$$\tan\left(-\frac{4\pi}{3}\right) = \tan\left(-\frac{4\pi}{3} + 2\pi\right) = \tan \frac{2\pi}{3}.$$

Then we will use reference angles to evaluate:
$$\tan \frac{2\pi}{3} = -\tan \frac{\pi}{3} = -\frac{\sqrt{3}}{3}.$$

Hence,
$$\tan\left(-\frac{4\pi}{3}\right) = -\frac{\sqrt{3}}{3}.$$

(c) Cosecant has period $360°$ when evaluated using degrees, and $585° \div 360° = 1\frac{5}{8}$. Subtracting $360°(1) = 360°$ from the input makes the computation more tractable:
$$\csc 585° = \csc(585° - 360°) = \csc 225°.$$

Using reference angles, we have

$$\csc 225° = -\csc 45°$$
$$= -\frac{1}{\sin 45°}$$
$$= -\frac{1}{\sqrt{2}/2}$$
$$= -\sqrt{2}.$$

Therefore,
$$\csc 585° = -\sqrt{2}.$$

■

Because trigonometric functions are periodic, if θ is a solution then so is θ plus or minus an integer multiple of the period. With this in mind, we are ready to handle general solutions of trigonometric equations.

Example 5.11 Suppose

$$\sec \theta = 2.$$

Find all values of θ.

Solution We know

$$\sec \theta = 2 \quad \text{implies} \quad \cos \theta = \frac{1}{2}.$$

Because $\cos(\pi/3) = 1/2$, the reference angle is $\pi/3$. Cosine is positive when θ lies in quadrant I or IV. Hence, the solutions for $0 \leq \theta < 2\pi$ are

$$\theta = \frac{\pi}{3} \quad \text{and} \quad \theta = \frac{5\pi}{3}.$$

Since secant is periodic with period 2π, any integer multiple of 2π added to either result will produce $1/2$. Thus, the solutions are

$$\theta = \frac{\pi}{3} + 2\pi n \quad \text{and} \quad \theta = \frac{5\pi}{3} + 2\pi n$$

for $n = 0, 1, -1, 2, -2, \ldots$.

■

Example 5.12 Solve for θ.

$$\sin(3\theta) = -\frac{\sqrt{2}}{2}.$$

Suppose $0 \leq \theta < 2\pi$.

Solution If $0 \leq \theta < 2\pi$, then $0 \leq 3\theta < 6\pi$. Using what we know about the unit circle, the solutions for 3θ in the interval $[0, 2\pi)$ are the values such that $3\theta = 5\pi/4$ and $3\theta = 7\pi/4$. To find the solutions in the interval $[2\pi, 6\pi)$ note that adding $2\pi = 8\pi/4$ to a previous solution produces another solution, as long as the sum remains in the interval. Using this insight, we have

$$3\theta = \frac{5\pi}{4}, \frac{7\pi}{4}, \frac{13\pi}{4}, \frac{15\pi}{4}, \frac{21\pi}{4}, \text{ and } \frac{23\pi}{4}$$

are the values of 3θ in the interval $[0, 6\pi)$ which satisfy the equation. Dividing by 3 yields our solutions:

$$\theta = \frac{5\pi}{12}, \frac{7\pi}{12}, \frac{13\pi}{12}, \frac{5\pi}{4}, \frac{7\pi}{4}, \text{ and } \frac{23\pi}{12}.$$

■

Definition 5.8

- The function f is **even** if

$$f(-x) = f(x).$$

- The function f is **odd** if

$$f(-x) = -f(x).$$

Proposition 5.4 (Even and Odd Identities)

(i) The function $\sin\theta$ is odd, so
$$\sin(-\theta) = -\sin\theta.$$

(ii) The function $\cos\theta$ is even, so
$$\cos(-\theta) = \cos\theta.$$

(iii) The function $\tan\theta$ is odd, so
$$\tan(-\theta) = -\tan\theta.$$

(iv) The function $\sec\theta$ is even, so
$$\sec(-\theta) = \sec\theta.$$

(v) The function $\csc\theta$ is odd, so
$$\csc(-\theta) = -\csc\theta.$$

(vi) The function $\cot\theta$ is odd, so
$$\cot(-\theta) = -\cot\theta.$$

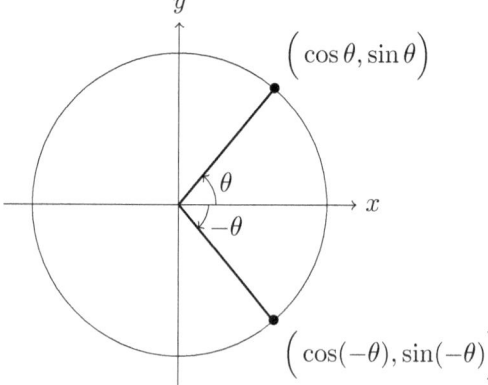

Proof Examination of the unit circle and the definitions of sine and cosine leads us to conclude that sine and cosine are odd and even, respectively. So, (i) and (ii) hold.

We will prove (iii) and (iv), and leave the rest as exercises.

(iii)
$$\tan(-x) = \frac{\sin(-x)}{\cos(-x)}$$
$$= \frac{-\sin x}{\cos x}$$
$$= -\tan x.$$

(iv)
$$\sec(-x) = \frac{1}{\cos(-x)}$$
$$= \frac{1}{\cos x}$$
$$= \sec x.$$

■

Example 5.13 Evaluate (a) $\sin(-30°)$ and (b) $\sec\left(-\dfrac{5\pi}{6}\right)$.

Solution

(a) Because sine is odd,
$$\sin(-30°) = -\sin 30° = -\frac{1}{2}.$$

(b) Since secant is even,
$$\sec\left(-\frac{5\pi}{6}\right) = \sec\frac{5\pi}{6}$$
$$= -\sec\frac{\pi}{6}$$
$$= -\frac{1}{\cos(\pi/6)}$$
$$= -\frac{1}{\sqrt{3}/2}$$
$$= -\frac{2\sqrt{3}}{3}.$$

■

5.4 Finding the Values of Trigonometric Functions

In this section, we will study how to use the value of one trigonometric function to find the values of the other five, e.g. we are given $\sin\theta$ and we will study how to find $\cos\theta$, $\tan\theta$, $\sec\theta$, etc. Surprisingly, a helpful approach to solving this type of problem is to consider where the terminal side of θ intersects a particular circle.

Our next theorem will be of great utility throughout the rest of this book. To understand the theorem, recall that a circle centered at $(0,0)$ and of radius $r > 0$ has equation

$$x^2 + y^2 = r^2.$$

Theorem 5.1 *Consider the circle with equation*

$$x^2 + y^2 = r^2$$

where $r > 0$. Let (x, y) be a point on the circle, and suppose θ is the standard position angle which includes (x, y) on its terminal side. Then

- $\sin\theta = \dfrac{y}{r}$
- $\cos\theta = \dfrac{x}{r}$
- $\tan\theta = \dfrac{y}{x}$
- $\sec\theta = \dfrac{r}{x}$
- $\csc\theta = \dfrac{r}{y}$
- $\cot\theta = \dfrac{x}{y}$

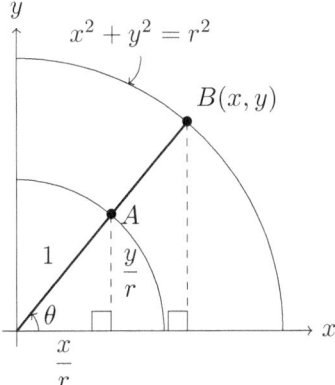

Proof A quick check shows that the identities hold for (x, y) on the x- or y-axis. Suppose the point (x, y) is contained in a quadrant. Then construct right triangles like in the diagram above. Using similar triangles, we have that the sides of the right triangle with a hypotenuse of length 1 are $1/r$ times the lengths of the sides of the right triangle with a hypotenuse of length r. Furthermore, the signs in each coordinate of A and B agree because both points lie in the same quadrant. We conclude that the terminal side of θ intersects the unit circle at $(x/r, y/r)$. Hence, the definitions of sine, cosine, and tangent tell us that

$$\sin\theta = \frac{y}{r}, \quad \cos\theta = \frac{x}{r}, \quad \text{and} \quad \tan\theta = \frac{y}{x}.$$

The other ratios follow from Proposition 5.1. ■

Example 5.14 Suppose

$$\sin\theta = \frac{3}{5} \quad \text{and} \quad \sec\theta < 0.$$

Find the values of the remaining five trigonometric functions.

Solution Because $\sin\theta > 0$ and $\sec\theta < 0$, the angle θ is in quadrant II. Let us suppose that we have a circle of radius 5.

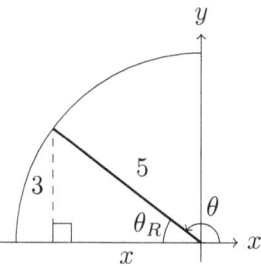

The Pythagorean Theorem (Theorem 2.5) tells us

$$x^2 + 3^2 = 5^2,$$

which implies $x = 4$ or $x = -4$. Since our triangle is in quadrant II, $x = -4$. From here, Theorem 5.1 tells us

$$\cos\theta = -\frac{4}{5}, \quad \tan\theta = -\frac{3}{4}, \quad \sec\theta = -\frac{5}{4},$$

$$\csc\theta = \frac{5}{3}, \quad \text{and} \quad \cot\theta = -\frac{4}{3}.$$

■

Example 5.15 Suppose
$$\tan \varphi = -\frac{12}{5} \quad \text{and} \quad \cos \varphi > 0.$$
Find the values of the remaining five trigonometric functions.

Solution Since $\tan \varphi < 0$ and $\cos \varphi > 0$, the terminal side of φ lies in quadrant IV. Suppose the signed length of the side opposite φ is -12. Then the side adjacent has length 5.

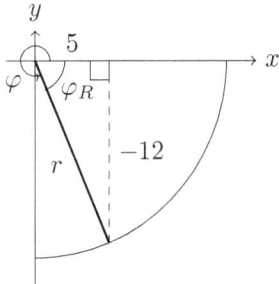

Using the Pythagorean Theorem (Theorem 2.5) we have
$$r^2 = 5^2 + (-12)^2 \quad \text{implies} \quad r = \pm 13.$$
The hypotenuse is always positive, so $r = 13$. Thus, the values of the five remaining trigonometric functions are the following.

$$\sin \theta = -\frac{12}{13}, \quad \cos \theta = \frac{5}{13}, \quad \sec \theta = \frac{13}{5},$$

$$\csc \theta = -\frac{13}{12}, \quad \text{and} \quad \cot \theta = -\frac{5}{12}.$$

∎

5.5 Pythagorean Identities

Let us introduce some important notation. When we write

$$\sin^2\theta, \quad \cos^2\theta, \quad \tan^2\theta, \quad \text{etc.},$$

we mean

$$(\sin\theta)^2, \quad (\cos\theta)^2, \quad (\tan\theta)^2, \quad \text{etc.},$$

respectively. That is, the notation $\sin^2\theta$, $\cos^2\theta$, $\tan^2\theta$, etc. means evaluate the trigonometric function at θ and then square the result. In contract, the notation $\sin\theta^2$, $\cos\theta^2$, $\tan\theta^2$, etc. means square θ and then evaluate the result.

Example 5.16 Evaluate $\tan^2\left(\dfrac{\pi}{3}\right)$ and $\tan\left(\dfrac{\pi}{3}\right)^2$. Use a calculator where necessary.

Solution

$$\tan^2\left(\frac{\pi}{3}\right) = \left(\tan\frac{\pi}{3}\right)^2$$
$$= \left(\sqrt{3}\right)^2$$
$$= 3$$

and

$$\tan\left(\frac{\pi}{3}\right)^2 = \tan\frac{\pi^2}{9} \approx 1.948.$$

∎

Theorem 5.2 (Pythagorean Identities)

For any real number θ the following equations hold whenever they are defined.

(i) $\cos^2\theta + \sin^2\theta = 1$

(ii) $1 + \tan^2\theta = \sec^2\theta$

(iii) $1 + \cot^2\theta = \csc^2\theta$

Proof

(i) Since the equation $x^2 + y^2 = 1$ is the unit circle, $x = \cos\theta$, and $y = \sin\theta$, the identity

$$\cos^2\theta + \sin^2\theta = 1$$

follows due to substitution.

(ii) Let us prove

$$1 + \tan^2\theta = \sec^2\theta.$$

Consider (i) and divide both sides by $\cos^2\theta$:

$$1 + \frac{\sin^2\theta}{\cos^2\theta} = \frac{1}{\cos^2\theta}$$

$$\Rightarrow \quad 1 + \left(\frac{\sin\theta}{\cos\theta}\right)^2 = \left(\frac{1}{\cos\theta}\right)^2$$

$$\Rightarrow \quad 1 + \tan^2\theta = \sec^2\theta.$$

(iii) Let us prove

$$1 + \cot^2\theta = \csc^2\theta.$$

Once again, consider (i). Divide both sides by $\sin^2\theta$:

$$\frac{\cos^2\theta}{\sin^2\theta} + 1 = \frac{1}{\sin^2\theta}$$

$$\Rightarrow \quad 1 + \left(\frac{\cos\theta}{\sin\theta}\right)^2 = \left(\frac{1}{\sin\theta}\right)^2$$

$$\Rightarrow \quad 1 + \cot^2\theta = \csc^2\theta.$$

Example 5.17 Solve
$$5\sin x - 2\cos^2 x = 1.$$
Find all values of x.

Solution Our goal is to rewrite the equation so that its only trigonometric function is $\sin\theta$. To achieve this goal, we will use Pythagorean Identity (i) to convert $\cos^2\theta$ into an expression of $\sin\theta$. In particular, we have
$$\cos^2\theta = 1 - \sin^2\theta.$$
Then we will use substitution to rewrite our equation:
$$5\sin x - 2\cos^2 x = 5\sin x - 2(1-\sin^2 x) = 2\sin^2 x + 5\sin x - 2.$$
It follows that
$$2\sin^2 x + 5\sin x - 3 = 0 \quad \text{implies} \quad (2\sin x - 1)(\sin x + 3) = 0$$
So,
$$\sin x = \frac{1}{2} \quad \text{or} \quad \sin x = -3.$$
The latter is an impossibility because sine is the y-coordinate on the unit circle and points on the unit circle have a y-coordinates between -1 and 1, inclusive. The equation
$$\sin x = \frac{1}{2}$$
has solutions $x = \pi/6$ and $x = 5\pi/6$ for $0 \leq x < 2\pi$. Since sine is periodic with period 2π any integer multiple of 2π added to either result is also a solution. Thus,
$$x = \frac{\pi}{6} + 2\pi n \quad \text{or} \quad x = \frac{5\pi}{6} + 2\pi n$$
for $n = 0, 1, -1, 2, -2, \ldots$. ∎

5.6 Verifying Identities

In this section, we will verify trigonometric identities. This requires knowledge of the identities we have already discussed. We suggest readers review the Reciprocal Identities (Proposition 5.1), the periods of the trigonometric functions (Proposition 5.3), the Even and Odd Identities (Proposition 5.4), and the Pythagorean Identities (Theorem 5.2). They will be used frequently within this section and the corresponding exercises.

When you are asked to verify an identity consider one side of the equation and perform operations on it until the expression is identical to the other side of the equation.

Example 5.18 Verify the identity.
$$\cos(-\theta)\tan(-\theta) = -\sin\theta.$$

Solution We know that cosine is even and tangent is odd, so $\cos(-\theta) = \cos\theta$ and $\tan(-\theta) = -\tan\theta$. Furthermore,
$$\tan\theta = \frac{\sin\theta}{\cos\theta}.$$

Let us start with the left side and work our way to the right.
$$\begin{aligned}\cos(-\theta)\tan(-\theta) &= \cos\theta\Big(-\tan\theta\Big) \\ &= -\cos\theta \cdot \frac{\sin\theta}{\cos\theta} \\ &= -\frac{\cos\theta}{1} \cdot \frac{\sin\theta}{\cos\theta} \\ &= -\frac{\sin\theta}{1} \\ &= -\sin\theta.\end{aligned}$$

∎

Example 5.19 Verify the identity.
$$\frac{\cos^2 x}{1 - \sin x} = 1 + \sin x.$$

Solution We will start on the left side, and use the Pythagorean Identity
$$\cos^2 x = 1 - \sin^2 x.$$
To do this we will multiply the top and bottom of the ratio by $1 + \sin x$ and use the difference of two squares formula:

$$\begin{aligned}
\frac{\cos^2 x}{1 - \sin x} &= \frac{\cos^2 x}{1 - \sin x} \cdot \frac{1 + \sin x}{1 + \sin x} \\
&= \frac{\cos^2 x (1 + \sin x)}{1 - \sin^2 x} \\
&= \frac{\cos^2 x (1 + \sin x)}{\cos^2 x} \\
&= 1 + \sin x.
\end{aligned}$$

■

Formulating an appropriate procedure to verify an identity is sometimes elusive. When this is the case, a good "rule-of-thumb" is to convert the trigonometric expressions into a more familiar form, e.g. converting the expression into one of sine and cosine.

Example 5.20 Verify the identity.
$$\cos\alpha + \sec\alpha \sin^2\alpha = \sec\alpha.$$

Solution

$$\begin{aligned}
\cos\alpha + \sec\alpha \sin^2\alpha &= \cos\alpha + \left(\frac{1}{\cos\alpha}\right)\sin^2\alpha \\
&= \cos\alpha + \frac{\sin^2\alpha}{\cos\alpha} \\
&= \frac{\cos^2\alpha}{\cos\alpha} + \frac{\sin^2\alpha}{\cos\alpha} \\
&= \frac{\cos^2\alpha + \sin^2\alpha}{\cos\alpha} \\
&= \frac{1}{\cos\alpha} \\
&= \sec\alpha.
\end{aligned}$$

■

5.7 Exercises

* Exercise 1

Find the quadrant or axis in which each point lies.

(a) $A(0, 4)$

(b) $B\left(5, -\dfrac{30}{7}\right)$

(c) $C(3, 1)$

(d) $D(-4, 0)$

(e) $E(1, -1.5)$

(f) $F(-\pi, 4)$

(g) $G(-1, \sqrt{3})$

(h) $H\left(-\dfrac{\sqrt{2}}{2}, -\dfrac{\sqrt{2}}{2}\right)$

* Exercise 2

Determine the quadrant or axis in which each standard position angle lies.

(a) $45°$

(b) $\dfrac{4\pi}{3}$

(c) $300°$

(d) $\dfrac{7\pi}{6}$

(e) $150°$

(f) $\dfrac{\pi}{2}$

(g) $315°$

(h) π

** Exercise 3

Find the point on the unit circle corresponding to $\theta =$

(a) $45°$

(b) $\dfrac{\pi}{3}$

(c) $210°$

(d) 0

(e) $\dfrac{7\pi}{4}$

(f) $120°$

(g) π

(h) $30°$

(i) $\dfrac{3\pi}{4}$

(j) $270°$

** Exercise 4

Determine the degree measure of the standard position angle α corresponding to each point on the unit circle. Assume that $0 \leq \alpha < 360°$.

(a) $\left(\dfrac{\sqrt{3}}{2}, \dfrac{1}{2}\right)$

(b) $\left(\dfrac{1}{2}, \dfrac{\sqrt{3}}{2}\right)$

(c) $(0, 1)$

(d) $\left(-\dfrac{\sqrt{2}}{2}, -\dfrac{\sqrt{2}}{2}\right)$

(e) $(0, -1)$

(f) $\left(\dfrac{\sqrt{3}}{2}, -\dfrac{1}{2}\right)$

** Exercise 5

Find the radian measure of the standard position angle β corresponding to each point on the unit circle. Suppose $0 \leq \beta < 2\pi$.

(a) $\left(\dfrac{\sqrt{2}}{2}, \dfrac{\sqrt{2}}{2}\right)$

(b) $(1, 0)$

(c) $(-1, 0)$

(d) $\left(-\dfrac{1}{2}, -\dfrac{\sqrt{3}}{2}\right)$

(e) $\left(-\dfrac{\sqrt{3}}{2}, \dfrac{1}{2}\right)$

(f) $\left(\dfrac{\sqrt{2}}{2}, -\dfrac{\sqrt{2}}{2}\right)$

* Exercise 6

Use the unit circle to find following. Some values are undefined.

(a) $\cos 45°$
(b) $\tan 30°$
(c) $\tan 90°$
(d) $\sec 180°$
(e) $\sin 270°$
(f) $\csc 60°$
(g) $\cot 90°$
(h) $\sin 0$

* Exercise 7

Use the unit circle to find the following. Some values are undefined.

(a) $\sin \dfrac{\pi}{3}$
(b) $\tan \dfrac{3\pi}{2}$
(c) $\cos \dfrac{\pi}{3}$
(d) $\sec \dfrac{\pi}{4}$
(e) $\tan \pi$
(f) $\sec \dfrac{\pi}{6}$
(g) $\csc \dfrac{\pi}{3}$
(h) $\cot \pi$

** Exercise 8

Solve for θ where $0 \leq \theta < 360°$.

(a) $\sin \theta = 1$

(b) $\cos \theta + 1 = 0$

(c) $\tan \theta - 1 = 0$

(d) $\cos \theta = \dfrac{\sqrt{3}}{2}$

(e) $\csc \theta = \dfrac{2\sqrt{3}}{3}$

(f) $\tan \theta = -\sqrt{3}$

** Exercise 9

Solve for φ where $0 \leq \varphi < 2\pi$.

(a) $\cos\varphi = 0$
(b) $2\sin\varphi - 1 = 0$
(c) $3\tan\varphi = -\sqrt{3}$
(d) $\frac{1}{2}\sec\varphi = 1$
(e) $\cot\varphi + 1 = 2$
(f) $3\csc\varphi = -6$

** Exercise 10

Determine the quadrant in which the terminal side of θ lies.

(a) $\sin\theta > 0$ and $\cos\theta > 0$
(b) $\cos\theta > 0$ and $\tan\theta < 0$
(c) $\sin\theta < 0$ and $\sec\theta < 0$
(d) $\csc\theta > 0$ and $\sec\theta < 0$
(e) $\sin\theta < 0$ and $\cot\theta < 0$
(f) $\sec\theta < 0$ and $\cot\theta > 0$

** Exercise 11

State whether the terminal side of φ lies in the positive x-axis, positive y-axis, negative x-axis, or negative y-axis.

(a) $\csc\varphi$ is undefined and $\cos\varphi > 0$.
(b) $\tan\varphi$ is undefined and $\csc\varphi > 0$.
(c) $\cot\varphi$ is undefined and $\sec\varphi < 0$.
(d) $\sec\varphi$ is undefined and $\sin\varphi < 0$.

** Exercise 12

Compute the reference angle.

(a) $147°$
(b) $314°$
(c) $29°$
(d) $307°$
(e) $217°$
(f) $201°$
(g) $316°$
(h) $118°$

** Exercise 13

Find the reference angle.

(a) $\dfrac{5\pi}{6}$
(b) $\dfrac{11\pi}{12}$
(c) $\dfrac{19\pi}{10}$
(d) $\dfrac{3\pi}{4}$
(e) $\dfrac{9\pi}{8}$
(f) $\dfrac{11\pi}{6}$
(g) $\dfrac{4\pi}{3}$
(h) $\dfrac{13\pi}{9}$
(i) 5
(j) 1

** Exercise 14

Evaluate.

(a) $\sin 30°$
(b) $\sin 150°$
(c) $\sin 210°$
(d) $\sin 330°$

** Exercise 15

Evaluate.

(a) $\cos \dfrac{\pi}{4}$

(b) $\cos \dfrac{3\pi}{4}$

(c) $\cos \dfrac{5\pi}{4}$

(d) $\cos \dfrac{7\pi}{4}$

** Exercise 16

Evaluate.

(a) $\tan 60°$

(b) $\tan 120°$

(c) $\tan 240°$

(d) $\tan 300°$

** Exercise 17

Use reference angles to evaluate. Some expressions are undefined.

(a) $\tan \dfrac{5\pi}{6}$

(b) $\cos \dfrac{3\pi}{2}$

(c) $\sin \dfrac{11\pi}{6}$

(d) $\tan \dfrac{\pi}{2}$

(e) $\csc \dfrac{7\pi}{4}$

(f) $\cot \dfrac{2\pi}{3}$

(g) $\csc \dfrac{7\pi}{6}$

(h) $\cot \dfrac{3\pi}{4}$

(i) $\sin \dfrac{5\pi}{4}$

(j) $\cos \dfrac{2\pi}{3}$

** Exercise 18

Use reference angles to evaluate. Some expressions are undefined.

(a) $\sin 120°$

(b) $\tan 315°$

(c) $\csc 315°$

(d) $\cos 270°$

(e) $\csc 240°$

(f) $\sec 210°$

(g) $\cos 150°$

(h) $\tan 330°$

(i) $\sin 225°$

(j) $\cos 120°$

** Exercise 19

Suppose α is an acute angle of degree measure such that the terminal side of α intersects the unit circle at the point $(4/5, 3/5)$.

(a) Find the values of the six trigonometric functions at α.

(b) What are the values of the six trigonometric functions at $360° - \alpha$?

(c) Compute the values of the six trigonometric functions at $\alpha + 180°$.

(d) Evaluate the six trigonometric functions at $180° - \alpha$.

** Exercise 20

Assume β is an acute angle of radian measure such that the terminal side of β intersects the unit circle at the point $(5/13, 12/13)$.

(a) Find the values of the six trigonometric functions at β.

(b) What are the values of the six trigonometric functions at $\pi - \beta$?

(c) Compute the values of the six trigonometric functions at $2\pi - \beta$.

(d) Evaluate the six trigonometric functions at $\beta + \pi$.

*** Exercise 21

The terminal side of θ lies in quadrant IV and intersects the unit circle at the point $(8/17, -15/17)$. Compute each of the following.

(a) $\sin\theta$

(b) $\cot\theta$

(c) $\cos(2\pi - \theta)$

(d) $\sec(\theta - \pi)$

(e) $\csc(\theta - \pi)$

(f) $\cos(3\pi - \theta)$

** Exercise 22

Compute each of the following. Some expressions are undefined.

(a) $\cot(-180°)$

(b) $\tan 870°$

(c) $\csc 585°$

(d) $\sin 630°$

(e) $\tan(-135°)$

(f) $\sin(-810°)$

(g) $\cot(-510°)$

(h) $\sec(-1050°)$

** Exercise 23

Calculate each of the following. Some expressions are undefined.

(a) $\cos 2\pi$

(b) $\cos\left(-\dfrac{9\pi}{2}\right)$

(c) $\csc\left(-\dfrac{11\pi}{3}\right)$

(d) $\sin\dfrac{7\pi}{2}$

(e) $\sin\dfrac{23\pi}{6}$

(f) $\csc\left(-\dfrac{13\pi}{4}\right)$

(g) $\sec\left(-\dfrac{5\pi}{6}\right)$

(h) $\cot\left(-\dfrac{9\pi}{4}\right)$

** Exercise 24

Prove Proposition 5.4 (v) and (vi).

** Exercise 25

The **domain** of a function is the set of inputs of a function, and the **range** of a function is the set of outputs of the function. Determine the domain and range of f.

(a) $f(x) = \sin x$

(b) $f(x) = \cos x$

(c) $f(x) = \tan x$

(d) $f(x) = \csc x$

(e) $f(x) = \sec x$

(f) $f(x) = \cot x$

** Exercise 26

Determine the values of θ which satisfy the equation.

(a) $\tan \theta = 0$

(b) $-2 \cos \theta = 1$

(c) $3 \cot^2 \theta = 1$

(d) $\sin^2 \theta - 2 \sin \theta = -1$

** Exercise 27

Solve for φ.

(a) $2 \sin 3\varphi = 1$

(b) $-\tan 2\varphi = 1$

(c) $3 \csc^2 \pi\varphi = 4$

(d) $\sec^2(5\varphi) - 3\sec(5\varphi) + 2 = 0$

** Exercise 28

Find all values of α within the interval $[0, 360°)$ which satisfy the equation.

(a) $\tan(-\alpha) = \sqrt{3}$

(b) $3 \csc 3\alpha = 3\sqrt{2}$

(c) $\cot^2 \dfrac{\alpha}{2} = 1$

(d) $2 \sin^2(2\alpha) - 9 \sin(2\alpha) = 5$

** Exercise 29

Solve for β. Assume $0 \leq \beta < 2\pi$.

(a) $2 \cos(-\beta) = \sqrt{3}$

(b) $\sec^2 \dfrac{\beta}{3} - 2 = 0$

(c) $\tan^2 \dfrac{\pi\beta}{3} = 3$

(d) $3 \csc^2 \pi\beta - 5 \csc \pi\beta + 2 = 0$

** Exercise 30

Find all values of θ in the interval $(-180°, 180°]$ which satisfy the equation.

(a) $6 \sin(-\theta) = 3\sqrt{2}$

(b) $\csc^2 \dfrac{\theta}{2} - 2 = 0$

(c) $\cot \dfrac{5\theta}{4} + \sqrt{3} = 0$

(d) $\tan^2 \dfrac{3\theta}{2} - 1 = 0$

** Exercise 31

Determine the values of the remaining five trigonometric functions.

(a) $\sin\alpha = -4/5$ and $\cos\alpha < 0$

(b) $\cot\beta = 15/8$ and $\sec\beta > 0$

(c) $\cos\gamma = 5/13$ and $\cot\gamma < 0$

(d) $\tan\theta = -24/7$ and $\sec\theta < 0$

(e) $\csc\varphi$ is undefined and $\sec\varphi < 0$

** Exercise 32

Solve for α.

(a) $2\cos^2\alpha - \sin\alpha = 1$

(b) $3\tan^2\alpha = 4\sec\alpha + 1$

(c) $5\csc^4\alpha - 9\cot^2\alpha - 11 = 0$

** Exercise 33

Find all values of β.

(a) $\cos^4(3\beta) + 3\sin^2(3\beta) = 1$

(b) $\tan^3(2\beta) + \sec^2(-2\beta) = 3\tan(2\beta) + 4$

(c) $\csc^2(5\beta)\cot(-5\beta) = \csc^2(5\beta)$

** Exercise 34

(i) 1 (iii) $-\cot x$

(ii) $\tan x$ (iv) $\sec x$

Match the expressions above with the equivalent expressions below. Some options may be used more than once.

(a) $\dfrac{\sec^2 x - 1}{\tan x}$

(b) $\sin(x)\sec(-x)$

(c) $\cos(-x)\csc(-x)$

(d) $\cos x \sec x$

(e) $\sin(-x)\csc(-x)$

(f) $\dfrac{\tan x}{\sin x}$

** Exercise 35

Verify each identity.

(a) $\sin(2\pi - \alpha) = -\sin\alpha$

(b) $\cos(360° - \beta) = \cos\beta$

(c) $\tan(2\pi - \gamma) = -\tan\gamma$

** Exercise 36

Verify the identity.

$$\dfrac{\tan(\theta + \pi)}{\sin\theta} = \sec\theta$$

** Exercise 37

Verify each identity.

(a) $\cos^2\alpha - \sin^2\alpha = 2\cos^2\alpha - 1$

(b) $\cos^2\alpha - \sin^2\alpha = 1 - 2\sin^2\alpha$

** Exercise 38

Verify each identity.

(a) $\dfrac{\sin^2 \theta}{1+\cos \theta} = 1 - \cos \theta$

(b) $\dfrac{\cos^2 \alpha}{1+\sin \alpha} = 1 - \sin \alpha$

(c) $\dfrac{\sec^2 \varphi - 1}{\tan \varphi} = \tan \varphi$

(d) $\dfrac{\tan \beta}{1-\cos \beta} = \sec \beta \csc \beta + \csc \beta$

** Exercise 39

Verify each identity.

(a) $\dfrac{\cos^2 x - \sin x(-\sin x)}{\cos^2 x} = \sec^2 x$

(b) $\dfrac{-\sin^2(-x) - \cos^2 x}{\sin^2 x} = -\csc^2 x$

** Exercise 40

Verify each identity.

(a) $\sec x + \tan x = \dfrac{1}{\sec x - \tan x}$

(b) $\csc x + \cot x = \dfrac{1}{\csc x - \cot x}$

** Exercise 41

Verify the identity.

$\sin \theta + \sin \theta \tan^2 \theta = \sec \theta \tan \theta$

Chapter 6

Graphing Trigonometric Functions

In this chapter, we will learn to graph the six trigonometric functions on the xy-plane. We assume a thorough understanding of Chapter 5. Some knowledge of Appendix C—which addresses shifts, stretches, and compressions—is helpful, but will not be directly utilized. We will not use calculators.

6.1 Graphing Sine and Cosine

We want to graph functions of the form

$$f(x) = A\sin(Bx + C) + D \quad \text{and} \quad g(x) = A\cos(Bx + C) + D.$$

Definition 6.1 A **parent function** is considered to be the most basic within a family of functions.

We like to think of $f(x) = \sin x$ and $g(x) = \cos x$ as the parent functions of their respective family of functions. From this perspective, the constants A, B, C, and D modify the parent functions. As a result, it is helpful to obtain an understanding of the graphs of $f(x) = \sin x$ and $g(x) = \cos x$.

$f(x) = \sin x$

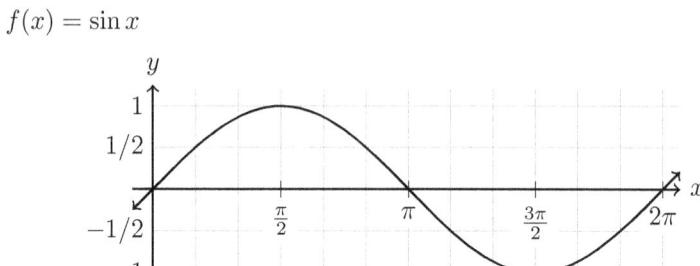

$g(x) = \cos x$

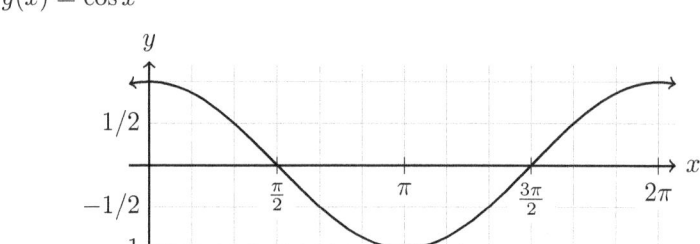

Proposition 5.3 tells us that sine and cosine are periodic with period 2π. This means that the behavior of $f(x) = \sin x$ and $g(x) = \cos x$, within the interval $[0, 2\pi]$, is repeated in subsequent intervals, so graphing more periods of either function is simply a matter of recognizing the pattern.

Definition 6.2

- The **amplitude** of f is
$$\frac{\max\{f(x)\} - \min\{f(x)\}}{2}.$$

- The **phase shift** of a periodic function f is how much its principal period is shifted left or right on a graph relative to the parent function of f.

Proposition 6.1 *Suppose we have a function of the form*

$$f(x) = A\sin(Bx + C) + D \quad \text{or} \quad g(x) = A\cos(Bx + C) + D,$$

where $B > 0$.

(i) *The amplitude of the function is $|A|$.*

(ii) *The graph has period*
$$\frac{2\pi}{B}.$$

(iii) *The neutral vertical position of its graph is $y = D$; this is called the vertical shift.*

(iv) *The phase shift is*
$$-\frac{C}{B}.$$

Note: a positive phase shift corresponds to a rightward shift and a negative phase shift corresponds to a leftward shift.

Proposition 6.1 will be used extensively to graph functions, but plotting a small number of points is also helpful. We will introduce a protocol to find five useful points.

Let $f(x) = A\sin(Bx + C) + D$ or $g(x) = A\cos(Bx + C) + D$, where $B > 0$.

- Start at an x-coordinate equal to the value of the phase shift, i.e. start at

$$x = -\frac{C}{B}.$$

Evaluate to find the y-coordinate.

- To find subsequent x-coordinates, add one-fourth the period to the previous x-coordinate. That is, add

$$\frac{2\pi/B}{4} = \frac{\pi}{2B}$$

to the previous x-coordinate. Evaluate the new x-value to find the corresponding y-value.

- Stop after you have found the point whose x-coordinate is equal to the phase shift plus the period. That is, stop after the point corresponding to

$$x = -\frac{C}{B} + \frac{2\pi}{B}.$$

This protocol gives five points. The y-coordinate of the last point should be the same as the first.

Example 6.1 Graph one period of

$$f(x) = 2\sin(3x - \pi) + 1.$$

Solution Proposition 6.1 gives us the table below.

Amplitude:	$\|2\| = 2$
Period:	$\dfrac{2\pi}{3}$
Vertical shift:	1
Phase shift:	$-\dfrac{(-\pi)}{3} = \dfrac{\pi}{3}$

Let us plot some points. The first x-coordinate is equal to the phase shift $\pi/3$ and we increase each subsequent x-coordinate by an increment of

$$\frac{2\pi/3}{4} = \frac{\pi}{6}.$$

150

x	$f(x)$
$\dfrac{\pi}{3}$	$2\sin(0)+1 \;=\; 1$
$\dfrac{3\pi}{6}=\dfrac{\pi}{2}$	$2\sin\left(\dfrac{\pi}{2}\right)+1 \;=\; 3$
$\dfrac{4\pi}{6}=\dfrac{2\pi}{3}$	$2\sin(\pi)+1 \;=\; 1$
$\dfrac{5\pi}{6}$	$2\sin\left(\dfrac{3\pi}{2}\right)+1 \;=\; -1$
π	$2\sin(2\pi)+1 \;=\; 1$

Hence, we have the following graph.

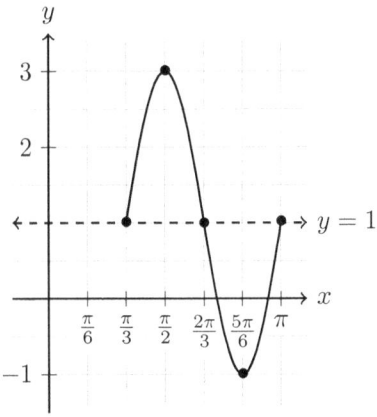

■

Example 6.2 Graph two periods of

$$g(x) = -2 - \frac{1}{2}\cos\left(\frac{3\pi x + \pi}{4}\right).$$

Solution Let us rewrite this into a more familiar form:

$$g(x) = -2 - \frac{1}{2}\cos\left(\frac{3\pi x + \pi}{4}\right) = -\frac{1}{2}\cos\left(\frac{3\pi}{4}x + \frac{\pi}{4}\right) - 2.$$

Proposition 6.1 tells us the key features of g's graph.

Amplitude:	$\left\lvert -\frac{1}{2} \right\rvert = \frac{1}{2}$
Period:	$\frac{2\pi}{3\pi/4} = \frac{8}{3}$
Vertical shift:	-2
Phase shift:	$-\frac{\pi/4}{3\pi/4} = -\frac{1}{3}$

Let us plot some points. The first point has an x-coordinate of $-1/3$ and subsequent x-values will be increased in increments of

$$\frac{8/3}{4} = \frac{2}{3}.$$

x	$g(x)$
$-\dfrac{1}{3}$	$-\dfrac{1}{2}\cos(0)-2 \;=\; -2\tfrac{1}{2}$
$\dfrac{1}{3}$	$-\dfrac{1}{2}\cos\left(\dfrac{\pi}{2}\right)-2 \;=\; -2$
$\dfrac{3}{3}=1$	$-\dfrac{1}{2}\cos(\pi)-2 \;=\; -1\tfrac{1}{2}$
$\dfrac{5}{3}$	$-\dfrac{1}{2}\cos\left(\dfrac{3\pi}{2}\right)-2 \;=\; -2$
$\dfrac{7}{3}$	$-\dfrac{1}{2}\cos(2\pi)-2 \;=\; -2\tfrac{1}{2}$

After we plot the points and graph the first period, we use the pattern to graph the second period.

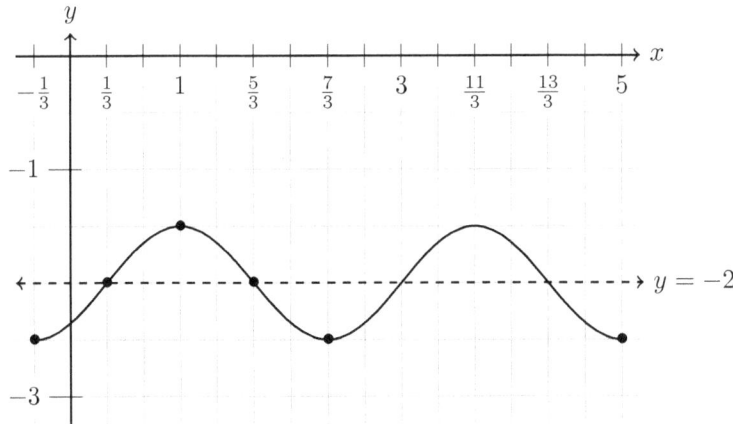

■

Proposition 6.1 assumes that $B > 0$. For $B < 0$, we can utilize Proposition 5.4 which says that sine is odd and cosine is even. This allows us to change the sign of the coefficient in front of x.

Example 6.3 Graph one period of

$$h(x) = 2\sin\left(-x + 45°\right).$$

Solution Sine is odd, which implies

$$2\sin\left(-x + 45°\right) = 2\sin\left(-(x - 45°)\right) = -2\sin\left(x - 45°\right).$$

Then we utilize Proposition 6.1.

Amplitude:	$|-2| = 2$	
Period:	$\dfrac{360°}{1} = 360°$	
Vertical shift:	0	
Phase shift:	$-\dfrac{-45°}{1} = 45°$	

The next step is to plot points. We begin at $x = 45°$ and add

$$\frac{360°}{4} = 90°$$

to the previous x-coordinate to find the next. We stop at $x = 405°$.

x	$h(x)$
$45°$	$-2\sin 0 = 0$
$135°$	$-2\sin 90° = -2$
$225°$	$-2\sin 180° = 0$
$315°$	$-2\sin 270° = 2$
$405°$	$-2\sin 360° = -2$

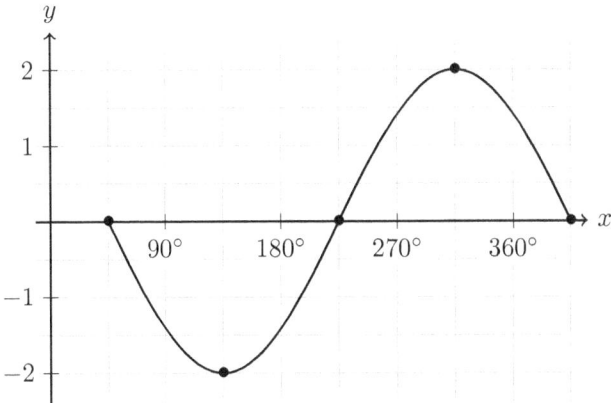

6.2 Graphing Tangent and Cotangent

The goal of this section is to graph functions of the form

$$f(x) = A\tan(Bx + C) + D \quad \text{and} \quad g(x) = A\cot(Bx + C) + D.$$

Much like sine and cosine graphs, we think of A, B, C, and D as modifying the graphs of the parent functions $f(x) = \tan x$ and $g(x) = \cot x$.

Tangent and cotangent are somewhat more difficult to graph because they contain vertical asymptotes.

Definition 6.3 The function h has a **vertical asymptote** of $x = a$ if $h(x)$ goes to $\pm\infty$ as x goes to a from the left or the right. Within sketches of graphs, asymptotes are usually denoted by dashed lines.

The graph of h to the right has a vertical asymptote of $x = 1$. This is because $h(x)$ goes to $-\infty$ as x goes to 1 from the left, and $h(x)$ goes to ∞ as x goes to 1 from the right.

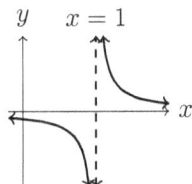

6.2.1 Graphing Tangent

Let us consider the graph of $f(x) = \tan x$.

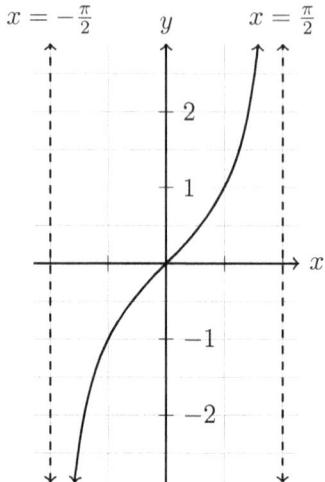

Graphing more of $f(x) = \tan x$ is not difficult, because Proposition 5.3 tells us tangent has period π. Hence, in subsequent intervals, the graph of tangent simply repeats its behavior.

Proposition 6.2 *Suppose*

$$f(x) = A\tan(Bx + C) + D,$$

where $B > 0$.

(i) The graph has period

$$\frac{\pi}{B}.$$

(ii) The neutral vertical position of the graph is $y = D$; this is called the vertical shift.

(iii) The phase shift of f is

$$-\frac{C}{B}.$$

Note: Being shifted a negative number of units right corresponds to being shifted left.

(iv) The function f has vertical asymptotes at the solutions of
$$Bx + C = -\frac{\pi}{2} \quad \text{and} \quad Bx + C = \frac{\pi}{2}.$$

Notice that we did not mentioned amplitude. The value of $|A|$ vertically compresses or stretches the graph of tangent, but it has no maximum or minimum value. This makes the concept of amplitude nonsensical.

Example 6.4 Determine the period, vertical shift, phase shift, and asymptotes of the function.
$$f(x) = -3\tan(15°x + 45°) - 7$$

Solution From Proposition 6.2, we know the period is
$$\frac{180°}{15°} = 12,$$
the vertical shift is -7, and the phase shift is
$$-\frac{45}{15} = -3.$$

One of the vertical asymptotes is the solution of
$$15°x + 45° = -90°.$$
Solving yields the vertical asymptote $x = -9$.

Since f has period 12, this means that all the asymptotes are of the form
$$x = -9 + 12n,$$
where $n = 0, 1, -1, 2, -2, \ldots$. ∎

Proposition 6.2 will be used extensively, but we will also plot some points when we graph. So, we will introduce a protocol to find three helpful points.

Suppose $f(x) = A\tan(Bx + C) + D$, where $B > 0$. Proposition 6.2 tells us that there are vertical asymptotes at
$$Bx + C = -\frac{\pi}{2} \quad \text{and} \quad Bx + C = \frac{\pi}{2}.$$

- Add one-fourth the period to the x-value of the left asymptote. That is, add

$$\frac{\pi}{4B}$$

to the solution of $Bx + C = -\pi/2$. This gives the x-coordinate of the first point. Evaluate f at the x-value to find the y-coordinate.

- To find the next x-coordinate, add one-fourth the period to the previous x-coordinate. Evaluate f at the x-value to find the y-coordinate.

- Stop before you reach the solution of

$$Bx + C = \frac{\pi}{2}.$$

This protocol should give three points. Their y-coordinates should be $-A + D$, D, and $A + D$, respectively.

Example 6.5 Graph one period of

$$g(x) = 2\tan\left(\frac{x}{2} + \frac{\pi}{3}\right) - 4.$$

Solution Let us compute the vertical asymptotes. Using Proposition 6.2, the vertical asymptotes are the solutions of

$$\frac{x}{2} + \frac{\pi}{3} = -\frac{\pi}{2} \quad \text{and} \quad \frac{x}{2} + \frac{\pi}{3} = \frac{\pi}{2}.$$

Hence, we have vertical asymptotes

$$x = -\frac{5\pi}{3} \quad \text{and} \quad x = \frac{\pi}{3}.$$

Using the above, and some other parts of Proposition 6.2 give us our table.

> Vertical shift: -4
>
> Period: $\dfrac{\pi}{1/2} = 2\pi$
>
> Phase shift: $-\dfrac{\pi/3}{1/2} = -\dfrac{2\pi}{3}$
>
> Asymptotes: $x = -\dfrac{5\pi}{3}$ and $x = \dfrac{\pi}{3}$

Let us plot some points. Consecutive x-coordinates will be increased in increments of
$$\frac{2\pi}{4} = \frac{\pi}{2}.$$
The first point has an x-coordinate of
$$-\frac{5\pi}{3} + \frac{\pi}{2} = -\frac{7\pi}{6}.$$

x	$g(x)$
$-\dfrac{7\pi}{6}$	$2\tan\left(-\dfrac{\pi}{4}\right) - 4 = -6$
$-\dfrac{4\pi}{6} = -\dfrac{2\pi}{3}$	$2\tan 0 - 4 = -4$
$-\dfrac{\pi}{6}$	$2\tan\left(\dfrac{\pi}{4}\right) - 4 = -2$

We have obtained enough information to graph one period of g.

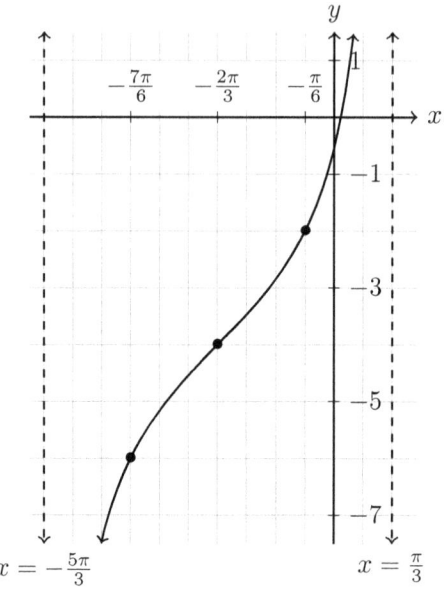

6.2.2 Graphing Cotangent

Consider the graph of $f(x) = \cot x$.

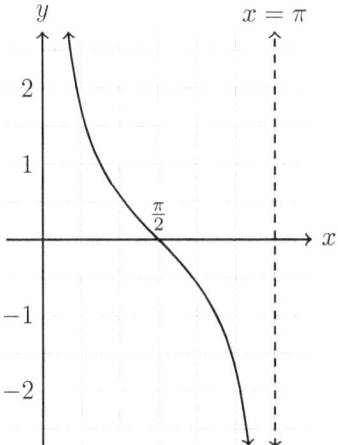

Proposition 5.3 tells us that cotangent has period π. Hence, the pattern within the interval $(0, \pi)$ is repeated in subsequent periods.

Proposition 6.3 *Suppose*

$$f(x) = A\cot(Bx + C) + D,$$

where $B > 0$.

(i) The graph has period
$$\frac{\pi}{B}.$$

(ii) The neutral vertical position of the graph is $y = D$; this is called the vertical shift.

(iii) The phase shift of f is
$$-\frac{C}{B}.$$
Note: being shifted a negative number of units right corresponds to being shifted left.

(iv) There are vertical asymptotes at
$$Bx + C = 0 \quad \text{and} \quad Bx + C = \pi.$$

We will use some point plotting, along with Proposition 6.3, to graph cotangent. This makes a protocol for finding points necessary. Ours gives three useful points.

Let $f(x) = A\cot(Bx + C) + D$, where $B > 0$. Proposition 6.3 tells us that f has vertical asymptotes at

$$Bx + C = 0 \quad \text{and} \quad Bx + C = \pi.$$

- Add one-fourth the period to the x-value of the left asymptote. That is, add

$$\frac{\pi}{4B}$$

to the solution of $Bx + C = 0$. This gives the x-coordinate of the first point. Evaluate f at the x-value to find the y-coordinate.

- To find the next x-coordinate, add one-fourth the period to the previous x-coordinate. Evaluate f at the x-value to find the y-coordinate.

- Stop before you reach the x-value of the right asymptote. That is, stop before the x-value is the solution of

$$Bx + C = \pi.$$

This procedure gives three points. Their y-coordinates should be $A + D$, D, and $-A + D$, respectively.

Example 6.6 Graph two periods of

$$f(x) = \frac{1}{2}\cot(180°x + 135°) - 1.$$

Solution Using Proposition 6.3, the asymptotes within the first period are the solutions of

$$180°x + 135° = 0 \quad \text{and} \quad 180°x + 135° = 180°.$$

Solving these yields

$$x = -\frac{3}{4} \quad \text{and} \quad x = \frac{1}{4}.$$

The period is
$$\frac{180°}{180°} = 1.$$
We need another asymptote, because we want to graph two periods of cotangent. Using the fact that cotangent has period 1, we conclude
$$x = \frac{1}{4} + 1 = \frac{5}{4}$$
is another asymptote.

We have the following table.

Period:	$\frac{180°}{180°} = 1$
Vertical shift:	-1
Phase shift:	$-\frac{135°}{180°} = -\frac{3}{4}$
Asymptotes:	$x = -\frac{3}{4}, \quad x = \frac{1}{4}, \quad \text{and} \quad x = \frac{5}{4}$

Next, we plot points within the first period. We increase subsequent x-coordinates by an increment $1/4$, and we start at
$$x = -\frac{3}{4} + \frac{1}{4} = -\frac{1}{2}.$$

x	$f(x)$		
$-\frac{2}{4} = -\frac{1}{2}$	$\frac{1}{2}\cot(45°) - 1$	$=$	$-\frac{1}{2}$
$-\frac{1}{4}$	$\frac{1}{2}\cot(90°) - 1$	$=$	-1
0	$\frac{1}{2}\cot(135°) - 1$	$=$	$-1\frac{1}{2}$

This enough to graph one period of f's graph. The next period is not difficult to draw, because we have the vertical asymptote $x =$

5/4 and we can determine the behavior of the graph by analyzing the previous period.

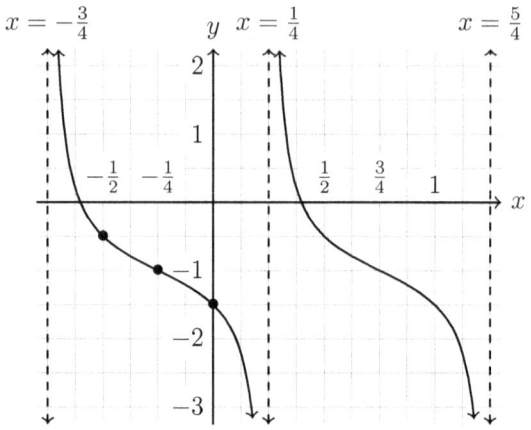

Example 6.7 A cotangent graph has vertical asymptotes $x = 1$ and $x = 7$, no vertical shift, and contains the point $(5/2, 4)$. Find the equation of the corresponding cotangent function.

Solution Suppose
$$g(x) = A \cot(Bx + C) + D$$
is the function. There is no vertical shift which implies $D = 0$. Since $x = 1$ and $x = 7$ are vertical asymptotes, the period is
$$7 - 1 = 6.$$
Hence, Proposition 6.3 tells us
$$\frac{\pi}{B} = 6 \quad \text{implies} \quad B = \frac{\pi}{6}.$$
For cotangent functions, the phase shift is the same as the left asymptote within the principal period. So, it must be 1. Using Proposition 6.3 again, it follows that
$$-\frac{C}{\pi/6} = 1 \quad \text{implies} \quad C = -\frac{\pi}{6}.$$

As of now, we have
$$g(x) = A\cot\left(\frac{\pi}{6}x - \frac{\pi}{6}\right).$$

To find A, we will plug in $5/2$, because we know $g(5/2) = 4$:
$$\begin{aligned}
g\left(\frac{5}{2}\right) &= A\cot\left(\frac{\pi}{6}\cdot\frac{5}{2} - \frac{\pi}{6}\right) \\
&= A\cot\frac{\pi}{4} \\
&= A(1) \\
&= A.
\end{aligned}$$

We conclude $A = 4$.

Thus, our function is
$$g(x) = 4\cot\left(\frac{\pi}{6}x - \frac{\pi}{6}\right).$$

■

6.3 Graphing Secant and Cosecant

In this section, we will learn to graph secant and cosecant. These graphs rely on the skills from Section 6.1, so we recommend that the reader masters those concepts before they continue. In particular, because
$$\sec x = \frac{1}{\cos x} \quad \text{and} \quad \csc x = \frac{1}{\sin x},$$
the reader must understand how to graph cosine to graph secant, and how to graph sine to graph cosecant.

6.3.1 Graphing Secant

Let us begin with an analysis of the graph of $f(x) = \sec x$.

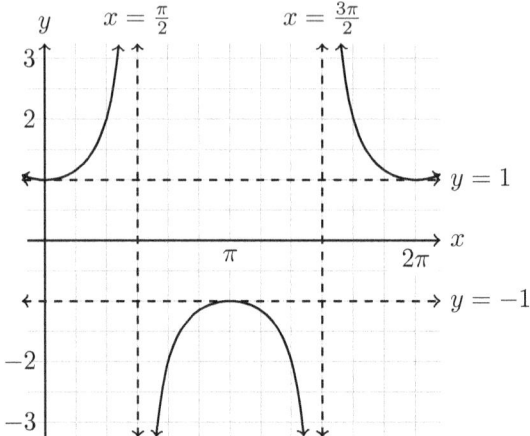

Each branch of secant is a U-shaped curve. Secant has asymptotes at $x = \pi/2$ and $x = 3\pi/2$. From Proposition 5.3, we know that secant has period 2π. As a result, we can graph more periods easily.

We now introduce a procedure to graph secant.

Let
$$f(x) = A \sec(Bx + C) + D.$$

1. Lightly sketch the graph of
$$y = A \cos(Bx + C) + D.$$

2. Draw vertical asymptotes where the cosine graph intersects the line $y = D$.

3. Draw horizontal dashed lines $y = A + D$ and $y = -A + D$. These values correspond to the maximum and minimum y-values of $y = A\cos(Bx + C) + D$.

4. Draw the graph of
$$f(x) = A\sec(Bx + C) + D.$$

Each U-shaped branch touches the cosine graph at its vertex and opens away from the cosine graph.

This strategy is based on the fact that secant is the reciprocal function of cosine.

Example 6.8 Graph one period of
$$f(x) = -2 - \frac{1}{2}\sec\left(\frac{3\pi x + \pi}{4}\right).$$

Solution The first step is to graph
$$y = -2 - \frac{1}{2}\cos\left(\frac{3\pi x + \pi}{4}\right) = -\frac{1}{2}\cos\left(\frac{3\pi}{4}x + \frac{\pi}{4}\right) - 2.$$

Using Section 6.1, and in particular Example 2, yields (a). Then we draw horizontal dashed lines at
$$y = -\frac{1}{2} - 2 = -\frac{5}{2} \quad \text{and} \quad y = \frac{1}{2} - 2 = -\frac{3}{2}.$$

Vertical asymptotes are drawn at
$$x = \frac{1}{3} \quad \text{and} \quad x = \frac{5}{3},$$

because this is where the cosine graph intersects $y = -2$. This gives (b).

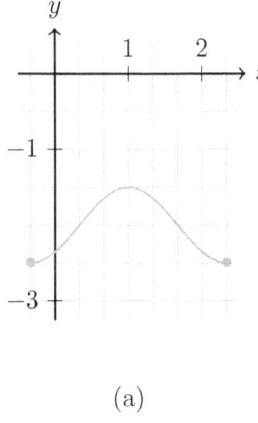

(a)

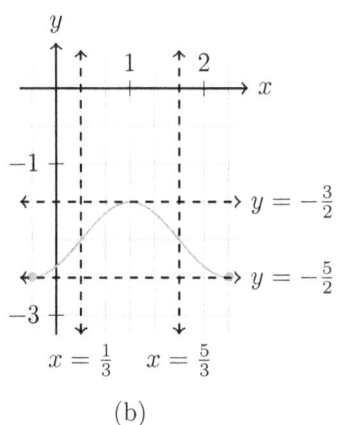

(b)

All that is left is to draw secant's branches. The branches have vertices of $(-1/3, -5/2)$, $(1, -3/2)$, and $(-7/3, -5/2)$, because those are the points where cosine intersects the dashed lines. The rest of each U-shape follows due to the position of the asymptotes.

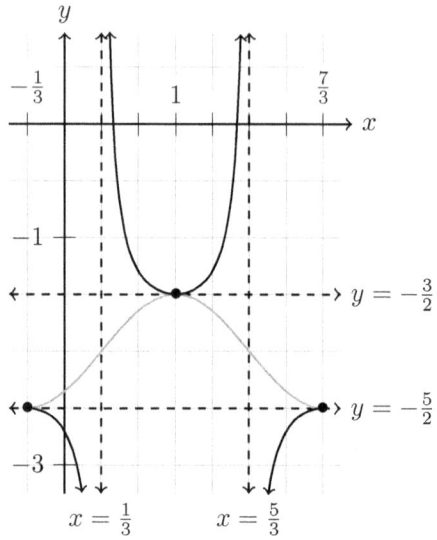

Note that only the darkened curves above is the secant graph; the rest is a graphing aid. So, plugging

$$f(x) = -2 - \frac{1}{2}\sec\left(\frac{3\pi x + \pi}{4}\right)$$

into a graphing calculator results in a graph like the one on the right.

6.3.2 Graphing Cosecant

Consider the graph of $f(x) = \csc x$.

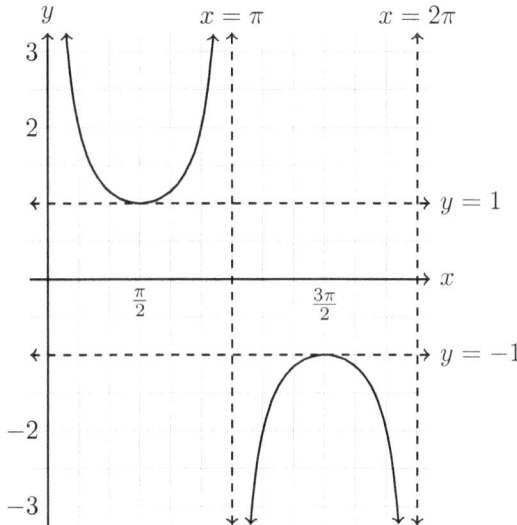

Notice that the principal period of cosecant has two U-shaped branches. Its vertical asymptotes occur at $x = 0$, $x = \pi$, and $x = 2\pi$. Graphing more branches of cosecant is a matter of emulating cosecant's behavior in the interval $(0, 2\pi)$. This is because of Proposition 5.3 which says cosecant's period is 2π.

Our procedure to graph

$$f(x) = A\csc(Bx + C) + D$$

is nearly identical to secant's which is described on page 166. Simply replace $y = A\cos(Bx + C) + D$ with $y = A\sin(Bx + C) + D$, and draw the U-shaped branches based on the latter's graph.

Example 6.9 Graph two periods of

$$g(x) = 2\csc(3x - \pi) + 1.$$

Solution In Example 1 of Section 6.1, we graphed

$$y = 2\sin(3x - \pi) + 1$$

for $\pi/3 \le x \le \pi$, which was one period.

We need two periods of the sine graph to obtain two periods of cosecant's graph. To keep the graph close to the y-axis, we will graph $y = 2\sin(3x - \pi) + 1$ for the period corresponding $-\pi/3 \le x \le \pi/3$. The result is (a).

We draw dashed horizontal lines at

$$y = 2 + 1 = 3 \quad \text{and} \quad y = -2 + 1 = -1.$$

Vertical asymptotes are drawn at

$$x = -\frac{\pi}{3}, \quad x = 0, \quad x = \frac{\pi}{3}, \quad x = \frac{2\pi}{3}, \quad \text{and} \quad x = \pi,$$

because this is where $y = 2\sin(3x - \pi) + 1$ intersects $y = 1$. The result is (b).

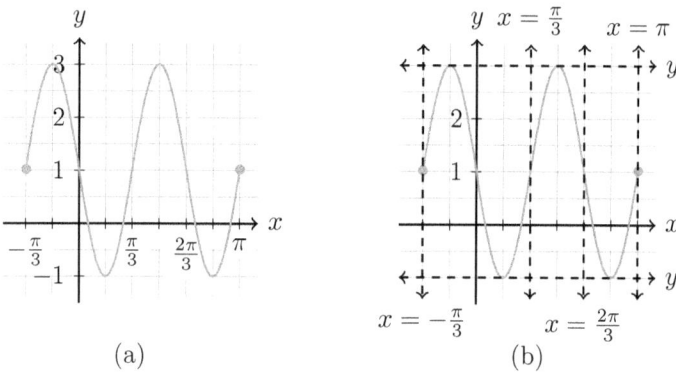

(a) (b)

We are ready to graph cosecant. The vertices for the branches are at $(-\pi/6, 3)$, $(\pi/6, -1)$, $(\pi/2, 3)$, and $(5\pi/6, -1)$. The U-shaped branches follow from the asymptotes.

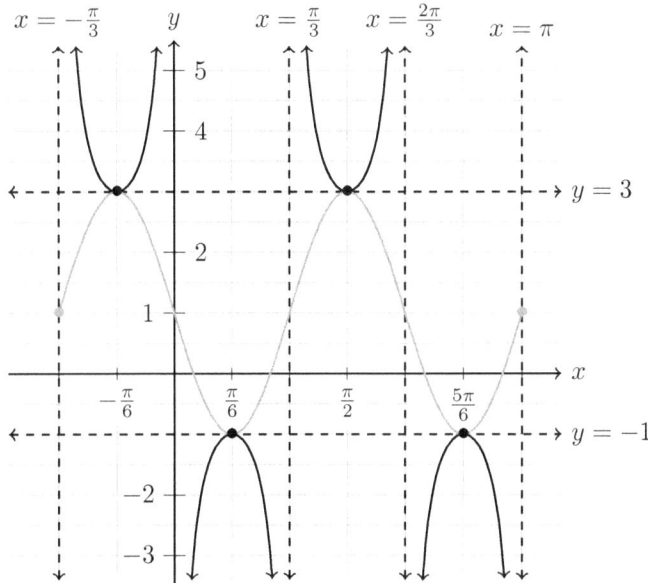

6.4 Miscellaneous Graphing Problems

In this section, we will examine some less essential graphing problems. We will study them for fun as well as the fact that working a few challenging problems is important for students' mathematical development.

Example 6.10 Graph
$$f(x) = x \sin x.$$

Solution Since
$$-1 \leq \sin x \leq 1 \quad \text{implies} \quad -|x| \leq x \sin x \leq |x|,$$
the graph of f oscillates between $y = -|x|$ and $y = |x|$.

Plotting points for x in the interval $[-\pi, \pi]$ is helpful because sine has period 2π and the sign change of x at $x = 0$ affects the look of the graph. We can then use the behavior we observe from our points to graph more of f.

Our first point will have an x-coordinate of $-\pi$ and subsequent x-coordinates will be $2\pi/4 = \pi/2$ greater than their previous x-coordinate.

x	$f(x)$		
$-\pi$	$-\pi \sin(-\pi)$	$=$	0
$-\dfrac{\pi}{2}$	$-\dfrac{\pi}{2} \sin\left(-\dfrac{\pi}{2}\right)$	$=$	$\dfrac{\pi}{2}$
0	$0 \sin 0$	$=$	0
$\dfrac{\pi}{2}$	$\dfrac{\pi}{2} \sin \dfrac{\pi}{2}$	$=$	$\dfrac{\pi}{2}$
π	$\pi \sin \pi$	$=$	0

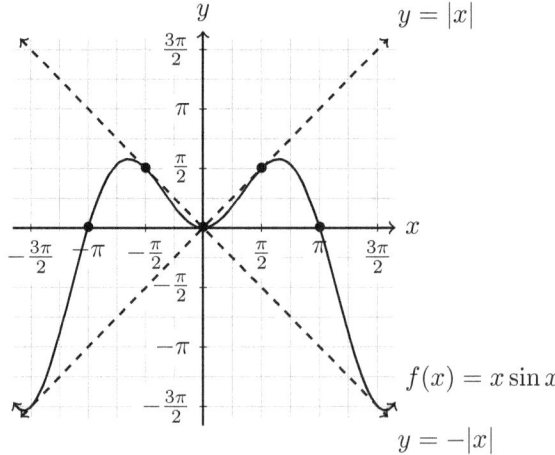

Example 6.11 Graph two periods of

$$g(x) = |\tan x|.$$

Solution We will use the graph of $y = \tan x$ as an aid. When $\tan x \geq 0$, the absolute value does nothing so we leave the graph of $y = \tan x$ unaltered. When $\tan x < 0$, the absolute value makes $g(x)$ positive so we reflect the graph of $y = \tan x$ about the x-axis.

The graph of $g(x) = |\tan x|$ is drawn in black, and the graph of $y = \tan x$ is drawn in gray.

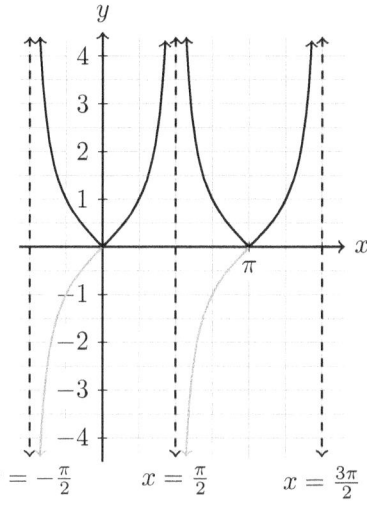

Example 6.12 Determine the number of times $y = x$ intersects $y = \cos 35x$ for $x > 0$.

Solution Let us examine one period, and make an inference about the general pattern.

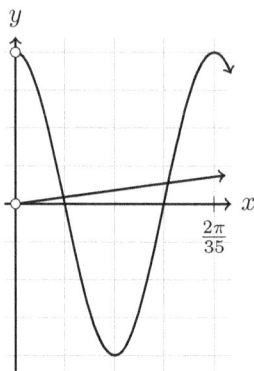

During the first quarter of each period cosine goes from 1 to 0, and in the fourth quarter cosine goes from 0 to 1. When $0 \leq x \leq 1$, this implies that the graphs of $y = \cos 35x$ and $y = x$ intersect once in the first and fourth quarter of each period. When $x > 1$, $y = x$ will never intersect $y = \cos 35x$, because $-1 \leq \cos 35x \leq 1$.

As a result, we can find the number of positive intersects by computing the number of periods of $y = \cos 35x$ within the interval $[0, 1]$. The period of $y = \cos 35x$ is $2\pi/35$. It follows that there are

$$\frac{1}{2\pi/35} = \frac{35}{2\pi} \approx 5.570$$

periods between $x = 0$ and $x = 1$. Since $5.25 < 5.570 < 5.75$, $y = \cos 35x$ and $y = x$ intersect

$$2(5) + 1 = 11$$

times. ∎

6.5 Exercises

* Exercise 1

Determine the amplitude, period, vertical shift, and phase shift.

(a) $y = 3\sin \pi x$

(b) $y = -\dfrac{1}{2}\cos(x - 30°) + 1$

(c) $y = -2\sin(2\pi(x-1)) - \pi$

(d) $y = 1 - \cos(x - \pi)$

(e) $y = -2\sin\left(\dfrac{x}{2} + \pi\right)$

(f) $y = \dfrac{3\cos\left(\dfrac{x+\pi}{3}\right) - 7}{4}$

** Exercise 2

Write a corresponding function.

(a) A sine function begins each period by decreasing from its neutral position.

Amplitude:	3
Period:	6
Vertical shift:	0
Phase shift:	2

(b) A cosine function begins each period at its maximum.

Amplitude:	π
Period:	$\pi/2$
Vertical shift:	-2
Phase shift:	$-\pi/6$

** Exercise 3

Graph one period.

(a) $y = 2\sin \pi x$

(b) $y = 1 - \cos(180° x - 30°)$.

(c) $y = -\dfrac{3}{5}\sin\left(\dfrac{\pi}{6}(x-2)\right) - \dfrac{3}{4}$

(d) $y = -6\cos\left(\dfrac{x}{4} + \dfrac{\pi}{10}\right) + 1$

(e) $y = -\sin(90° + x) + 1$

(f) $y = 2\cos\left(\dfrac{\pi - x}{6}\right)$

(g) $y = \sin\left(\dfrac{\pi - 2x}{3}\right) + 1$

(h) $y = 7 - \cos(-\pi x)$

** Exercise 4

Graph two periods.

(a) $y = -3\sin 60° x$

(b) $y = 5 - \tfrac{1}{2}\cos 5x$

(c) $y = 2 - \sin\left(\dfrac{x-\pi}{3}\right)$

(d) $y = 6\cos\left(\dfrac{x + 18°}{5}\right)$

(e) $y = -\sin\left(-x + \dfrac{\pi}{4}\right) + 1$

(f) $y = -\dfrac{3}{4}\cos\left(\dfrac{\pi - x}{4}\right)$

** Exercise 5

Find A, B, and C such that ...

(a) $\cos x = A\sin(Bx + C)$.

(b) $\sin x = A\cos(Bx + C)$.

* Exercise 6

Determine the period, vertical shift, phase shift, and asymptotes.

(a) $y = -3\tan\left(\dfrac{x}{2} + 45°\right)$

(b) $y = \dfrac{3\pi}{2} - \cot(17°x)$

(c) $y = \dfrac{5\tan\left(\dfrac{x+\pi}{3}\right) + 2}{\pi}$

(d) $y = -3\cot(18°x + 15°) + 1$

(e) $y = \dfrac{3}{4} - \tan(x - \pi)$

(f) $y = -4\cot\left(\dfrac{\pi x - 3}{2}\right) + 1$

** Exercise 7

Write a corresponding function.

(a) A tangent function has vertical asymptotes of $x = -1$ and $x = 9$. Its vertical shift is 1, and it contains the point $(3/2, -2)$.

(b) A cotangent function with vertical asymptotes $x = 0$ and $x = 5\pi$. Its vertical shift is -3, and it contains the point $(5\pi/4, -7)$.

** Exercise 8

Graph one period.

(a) $y = -2\tan 60°x$

(b) $y = \dfrac{\pi}{2} - \pi\cot 2x$

(c) $y = -3\tan\left(\dfrac{\pi}{10}(x - 7)\right) - 5$

(d) $y = -3\cot\left(\dfrac{x}{5} - 20°\right) + 1$

(e) $y = -\tan(60° - x) + 1$

(f) $y = \pi\cot\left(\dfrac{\pi - x}{3}\right)$

(g) $y = \tan(15° - 3x) - 1$

(h) $y = -\cot\left(\dfrac{2\pi}{3}(2 - x)\right) + 2$

** Exercise 9

Graph two periods.

(a) $y = -\pi \tan \dfrac{\pi x}{6}$

(b) $y = 2 - \dfrac{3}{4} \cot 10° x$

(c) $y = 1 - \dfrac{1}{2} \tan\left(\dfrac{2x - \pi}{4}\right)$

(d) $y = -\cot(20°(x+2)) + 3$

(e) $y = -\tan\left(-x + \dfrac{\pi}{8}\right) + 2$

(f) $y = \cot\left(\dfrac{\pi - x}{6}\right) + 1$

** Exercise 10

Find A, B, and C such that ...

(a) $\cot x = A\tan(Bx + C)$.

(b) $\tan x = A\cot(Bx + C)$.

* Exercise 11

Determine the period, vertical shift, phase shift, and asymptotes.

(a) $y = 2\sec(120° x) - 3$

(b) $y = -\pi \csc\left(x + \dfrac{\pi}{3}\right) + \dfrac{\pi}{6}$

(c) $y = 1 - \csc(4(x - \pi))$

(d) $y = -2\sec\left(\dfrac{x}{2} + \pi\right)$

** Exercise 12

Write a corresponding function.

(a) A secant function has asymptotes $x = 1$, $x = 3$, and $x = 5$. Its vertical shift is 2. The vertex of a downward opening branch is $(2, -1)$.

(b) A cosecant function has asymptotes $x = \pi$, $x = 3\pi$, and $x = 5\pi$. Its vertical shift is -11π. The vertex of an upward opening branch is $(4\pi, \pi/2)$.

** Exercise 13

Graph one period.

(a) $y = -5\sec 9° x$

(b) $y = \csc\left(x + \dfrac{\pi}{2}\right) - 5$

(c) $y = 2 - \sec\left(\dfrac{x - 2\pi}{3}\right)$

(d) $y = \csc\left(\dfrac{\pi}{12}(x - 4)\right) + 1$.

(e) $y = 2\pi + \pi \sec(20° x + 15°)$

(f) $y = -2\csc\left(\dfrac{x}{6} - \dfrac{\pi}{18}\right) + 3$

(g) $y = -\sec\left(\dfrac{\pi}{4} - x\right) + 1$

(h) $y = 3\csc\left(\dfrac{\pi + x}{4}\right)$

** Exercise 14

Graph two periods.

(a) $y = -2\sec\dfrac{2x}{3}$

(b) $y = \pi - \dfrac{\pi}{6}\csc 5x$

(c) $y = 3 - \sec\left(\dfrac{x+\pi}{8}\right)$

(d) $y = -\csc\left(\dfrac{\pi}{6}(2x+6)\right) + 2$.

(e) $y = -2\sec\left(-x + \dfrac{\pi}{3}\right)$

(f) $y = -\dfrac{\pi}{8}\csc\left(\dfrac{3\pi - \pi x}{4}\right)$

** Exercise 15

Graph one period.

(a) $y = \dfrac{3\csc(x + 17°) - 4}{5}$

(b) $y = 4\sin\left(\dfrac{x - \pi}{2}\right)$

(c) $y = \dfrac{3}{4}\tan\left(\dfrac{x - \pi}{3}\right)$

(d) $y = 3 - \cos 3x$

(e) $y = 1 - \cot\left(\dfrac{\pi}{4}(x - 3)\right)$.

(f) $y = 2\sec(2\pi(x - 1)) - 1$

** Exercise 16

Graph two periods.

(a) $y = 1 - \sin(20°(x - 2))$

(b) $y = -4\csc\left(\dfrac{x + \pi}{6}\right)$

(c) $y = 3 - \tan\left(\dfrac{x - 2\pi}{3}\right)$

(d) $y = 1 - \sec(x - \pi)$

(e) $y = \dfrac{1}{2}\cot\left(\dfrac{x + 25°}{10}\right)$

(f) $y = \cos(20°x + 30°) - 1$

** Exercise 17

Write corresponding equations for the graphs on pages 180 and 181.

*** Exercise 18

Graph each of the following.

(a) $y = x\cos x$

(b) $y = x\sin \pi x$

(c) $y = \cos x + x$

(d) $y = x\sec x$

*** Exercise 19

Graph each of the following.

(a) $y = |\cos x|$

(b) $y = |\cot x|$

(c) $y = |\csc x|$

*** Exercise 20

Suppose $x > 0$. Determine the number of times $y = x$ intersects each graph.

(a) $y = \sin 3x$

(b) $y = \cos 15x$

(c) $y = \sin 40x$

(d) $y = \cos 85x$

(e) $y = \tan x$

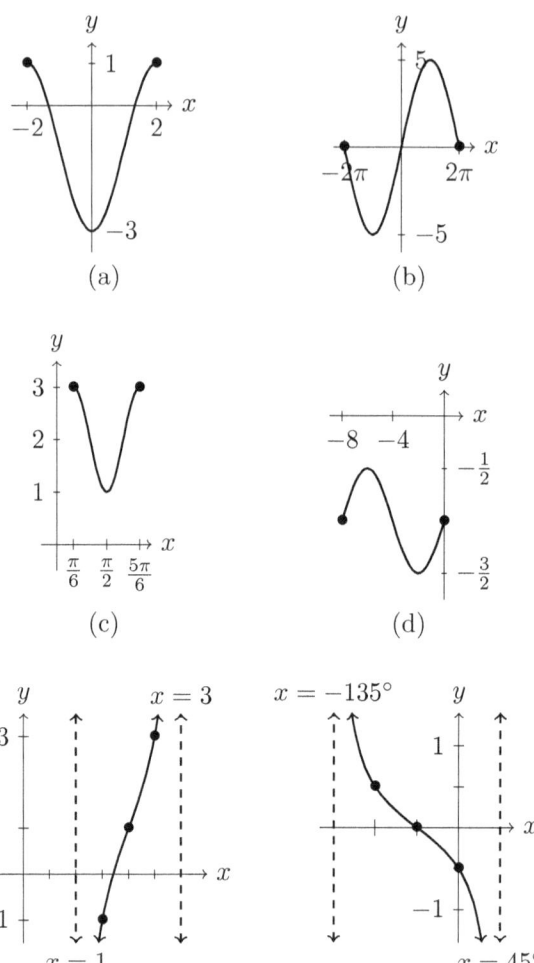

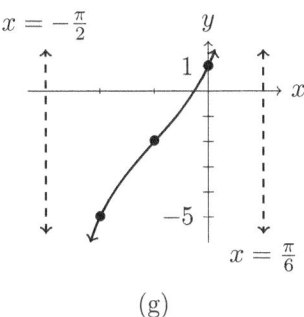
(g)

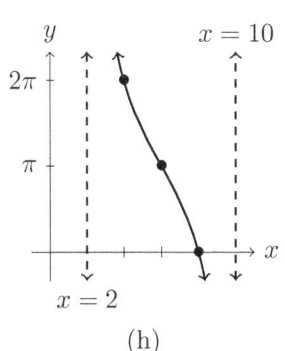
(h)

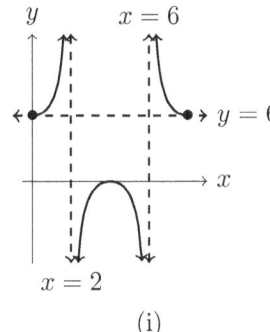
(i)

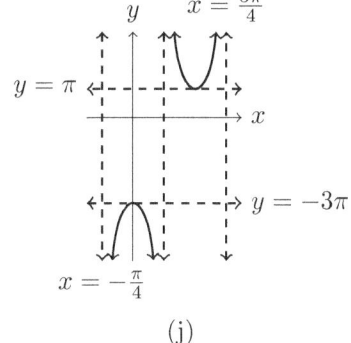
(j)

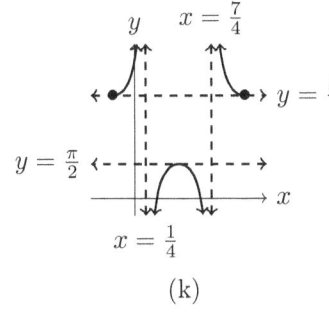
(k)

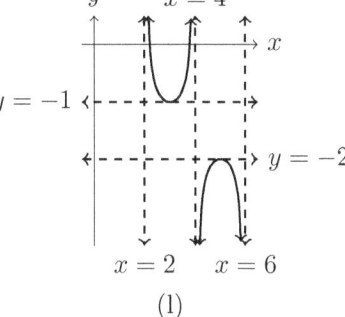
(l)

Chapter 7

Using Identities

In this chapter, we will examine some popular identities. We assume a solid command of Chapter 5. A modest amount of information from Chapter 6 will also be used. Calculators are not required. Indeed, many techniques discussed further expand the set of angles at which we can evaluate the trigonometric functions exactly by hand.

7.1 Sum and Difference Identities

Theorem 7.1 (Sum and Difference Identities) *Suppose α and β are standard position angles.*

(i) $\sin(\alpha \pm \beta) = \sin\alpha\cos\beta \pm \cos\alpha\sin\beta$

(ii) $\cos(\alpha \pm \beta) = \cos\alpha\cos\beta \mp \sin\alpha\sin\beta$

(iii) $\tan(\alpha \pm \beta) = \dfrac{\tan\alpha \pm \tan\beta}{1 \mp \tan\alpha\tan\beta}$

The proof of Theorem 7.1 is difficult. As a result, it is broken up into pieces within this chapter. The proof of (ii) is in Subsection 7.1.1, (i) is proven in Subsection 7.2.1, and (iii) is left as an exercise for the reader.

Example 7.1 Evaluate
$$\sin \frac{\pi}{12}.$$

Solution Since
$$\frac{\pi}{12} = \frac{\pi}{3} - \frac{\pi}{4},$$
Theorem 7.1 (i) tells us
$$\begin{aligned}\sin \frac{\pi}{12} &= \sin\left(\frac{\pi}{3} - \frac{\pi}{4}\right) \\ &= \sin\frac{\pi}{3}\cos\frac{\pi}{4} - \cos\frac{\pi}{3}\sin\frac{\pi}{4} \\ &= \frac{\sqrt{3}}{2} \cdot \frac{\sqrt{2}}{2} - \frac{1}{2} \cdot \frac{\sqrt{2}}{2} \\ &= \frac{\sqrt{6}}{4} - \frac{\sqrt{2}}{4} \\ &= \frac{\sqrt{6} - \sqrt{2}}{4}.\end{aligned}$$

∎

Example 7.2 Evaluate
$$\cos 195° \cos 15° + \sin 195° \sin 15°.$$

Solution Using Theorem 7.1 (ii),
$$\begin{aligned}\cos 195° \cos 15° + \sin 195° \sin 15° &= \cos(195° - 15°) \\ &= \cos 180° \\ &= -1.\end{aligned}$$

∎

Example 7.3 Suppose the terminal side of α is in quadrant II and the terminal side of β is in quadrant III. Assume
$$\sin\alpha = \frac{3}{5} \quad \text{and} \quad \tan\beta = \frac{5}{12}.$$
Compute (a) $\cos(\alpha + \beta)$ and (b) $\tan(\alpha - \beta)$.

Solution This will require Theorem 7.1 (ii) and (iii). However, in addition to the given information, the identities require $\cos\alpha$, $\tan\alpha$, $\sin\beta$, and $\cos\beta$. To find these values we will build triangles using the techniques outlined in Section 5.4.

Say the side opposite the reference angle of α is 3. Then the hypotenuse must be 5. The side adjacent the reference angle has signed length -4 due to the Pythagorean Theorem and the fact that the terminal side of α lies in quadrant II.

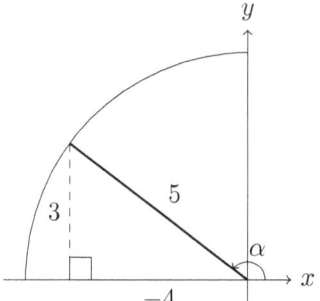

Let us say the side opposite β's reference angle has length 5. Then the side adjacent must have length 12. Because terminal side of β lies in quadrant III, these sides' signed lengths are -5 and -12, respectively. The Pythagorean Theorem tells us the length of the hypotenuse is 13.

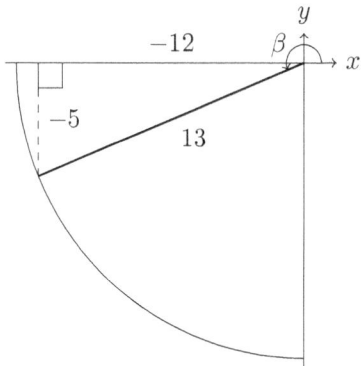

We are ready to answer the questions.

(a)
$$\cos(\alpha + \beta) = \cos\alpha\cos\beta - \sin\alpha\sin\beta$$
$$= \left(-\frac{4}{5}\right)\left(-\frac{12}{13}\right) - \left(\frac{3}{5}\right)\left(-\frac{5}{13}\right)$$
$$= \frac{48}{65} + \frac{15}{65}$$
$$= \frac{63}{65}.$$

(b)
$$\tan(\alpha - \beta) = \frac{\tan\alpha - \tan\beta}{1 + \tan\alpha\tan\beta}$$
$$= \frac{-3/4 - 5/12}{1 + (-3/4)(5/12)}$$
$$= \frac{-14/12}{11/16}$$
$$= -\frac{56}{33}.$$

■

Example 7.4 Graph
$$y = 2\sqrt{3}\sin x - 2\cos x.$$

Solution Our goal is to use Theorem 7.1 (i) to rewrite the expression into the form
$$y = A\sin(Bx + C) + D,$$
and then use the techniques discussed in Section 6.1 to graph the function. To do this, we will find a length r and a standard position angle θ such that
$$2\sqrt{3}\sin x - 2\cos x = r(\cos\theta\sin x - \sin\theta\cos x) = r\sin(x - \theta).$$

The terminal side of θ contains the point $(2\sqrt{3}, 2)$, because
$$r\cos\theta = 2\sqrt{3} \quad \text{and} \quad r\sin\theta = 2.$$

This allows us to build a triangle.

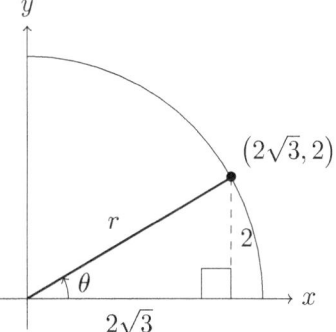

Using the Pythagorean Theorem,

$$\left(2\sqrt{3}\right)^2 + 2^2 = r^2 \quad \text{implies} \quad r = 4.$$

It follow that

$$\cos\theta = \frac{2\sqrt{3}}{4} = \frac{\sqrt{3}}{2} \quad \text{and} \quad \sin\theta = \frac{2}{4} = \frac{1}{2}.$$

From here, we see that $\theta = \pi/6$ satisfies the necessary criteria. Then Theorem 7.1 (i) allows us to rewrite $y = 2\sqrt{3}\sin x - 2\cos x$:

$$y = 2\sqrt{3}\sin x - 2\cos x$$
$$= 4\left(\frac{\sqrt{3}}{2}\sin x - \frac{1}{2}\cos x\right)$$
$$= 4\left(\cos\frac{\pi}{6}\sin x - \sin\frac{\pi}{6}\cos x\right)$$
$$= 4\left(\sin x\cos\frac{\pi}{6} - \cos x\sin\frac{\pi}{6}\right)$$
$$= 4\sin\left(x - \frac{\pi}{6}\right).$$

All that is left is to graph the result.

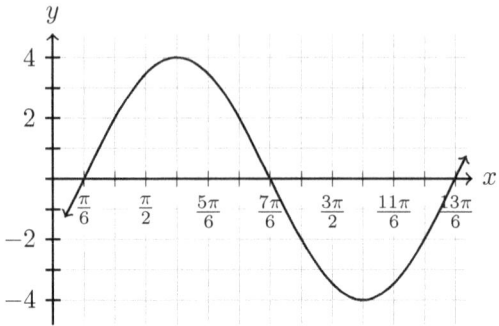

7.1.1 Proof of Theorem 7.1 (ii)

Theorem 7.1 (ii) says

$$\cos(\alpha \pm \beta) = \cos\alpha\cos\beta \mp \sin\alpha\sin\beta.$$

Proof Suppose α and β are angles in standard position. Let A and B be the points

$$(\cos\alpha, \sin\alpha) \quad \text{and} \quad (\cos\beta, \sin\beta),$$

respectively.

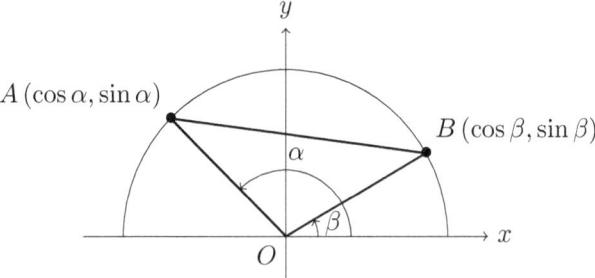

Using the distance formula,

$$AB = \sqrt{(\cos\alpha - \cos\beta)^2 + (\sin\alpha - \sin\beta)^2}.$$

Rotate $\triangle OBA$ measure β clockwise, so $\overline{OB}$ lies on the positive x-axis. Call the images of A and B under rotation A' and B'

respectively. Then A' has coordinates $\Big(\cos{(\alpha-\beta)},\sin{(\alpha-\beta)}\Big)$ and B' has coordinates $(1,0)$.

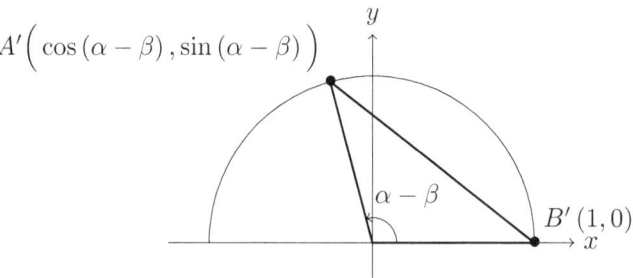

Using the distance formula,

$$A'B' = \sqrt{\Big(\cos(\alpha-\beta)-1\Big)^2 + \sin^2{(\alpha-\beta)}}.$$

Rotations do not change lengths, so

$$A'B' = AB.$$

It follows that

$$\sqrt{\Big(\cos(\alpha-\beta)-1\Big)^2 + \sin^2{(\alpha-\beta)}} = \sqrt{\Big(\cos\alpha-\cos\beta\Big)^2 + (\sin\alpha-\sin\beta)^2}.$$

Squaring both sides yields

$$\Big(\cos(\alpha-\beta)-1\Big)^2+\sin^2{(\alpha-\beta)} = \Big(\cos\alpha-\cos\beta\Big)^2+\Big(\sin\alpha-\sin\beta\Big)^2.$$

Let us simplify each side separately. On the left side of the equation, we have

$$\Big(\cos(\alpha-\beta)-1\Big)^2 + \sin^2{(\alpha-\beta)}$$
$$= \cos^2{(\alpha-\beta)} - 2\cos(\alpha-\beta) + 1 + \sin^2{(\alpha-\beta)}$$
$$= \underbrace{\cos^2{(\alpha-\beta)} + \sin^2{(\alpha-\beta)}}_{1} - 2\cos(\alpha-\beta) + 1$$
$$= 2 - 2\cos(\alpha-\beta).$$

On the right side, we have

$$\left(\cos\alpha - \cos\beta\right)^2 + \left(\sin\alpha - \sin\beta\right)^2$$
$$= \cos^2\alpha - 2\cos\alpha\cos\beta + \cos^2\beta + \sin^2\alpha - 2\sin\alpha\sin\beta + \sin^2\beta$$
$$= \underbrace{\cos^2\alpha + \sin^2\alpha}_{1} - 2\cos\alpha\cos\beta - 2\sin\alpha\sin\beta + \underbrace{\cos^2\beta + \sin^2\beta}_{1}$$
$$= 2 - 2\cos\alpha\cos\beta - 2\sin\alpha\sin\beta.$$

Hence,

$$\begin{aligned} 2 - 2\cos(\alpha - \beta) &= 2 - 2\cos\alpha\cos\beta - 2\sin\alpha\sin\beta \\ \Rightarrow \quad -2\cos(\alpha - \beta) &= -2\cos\alpha\cos\beta - 2\sin\alpha\sin\beta \\ \Rightarrow \quad \cos(\alpha - \beta) &= \cos\alpha\cos\beta + \sin\alpha\sin\beta \end{aligned}$$

Now to prove

$$\cos(\alpha + \beta) = \cos\alpha\cos\beta - \sin\alpha\sin\beta.$$

Using Proposition 5.4, we know sine and cosine are odd and even, respectively. Thus,

$$\begin{aligned} \cos(\alpha + \beta) &= \cos\left(\alpha - (-\beta)\right) \\ &= \cos\alpha\cos(-\beta) + \sin\alpha\sin(-\beta) \\ &= \cos\alpha\cos\beta + \sin\alpha\left(-\sin\beta\right) \\ &= \cos\alpha\cos\beta - \sin\alpha\sin\beta. \end{aligned}$$

∎

7.2 Other Identities

In this section, we will examine a few other popular trigonometric identities. Their proofs all follow from Theorem 7.1, albeit indirectly in some cases.

7.2.1 The Cofunction Identities

Proposition 7.1 (Cofunction Identities) *Let θ be a standard position angle.*

(i) $\sin(90° - \theta) = \cos\theta$
(ii) $\cos(90° - \theta) = \sin\theta$
(iii) $\tan(90° - \theta) = \cot\theta$
(iv) $\cot(90° - \theta) = \tan\theta$
(v) $\sec(90° - \theta) = \csc\theta$
(vi) $\csc(90° - \theta) = \sec\theta$

Proof The reader is given the opportunity to prove (i) in Exercise 12. We will prove (ii) and (iii).

(ii) Using Theorem 7.1 (ii),

$$\cos(90° - \theta) = \cos 90° \cos\theta + \sin 90° \sin\theta$$
$$= 0 \cdot \cos\theta + 1 \cdot \sin\theta$$
$$= \sin\theta.$$

(iii) Assume (i) and (ii) hold. Then

$$\tan(90° - \theta) = \frac{\sin(90° - \theta)}{\cos(90° - \theta)}$$
$$= \frac{\cos\theta}{\sin\theta}$$
$$= \cot\theta.$$

■

Example 7.5 Suppose $\tan 40° \approx 0.839$. Without using a calculator, approximately what is the value of $\cot 50°$?

Solution Since

$$\cot 50° = \tan(90° - 40°) = \tan 40°,$$

we conclude that
$$\cot 50° \approx 0.839.$$

■

Example 7.6 Use the Cofunction Identities (i) and (ii) as well as Theorem 7.1 (ii) to prove

$$\sin(\alpha \pm \beta) = \sin\alpha\cos\beta \pm \sin\beta\cos\alpha.$$

Solution Theorem 7.1 (ii) tells us

$$\cos(\alpha \pm \beta) = \cos\alpha\cos\beta \mp \sin\alpha\sin\beta,$$

and Cofunction Identity (ii) says $\cos(90° - \theta) = \sin\theta$. Therefore,

$$\begin{aligned}\sin(\alpha \pm \beta) &= \cos\left(90° - (\alpha \pm \beta)\right) \\ &= \cos\left((90° - \alpha) \mp \beta\right) \\ &= \cos(90° - \alpha)\cos\beta \pm \sin(90° - \alpha)\sin\beta \\ &= \sin\alpha\cos\beta \pm \cos\alpha\sin\beta.\end{aligned}$$

∎

7.2.2 Double Angle Identities

Proposition 7.2 (Double Angle Identities) *Suppose θ is a standard position angle.*

(i) $\sin 2\theta = 2\sin\theta\cos\theta$

(ii) $\cos 2\theta = \begin{cases} \cos^2\theta - \sin^2\theta \\ 2\cos^2\theta - 1 \\ 1 - 2\sin^2\theta \end{cases}$

(iii) $\tan 2\theta = \dfrac{2\tan\theta}{1 - \tan^2\theta}$

Proof The proofs for these properties are an application of Theorem 7.1.

(i) Using Theorem 7.1 (i),

$$\begin{aligned}\sin 2\theta &= \sin(\theta + \theta) \\ &= \sin\theta\cos\theta + \sin\theta\cos\theta \\ &= 2\sin\theta\cos\theta.\end{aligned}$$

(ii) Due to Theorem 7.1 (ii),
$$\begin{aligned}\cos 2\theta &= \cos(\theta + \theta) \\ &= \cos\theta\cos\theta - \sin\theta\sin\theta \\ &= \cos^2\theta - \sin^2\theta.\end{aligned}$$

The other two variations of Proposition 7.2 (ii) following from the Pythagorean Identities
$$\sin^2\theta = 1 - \cos^2\theta \quad \text{and} \quad \cos^2\theta = 1 - \sin^2\theta.$$

We have
$$\begin{aligned}\cos 2\theta &= \cos^2\theta - \sin^2\theta \\ &= \cos^2\theta - (1 - \cos^2\theta) \\ &= 2\cos^2\theta - 1\end{aligned}$$

and
$$\begin{aligned}\cos 2\theta &= \cos^2\theta - \sin^2\theta \\ &= (1 - \sin^2\theta) - \sin^2\theta \\ &= 1 - 2\sin^2\theta.\end{aligned}$$

(iii) From Theorem 7.1 (iii),
$$\begin{aligned}\tan 2\theta &= \tan(\theta + \theta) \\ &= \frac{\tan\theta + \tan\theta}{1 - \tan^2\theta} \\ &= \frac{2\tan\theta}{1 - \tan^2\theta}.\end{aligned}$$

■

Example 7.7 Assume
$$\cos^2 2x = 2 - 5\cos x.$$
Solve for x.

Solution The second case of Proposition 7.2 (ii) tells us
$$\cos 2x = 2\cos^2 x - 1.$$
So,
$$\begin{aligned}
& & \cos^2 2x &= 2 - 5\cos x \\
&\Rightarrow & 2\cos^2 x - 1 &= 2 - 5\cos x \\
&\Rightarrow & 2\cos^2 x + 5\cos x - 3 &= 0 \\
&\Rightarrow & (2\cos x - 1)(\cos x + 3) &= 0
\end{aligned}$$

It follows that
$$\cos x = \frac{1}{2} \quad \text{or} \quad \cos x = -3.$$
The latter case is impossible. If $\cos x = 1/2$, then
$$x = \frac{\pi}{3} + 2\pi n \quad \text{or} \quad x = \frac{5\pi}{3} + 2\pi n$$
for $n = 0, 1, -1, 2, -2, \ldots$. ■

Example 7.8 Suppose
$$\tan \alpha = -\frac{3}{2} \quad \text{and} \quad \sec \alpha > 0.$$
Find (a) $\sin 2\alpha$, (b) $\cos 2\alpha$, and (c) $\tan 2\alpha$.

Solution The first step is to build a triangle using the techniques outlined in Section 5.4. Since $\tan \alpha < 0$ and $\sec \alpha > 0$, the terminal side of α is in quadrant IV. Say, the side opposite the reference angle has signed length of -3. Then the adjacent side has length 2. Due to the Pythagorean Theorem the hypotenuse has length $\sqrt{13}$.

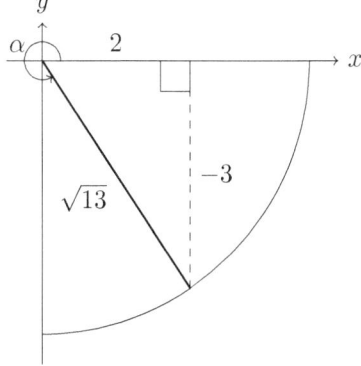

(a) Using Proposition 7.2 (i),

$$\sin 2\alpha = 2 \sin \alpha \cos \alpha$$
$$= 2\left(-\frac{3}{\sqrt{13}}\right)\left(\frac{2}{\sqrt{13}}\right)$$
$$= -\frac{12}{13}.$$

(b) Proposition 7.2 (ii) tells us

$$\cos 2\alpha = 2\cos^2 \alpha - 1$$
$$= 2\left(\frac{2}{\sqrt{13}}\right)^2 - 1$$
$$= -\frac{5}{13}.$$

(c) Because of Proposition 7.2 (iii),

$$\tan 2\alpha = \frac{2 \tan \alpha}{1 - \tan^2 \alpha}$$
$$= \frac{2(-3/2)}{1 - (-3/2)^2}$$
$$= \frac{12}{5}.$$

■

The triangle we constructed in Example 8 was not required for (c). This is because we were given tangent.

7.2.3 Half Angle Identities

Proposition 7.3 (Half Angle Identities) *Assume θ is a standard position angle.*

(i) $\sin \dfrac{\theta}{2} = \pm\sqrt{\dfrac{1-\cos\theta}{2}}$

(ii) $\cos \dfrac{\theta}{2} = \pm\sqrt{\dfrac{1+\cos\theta}{2}}$

(iii) $\tan \dfrac{\theta}{2} = \begin{cases} \pm\sqrt{\dfrac{1-\cos\theta}{1+\cos\theta}} \\[2mm] \dfrac{1-\cos\theta}{\sin\theta} \\[2mm] \dfrac{\sin\theta}{1+\cos\theta} \end{cases}$

Proof

(i) Proposition 7.2 (ii) gives

$$\cos 2\alpha = 1 - 2\sin^2\alpha.$$

Solving for $\sin^2\alpha$ yields

$$\sin^2\alpha = \frac{1-\cos(2\alpha)}{2}.$$

Taking square roots and substituting $\theta/2$ for α gives

$$\sin \frac{\theta}{2} = \pm\sqrt{\frac{1-\cos\theta}{2}}.$$

(ii) From Proposition 7.2 (ii),

$$\cos(2\alpha) = 2\cos^2\alpha - 1.$$

Solving for $\cos^2 \alpha$ gives

$$\cos^2 \alpha = \frac{1+\cos(2\alpha)}{2}.$$

Then, after taking square roots and replacing α with $\theta/2$, we have

$$\cos\frac{\theta}{2} = \pm\sqrt{\frac{1+\cos\theta}{2}}.$$

(iii) The proof for (iii) follows from (i) and (ii):

$$\begin{aligned}\tan\frac{\theta}{2} &= \frac{\pm\sqrt{\frac{1-\cos\theta}{2}}}{\pm\sqrt{\frac{1+\cos\theta}{2}}} = \\ &= \pm\sqrt{\frac{1-\cos\theta}{2}\cdot\frac{2}{1+\cos\theta}} \\ &= \pm\sqrt{\frac{1-\cos\theta}{1+\cos\theta}}.\end{aligned}$$

To prove the second and third cases of (iii), we will use the identity

$$1-\cos^2\theta = \sin^2\theta.$$

We have

$$\begin{aligned}\tan\frac{\theta}{2} &= \pm\sqrt{\frac{1-\cos\theta}{1+\cos\theta}} \\ &= \pm\sqrt{\frac{1-\cos\theta}{1+\cos\theta}\cdot\frac{1-\cos\theta}{1-\cos\theta}} \\ &= \pm\sqrt{\frac{(1-\cos\theta)^2}{1-\cos^2\theta}} \\ &= \pm\sqrt{\frac{(1-\cos\theta)^2}{\sin^2\theta}} \\ &= \pm\left|\frac{1-\cos\theta}{\sin\theta}\right|\end{aligned}$$

$$\begin{aligned}\tan\frac{\theta}{2} &= \pm\sqrt{\frac{1-\cos\theta}{1+\cos\theta}} \\ &= \pm\sqrt{\frac{1-\cos\theta}{1+\cos\theta}\cdot\frac{1+\cos\theta}{1+\cos\theta}} \\ &= \pm\sqrt{\frac{1-\cos^2\theta}{(1-\cos\theta)^2}} \\ &= \pm\sqrt{\frac{\sin^2\theta}{(1-\cos\theta)^2}} \\ &= \pm\left|\frac{\sin\theta}{1-\cos\theta}\right|\end{aligned}$$

Then a careful analysis of signs reveals

$$\tan\frac{\theta}{2} = \frac{1-\cos\theta}{\sin\theta} \quad \text{and} \quad \tan\frac{\theta}{2} = \frac{\sin\theta}{1-\cos\theta}.$$

Example 7.9 Use the Half Angle Identities to compute (a) $\sin 157.5°$, (b) $\cos 157.5°$, and (c) $\tan 157.5°$.

Solution

(a) Using Half Angle Identity (i),

$$\sin 157.5° = \sin\left(\frac{1}{2} \cdot 315°\right) = \pm\sqrt{\frac{1 - \cos 315°}{2}}.$$

Because $315°$ is in quadrant IV and its reference angle is $45°$, we know

$$\cos 315° = \cos 45° = \frac{\sqrt{2}}{2}.$$

It follows that

$$\sin 157.5° = \pm\sqrt{\frac{1 - \cos 315°}{2}}$$

$$= \pm\sqrt{\frac{1 - \sqrt{2}/2}{2}}$$

$$= \pm\sqrt{\frac{2 - \sqrt{2}}{4}}$$

$$= \pm\frac{\sqrt{2 - \sqrt{2}}}{2}.$$

Since $157.5°$ is in quadrant II, sine is positive. Hence,

$$\sin 157.5° = \frac{\sqrt{2 - \sqrt{2}}}{4}.$$

(b) Using Half Angle Identity (ii),

$$\cos 157.5° = \cos\left(\frac{1}{2} \cdot 315°\right) = \pm\sqrt{\frac{1 + \cos 315°}{2}}.$$

From (a), $\cos 315° = \sqrt{2}/2$. It follows that

$$\cos 157.5° = \pm\sqrt{\frac{1 + \cos 315°}{2}}$$

$$= \pm\sqrt{\frac{1 + \sqrt{2}/2}{2}}$$

$$= \pm\sqrt{\frac{2 + \sqrt{2}}{4}}$$

$$= \pm\frac{\sqrt{2 + \sqrt{2}}}{2}.$$

Since $157.5°$ is in the quadrant II, cosine is negative. Thus,

$$\cos 157.5° = -\frac{\sqrt{2 + \sqrt{2}}}{2}.$$

(c) Using the second Half Angle Identity (iii),

$$\tan 157.5° = \tan\left(\frac{1}{2} \cdot 315°\right) = \frac{1 - \cos 315°}{\sin 315°}.$$

All is left is to plug in the appropriate values for sine and cosine. From our previous work, we know $\cos 315° = \sqrt{2}/2$. Because $315°$ is in quadrant IV and its reference angle is $45°$,

$$\sin 315° = -\sin 45° - \frac{\sqrt{2}}{2}.$$

Ergo,

$$\tan 157.5° = \frac{1 - \cos 315°}{\sin 315°}$$

$$= \frac{1 - \sqrt{2}/2}{-\sqrt{2}/2}$$

$$= \frac{1 - \sqrt{2}/2}{-\sqrt{2}/2} \cdot \frac{2\sqrt{2}}{2\sqrt{2}}$$

$$= \frac{2\sqrt{2} - 2}{-2}$$

$$= 1 - \sqrt{2}.$$

∎

Example 7.10 Suppose θ is in $[0, 2\pi)$, and

$$\cot\theta = -\frac{\sqrt{7}}{3} \quad \text{and} \quad \cos\theta < 0.$$

What are the exact values of (a) $\sin(\theta/2)$, (b) $\cos(\theta/2)$, and (c) $\tan(\theta/2)$?

Solution The first step is to build a triangle. Since $\cot\theta < 0$ and $\cos\theta < 0$, the terminal side of θ lies in quadrant II. Suppose the signed length of the side adjacent the reference angle of θ is $-\sqrt{7}$. Then the side opposite has length 3. Using the Pythagorean Theorem, the hypotenuse must have length 4.

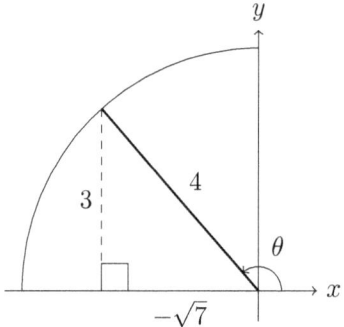

We are ready to answer the questions.

(a) Using Half Angle Identity (i),

$$\sin\frac{\theta}{2} = \pm\sqrt{\frac{1-\cos\theta}{2}}$$

$$= \pm\sqrt{\frac{1-\left(-\sqrt{7}/4\right)}{2}}$$

$$= \pm\sqrt{\frac{4+\sqrt{7}}{8}}$$

$$= \pm\sqrt{\frac{8+2\sqrt{7}}{16}}$$

$$= \pm\frac{\sqrt{8+2\sqrt{7}}}{4}.$$

Since $\pi/2 < \theta < \pi$ implies $\pi/4 < \theta/2 < \pi/2$, we conclude $\sin(\theta/2)$ is positive. Thus,

$$\sin\frac{\theta}{2} = \frac{\sqrt{8+2\sqrt{7}}}{4}.$$

(b) Using Half Angle Identity (ii),

$$\cos\frac{\theta}{2} = \pm\sqrt{\frac{1+\cos\theta}{2}}$$

$$= \pm\sqrt{\frac{1+(-\sqrt{7}/4)}{2}}$$

$$= \pm\sqrt{\frac{4-\sqrt{7}}{8}}$$

$$= \pm\sqrt{\frac{8-2\sqrt{7}}{16}}$$

$$= \pm\frac{\sqrt{8-2\sqrt{7}}}{4}.$$

Since $\pi/2 < \theta < \pi$ implies $\pi/4 < \theta/2 < \pi/2$, it follows that $\cos(\theta/2)$ is positive. Therefore,

$$\cos\frac{\theta}{2} = \frac{\sqrt{8-2\sqrt{7}}}{4}.$$

(c) Lastly, using the second case of Half Angle Identity (iii),

$$\tan\frac{\theta}{2} = \frac{1-\cos\theta}{\sin\theta}$$

$$= \frac{1+\sqrt{7}/4}{3/4}$$

$$= \frac{4+\sqrt{7}}{3}.$$

■

Another set of identities, which can be used to solve the same type of problems, are the Power Reducing Identities.

Corollary 7.1 (Power Reducing Identities) *Say that θ is a standard position angle.*

(i) $\sin^2 \theta = \dfrac{1 - \cos 2\theta}{2}$

(ii) $\cos^2 \theta = \dfrac{1 + \cos 2\theta}{2}$

(iii) $\tan^2 \theta = \dfrac{1 - \cos 2\theta}{1 + \cos 2\theta}$

Readers that prefer these identities to the Half Angle Identities are welcome to use them instead.

7.2.4 Product to Sum and Difference Identities

Proposition 7.4 (Product to Sum and Difference Identities) *Suppose α and β are in $\mathbb{R}$.*

(i) $\sin \alpha \sin \beta = \dfrac{1}{2}\Big(\cos(\alpha - \beta) - \cos(\alpha + \beta) \Big)$

(ii) $\cos \alpha \cos \beta = \dfrac{1}{2}\Big(\cos(\alpha + \beta) + \cos(\alpha - \beta) \Big)$

(iii) $\sin \alpha \cos \beta = \dfrac{1}{2}\Big(\sin(\alpha + \beta) + \sin(\alpha - \beta) \Big)$

(iv) $\cos \alpha \sin \beta = \dfrac{1}{2}\Big(\sin(\alpha + \beta) - \sin(\alpha - \beta) \Big)$

These identities were invaluable for non-exact evaluation before students had access to calculators. In those days, students used tables to evaluate trigonometric expressions. As a result, these identities made it less cumbersome for students to evaluate products, because they were able to convert them into sums or differences which are easier to compute. Some programmers are still interested in the identities for the same reason.

There are other modern applications. The Product to Sum and Difference Identities allow students to find exact values of trigonometric functions at a slightly larger set of angles. The identities have applications in Calculus as well.

Proof

(i) Theorem 7.1 (ii) gives

$$\begin{array}{rcl} \cos\alpha\cos\beta + \sin\alpha\sin\beta & = & \cos(\alpha-\beta) \\ -\left(\cos\alpha\cos\beta - \sin\alpha\sin\beta\right. & = & \left.\cos(\alpha+\beta)\right) \\ \hline 2\sin\alpha\sin\beta & = & \cos(\alpha-\beta) - \cos(\alpha+\beta) \end{array}$$

Then dividing by 2 yields

$$\sin\alpha\sin\beta = \frac{1}{2}\Big(\cos(\alpha+\beta) - \cos(\alpha-\beta)\Big).$$

(ii) Using Theorem 7.1 (ii),

$$\begin{array}{rcl} \cos\alpha\cos\beta - \sin\alpha\sin\beta & = & \cos(\alpha+\beta) \\ +\ \cos\alpha\cos\beta + \sin\alpha\sin\beta & = & \cos(\alpha-\beta) \\ \hline 2\cos\alpha\cos\beta & = & \cos(\alpha+\beta) + \cos(\alpha-\beta). \end{array}$$

Then dividing by 2 yields

$$\cos\alpha\cos\beta = \frac{1}{2}\Big(\cos(\alpha+\beta) + \cos(\alpha-\beta)\Big).$$

(iii) From Theorem 7.1 (i),

$$\begin{array}{rcl} \sin\alpha\cos\beta + \cos\alpha\sin\beta & = & \sin(\alpha+\beta) \\ +\ \sin\alpha\cos\beta - \cos\alpha\sin\beta & = & \sin(\alpha-\beta) \\ \hline 2\sin\alpha\cos\beta & = & \sin(\alpha+\beta) + \sin(\alpha-\beta). \end{array}$$

Then dividing by 2 yields

$$\sin\alpha\cos\beta = \frac{1}{2}\Big(\sin(\alpha+\beta) + \sin(\alpha-\beta)\Big).$$

(iv) Because of Theorem 7.1 (i),

$$\begin{array}{rcl} \sin\alpha\cos\beta + \cos\alpha\sin\beta & = & \sin(\alpha+\beta) \\ -\left(\sin\alpha\cos\beta - \cos\alpha\sin\beta\right. & = & \left.\sin(\alpha-\beta)\right) \\ \hline 2\cos\alpha\sin\beta & = & \sin(\alpha+\beta) - \sin(\alpha-\beta). \end{array}$$

Then dividing by 2 yields

$$\cos\alpha\cos\beta = \frac{1}{2}\Big(\sin(\alpha+\beta) - \sin(\alpha-\beta)\Big).$$

Example 7.11 Compute (a) $\cos 45° \cos 15°$ and (b) $\sin 22.5° \cos 22.5°$.

Solution

(a) Using Proposition 7.4 (ii),

$$\begin{aligned}
\cos 45° \cos 15° &= \frac{1}{2}\Big(\cos(45° + 15°) + \cos(45° - 15°)\Big) \\
&= \frac{1}{2}\Big(\cos 60° + \cos 30°\Big) \\
&= \frac{1}{2}\left(\frac{1}{2} + \frac{\sqrt{3}}{2}\right) \\
&= \frac{1}{2}\left(\frac{1 + \sqrt{3}}{2}\right) \\
&= \frac{1 + \sqrt{3}}{4}.
\end{aligned}$$

(b) Due to Proposition 7.4 (iii),

$$\begin{aligned}
\sin 22.5° \cos 22.5° &= \frac{1}{2}\Big(\sin(22.5° + 22.5°) + \sin(22.5° - 22.5°)\Big) \\
&= \frac{1}{2}\Big(\sin 45° + \sin 0\Big) \\
&= \frac{1}{2}\left(\frac{\sqrt{2}}{2} + 0\right) \\
&= \frac{\sqrt{2}}{4}.
\end{aligned}$$

7.3 Verifying Identities

We will verify identities in the final section of this chapter. The key ideas of the verification process were introduced in Section 5.6. However, this chapter has added more identities to our knowledge base, and the reader will be expected to utilize them.

Example 7.12 Verify

$$\frac{\csc(\pi/2 - \alpha)}{1 + \tan^2 \alpha} = \cos \alpha.$$

Solution We need three identities: Pythagorean Identity (ii), Proposition 7.1 (iv), and a Reciprocal Identity. They say

$$1 + \tan^2 \alpha = \sec^2 \alpha, \quad \csc\left(\frac{\pi}{2} - \alpha\right) = \sec \alpha, \quad \text{and} \quad \frac{1}{\sec \alpha} = \cos \alpha,$$

respectively. So,

$$\frac{\csc(\pi/2 - \alpha)}{1 + \tan^2 \alpha} = \frac{\csc(\pi/2 - \alpha)}{\sec^2 \alpha}$$
$$= \frac{\sec \alpha}{\sec^2 \alpha}$$
$$= \frac{1}{\sec \alpha}$$
$$= \cos \alpha.$$

∎

Example 7.13 Verify that the equation is an identity.

$$\frac{\tan 2\beta}{\tan \beta} = \frac{2\cos^2 \beta}{\cos^2 \beta - \sin^2 \beta}$$

Solution Recall that
$$\tan \beta = \frac{\sin \beta}{\cos \beta}.$$
We also need Proposition 7.2 (iii) which says
$$\tan 2\beta = \frac{2\tan \beta}{1 - \tan^2 \beta}.$$
Utilizing these identities, we have

$$\begin{aligned}
\frac{\tan 2\beta}{\tan \beta} &= \tan(2\beta)\frac{1}{\tan \beta} \\
&= \frac{2\tan \beta}{1 - \tan^2 \beta} \cdot \frac{1}{\tan \beta} \\
&= \frac{2}{1 - \tan^2 \beta} \\
&= \frac{2}{1 - \frac{\sin^2 \beta}{\cos^2 \beta}} \\
&= \frac{2/1}{\frac{\cos^2 \beta - \sin^2 \beta}{\cos^2 \beta}} \\
&= \frac{2}{1} \cdot \frac{\cos^2 \beta}{\cos^2 \beta - \sin^2 \beta} \\
&= \frac{2\cos^2 \beta}{\cos^2 \beta - \sin^2 \beta}.
\end{aligned}$$
∎

Example 7.14 Verify the identity.
$$-\frac{1}{2} + \frac{1}{2}\sin\theta \cot\frac{\theta}{2} + \frac{1}{2}\cos\theta = \cos\theta.$$

Solution We need a Reciprocal Identity and the third case of Proposition 7.3 (iii), which say
$$\cot\frac{\theta}{2} = \frac{1}{\tan(\theta/2)} \quad \text{and} \quad \tan\frac{\theta}{2} = \frac{\sin\theta}{1+\cos\theta}.$$

With our identities in mind, we proceed as follows:
$$\begin{aligned}
-\frac{1}{2} + \frac{1}{2}\sin\theta\cot\frac{\theta}{2} + \frac{1}{2}\cos\theta &= -\frac{1}{2} + \frac{\sin\theta}{2}\cdot\frac{1}{\tan(\theta/2)} + \frac{1}{2}\cos\theta \\
&= -\frac{1}{2} + \frac{\sin\theta}{2}\cdot\frac{1+\cos\theta}{\sin\theta} + \frac{1}{2}\cos\theta \\
&= -\frac{1}{2} + \frac{1}{2}(1+\cos\theta) + \frac{1}{2}\cos\theta \\
&= -\frac{1}{2} + \frac{1}{2} + \frac{1}{2}\cos\theta + \frac{1}{2}\cos\theta \\
&= \cos\theta.
\end{aligned}$$

∎

7.4 Exercises

** Exercise 1

Find the exact value.

(a) $\cos 285°$
(b) $\tan(-165°)$
(c) $\sin 375°$
(d) $\tan 255°$
(e) $\cos(-525°)$
(f) $\csc 345°$
(g) $\sec 105°$
(h) $\cot 15°$

** Exercise 2

Calculate the exact value.

(a) $\tan\left(-\frac{\pi}{12}\right)$
(b) $\sin\frac{19\pi}{12}$
(c) $\cot\frac{11\pi}{12}$
(d) $\sin\left(-\frac{35\pi}{12}\right)$
(e) $\sec\frac{23\pi}{12}$
(f) $\csc\left(-\frac{7\pi}{12}\right)$
(g) $\cos\frac{25\pi}{12}$
(h) $\cot\left(-\frac{85\pi}{12}\right)$

** Exercise 3

What is the exact value?

(a) $\cos 408° \cos 198° + \sin 408° \sin 198°$
(b) $\dfrac{\tan 57° + \tan 78°}{1 - \tan 57° \tan 78°}$
(c) $\sin 575° \cos 275° - \sin 275° \cos 575°$
(d) $\dfrac{\tan 312° - \tan 192°}{1 + \tan 312° \tan 197°}$
(e) $\cos 40° \cos 20° - \sin 40° \sin 20°$
(f) $\dfrac{\tan 286° - \tan 136°}{1 + \tan 286° \tan 136°}$

** Exercise 4

Compute the exact value.

(a) $\dfrac{\tan\frac{5\pi}{9} + \tan\frac{43\pi}{36}}{1 - \tan\frac{5\pi}{9} \tan\frac{43\pi}{36}}$
(b) $\sin\frac{19\pi}{9}\cos\frac{7\pi}{9} - \sin\frac{7\pi}{9}\cos\frac{19\pi}{9}$
(c) $\dfrac{\tan\frac{17\pi}{18} - \tan\frac{\pi}{9}}{1 + \tan\frac{17\pi}{18}\tan\frac{\pi}{9}}$
(d) $\cos\frac{5\pi}{24}\cos\frac{\pi}{24} - \sin\frac{5\pi}{24}\sin\frac{\pi}{24}$
(e) $\dfrac{\tan\frac{31\pi}{18} - \tan\frac{17\pi}{36}}{1 + \tan\frac{31\pi}{18}\tan\frac{17\pi}{36}}$
(f) $\cos\frac{\pi}{8}\cos\frac{23\pi}{24} + \sin\frac{\pi}{8}\sin\frac{23\pi}{24}$

** Exercise 5

Solve for θ.

(a) $\cos\left(\theta + \dfrac{\pi}{6}\right) = \sin\theta$
(b) $\sin\left(\theta + \frac{\pi}{4}\right) + \cos\left(\theta + \frac{\pi}{4}\right) = -1$
(c) $\tan\left(\theta + \dfrac{\pi}{6}\right) = -\sqrt{3}$

** Exercise 6

Suppose $\sin\alpha = 12/13$, $\tan\beta = -\sqrt{17}/8$, $\cos\alpha < 0$, and $\cos\beta > 0$.

(a) Find $\sin(\alpha + \beta)$.
(b) What is $\cos(\alpha - \beta)$?
(c) Evaluate $\tan(\alpha + \beta)$.
(d) Calculate $\csc(\alpha - \beta)$.

** Exercise 7

Assume $\cos\alpha = -4/5$, $\cot\beta = 15/8$, $\tan\alpha > 0$, and $\sin\beta > 0$.

(a) Find $\cos(\alpha - \beta)$.

(b) What is $\tan(\alpha - \beta)$?

(c) Evaluate $\sec(\alpha + \beta)$.

(d) Calculate $\cot(\alpha + \beta)$.

** Exercise 8

Write each expression in the form

$$y = A\sin(Bx + C) + D$$

for some A, B, C, and D.

(a) $y = 5\sin \pi x + 5\cos \pi x$

(b) $y = -3\sin x - 3\sqrt{3}\cos x$

(c) $y = \cos x - \sin x$

(d) $y = \sqrt{3}\sin 2x - \cos 2x$

** Exercise 9

Graph each expression.

(a) $y = 3\sqrt{2}\sin\dfrac{x}{2} + 3\sqrt{2}\cos\dfrac{x}{2}$

(b) $y = \sqrt{3}\sin x - \cos x$

(c) $y = -\dfrac{1}{4}\sin x - \dfrac{\sqrt{3}}{4}\cos x$

(d) $y = \sqrt{2}\cos\dfrac{\pi x}{2} - \sqrt{2}\sin\dfrac{\pi x}{2}$

* Exercise 10

$\sin 17° \approx 0.292$ $\csc 43° \approx 1.466$

$\cos 22° \approx 0.927$ $\sec 55° \approx 1.743$

$\tan 77° \approx 4.331$ $\cot 25° \approx 2.145$

Find the approximate value without a calculator.

(a) $\cot 13°$ (d) $\tan 65°$

(b) $\cos 73°$ (e) $\sin 68°$

(c) $\csc 35°$ (f) $\sec 47°$

* Exercise 11

$\sin\dfrac{5\pi}{16} \approx 0.831$ $\csc\dfrac{\pi}{7} \approx 2.305$

$\cos\dfrac{2\pi}{5} \approx 0.309$ $\sec\dfrac{5\pi}{18} \approx 1.556$

$\tan\dfrac{7\pi}{11} \approx -2.190$ $\cot\dfrac{11\pi}{16} \approx -0.668$

Compute the exact value without a calculator.

(a) $\tan\left(-\dfrac{3\pi}{16}\right)$ (d) $\sin\dfrac{\pi}{10}$

(b) $\cos\dfrac{3\pi}{16}$ (e) $\sec\dfrac{5\pi}{14}$

(c) $\csc\dfrac{2\pi}{9}$ (f) $\cot\left(-\dfrac{3\pi}{22}\right)$

** Exercise 12

Use Proposition 7.1 (ii) and the substitution

$$\theta = 90° - \varphi$$

to prove Proposition 7.1 (i).

** Exercise 13

Use parts (i) and (ii) of Proposition 7.1 to prove parts (iv), (v), and (vi).

** Exercise 14

Suppose $\csc\theta = -17/15$ and $\cos\theta < 0$. Find the six trigonometric functions evaluated at 2θ.

** Exercise 15

Say $\sec\varphi = 5/4$ and $\tan\varphi > 0$. What are the six trigonometric functions evaluated at 2φ?

** Exercise 16

Solve for θ.

(a) $\sin 2\theta + \cos\theta = 0$

(b) $\sin\theta = 1 - \cos 2\theta$

(c) $\cos 2\theta = 3\cos\theta + 4$

(d) $\tan 2\theta + 7 = 7 - \tan\theta$

** Exercise 17

Evaluate without a calculator.

(a) $\cos 165°$

(b) $\tan 285°$

(c) $\sin 22.5°$

(d) $\csc 15°$

(e) $\cot 255°$

(f) $\sec 202.5°$

(g) $\sin 7.5°$

(h) $\cos 191.25°$

** Exercise 18

Compute without a calculator.

(a) $\cos\dfrac{\pi}{8}$

(b) $\tan\dfrac{11\pi}{12}$

(c) $\sin\dfrac{19\pi}{8}$

(d) $\sec\dfrac{17\pi}{12}$

(e) $\cot\dfrac{9\pi}{8}$

(f) $\csc\dfrac{\pi}{12}$

(g) $\tan\dfrac{\pi}{24}$

(h) $\sin\dfrac{17\pi}{16}$

** Exercise 19

Say $\tan\theta = -7/24$, $\cos\theta > 0$, and $0 \leq \theta < 360°$. Compute the values of the six trigonometric functions at $\theta/2$.

** Exercise 20

Assume $\sec\varphi = -17/8$, $\csc\varphi > 0$, and $0 \leq \varphi < 2\pi$. What are the six trigonometric functions at $\varphi/2$?

** Exercise 21

Calculate.

(a) $\cos 105° \cos 45°$

(b) $\sin 30° \sin 15°$

(c) $\sin 105° \cos 105°$

(d) $\cos\dfrac{435°}{2} \sin\dfrac{375°}{2}$

** Exercise 22

Evaluate.

(a) $\cos\dfrac{7\pi}{12}\sin\dfrac{\pi}{4}$

(b) $\sin\dfrac{\pi}{6}\cos\dfrac{\pi}{12}$

(c) $\sin\dfrac{29\pi}{24}\sin\dfrac{25\pi}{24}$

(d) $\cos\dfrac{7\pi}{12}\cos\dfrac{\pi}{12}$

** Exercise 23

(i) 1   (iii) $-\cot x$

(ii) $\tan x$ (iv) $\sec x$

Match the above with the expressions below. Some options may be used more than once.

(a) $\dfrac{1+\tan^2 x}{\csc(90°-x)}$

(b) $\cos(90°-x)\csc(x)$

(c) $\cos(-x)\csc(-x)$

(d) $\dfrac{\sin(90°-x)}{\sin(-x)}$

(e) $\cos x \csc(90°-x)$

(f) $\dfrac{\sec x}{\csc x}$

(g) $\dfrac{\sin 2x}{2\cos^2 x}$

(h) $\dfrac{\sin(90°-2x)}{1-2\sin^2 x}$

** Exercise 24

Verify the identity.

(a) $\dfrac{2\sin(\alpha+\pi/4)}{\sqrt{2}}=\sin\alpha+\cos\alpha$

(b) $\dfrac{2\cos(\theta-45°)}{\sqrt{2}}=\sin\beta+\cos\beta$

(c) $\tan\left(\gamma+\dfrac{\pi}{4}\right)=\dfrac{1+\tan\gamma}{1-\tan\gamma}$

** Exercise 25

Verify.

(a) $\dfrac{\cos(90°-\alpha)}{1-\cos^2\alpha}=\csc\alpha$

(b) $\dfrac{\sec(\pi/2-\beta)}{1+\cot^2\beta}=\sin\beta$

(c) $\dfrac{\cot(90°-\gamma)}{\sec^2\gamma-1}=\cot\gamma$

** Exercise 26

Verify the identities.

(a) $\dfrac{\tan\alpha}{\tan 2\alpha}=\dfrac{2-\sec^2\alpha}{2}$

(b) $\dfrac{\tan 2\beta}{\sin\beta}=\dfrac{2}{2\cos\beta-\sec\beta}$

** Exercise 27

Verify.

(a) $\sec 2\alpha = \dfrac{\sec^2 \alpha}{1 - \tan^2 \alpha}$

(b) $\csc 2\beta = \dfrac{1}{2} \sec \beta \csc \beta$

(c) $\cot 2\gamma = \dfrac{\cot^2 \gamma - 1}{2 \cot \gamma}$

** Exercise 28

Verify the identities.

(a) $\dfrac{\sin 2\alpha}{1 - \cos^2 \alpha} = 2 \cot \alpha$

(b) $\dfrac{1 - \sin^2 \beta}{\sin 2\beta} = \dfrac{\cot \beta}{2}$

(c) $\dfrac{\sin 2\gamma}{2 - 2\sin^2 \gamma} = \tan \gamma$

** Exercise 29

Verify.

(a) $\dfrac{\cos \alpha + \sin \alpha}{\cos 2\alpha} = \dfrac{\sec \alpha}{1 - \tan \alpha}$

(b) $\dfrac{\cos \beta - \sin \beta}{\cos 2\beta} = \dfrac{\csc \beta}{1 + \cot \beta}$

(c) $\dfrac{\cos 2\gamma + \cos \gamma}{2\cos \gamma - 1} = \cos \gamma + 1$

** Exercise 30

Verify the identities.

(a) $\dfrac{2\sin^2(\alpha/2)}{1 - \cos^2 \alpha} = \dfrac{\sec \alpha}{1 + \sec \alpha}$

(b) $\dfrac{\cos^2(\beta/2)}{\sin^2 \beta} = \dfrac{\csc^2 \beta + \csc \beta \cot \beta}{2}$

(c) $\dfrac{\tan(\gamma/2)}{1 - \cos \gamma} = \csc \gamma$

*** Exercise 31

Verify.

$$\sin \theta = \cot \dfrac{\theta}{2} - \cos \theta \cot \dfrac{\theta}{2}.$$

*** Exercise 32

Use Theorem 7.1 (i) and (ii) to prove

(a) $\tan(\alpha+\beta) = \dfrac{\tan \alpha + \tan \beta}{1 - \tan \alpha \tan \beta}$

(b) $\tan(\alpha-\beta) = \dfrac{\tan \alpha - \tan \beta}{1 + \tan \alpha \tan \beta}$

** Exercise 33

Use Theorem 7.1 to prove (a) sine is odd and (b) cosine is even. This exercise is only for didactic purposes; we used that sine and cosine are odd and even, respectively, when we proved Theorem 7.1.

Chapter 8

Inverse Trigonometric Functions

This chapter will analyze inverses of trigonometric functions. We studied inverse trigonometric functions in Section 4.2, but only to find acute angle measures. Inverse trigonometric functions require more care when utilized to find angle measures more generally.

We assume thorough knowledge of Chapters 5 and 6. Some understanding of Chapter 7 is also helpful. Scientific calculators are necessary.

8.1 Inverses

Definition 8.1 The function g is the **inverse** of f if

$$f\big(g(x)\big) = x \quad \text{and} \quad g\big(f(x)\big) = x.$$

Inverse functions "undo" the original function. So, if $f(x) = y$ and g is the inverse of f, then $g(y) = x$.

x	-2	-1	0	1	2
$f(x)$	1	2	0	7	-4

Example 8.1 Suppose f is defined via the table above, and g is the inverse of f. Compute (a) $g(7)$, (b) $g\bigl(f(2)\bigr)$, (c) $f\bigl(g(1)\bigr)$, and (d) $g\bigl(g(-4)\bigr)$.

Solution

(a) Since $f(1) = 7$, we know $g(7) = 1$.

(b) Inverses "undo" the original functions, so we immediately know $g\bigl(f(2)\bigr) = 2$.

(c) The original function also "undoes" the inverse, which means $f\bigl(g(1)\bigr) = 1$.

(d) We know $f(2) = -4$ and $f(-1) = 2$. Hence,
$$g\bigl(g(-4)\bigr) = g(2) = -1.$$

■

Some functions are not invertible, i.e. they do not have an inverse. For example, $f(x) = x^2$ has no inverse. This is because f sends more than one input to the same output, e.g.
$$f(-2) = (-2)^2 = 4 \quad \text{and} \quad f(2) = 2^2 = 4.$$
As a result, if g were the inverse of $f(x) = x^2$, then $g(4) = -2$ and $g(4) = 2$. This contradicts the definition of a function, so no such function g exists.

Definition 8.2 A function f is **one-to-one** if
$$f(u) = f(v) \quad \text{implies} \quad u = v.$$

Example 8.2 Determine which functions are one-to-one.
(a) $f(x) = \sin x$, (b) $g(x) = 2x - 3$, and (c) $h(x) = x^2 - 2x + 1$

(a) The function $f(x) = \sin x$ is *not* one-to-one. For example,
$$f(0) = \sin 0 = 0 \quad \text{and} \quad f(2\pi) = \sin 2\pi = 0.$$

(b) The function $g(x) = 2x - 3$ is one-to-one, because each output corresponds to exactly one input.

(c) The function $h(x) = x^2 - 2x + 1$ is *not* one-to-one. For example,
$$h(0) = 1 \quad \text{and} \quad h(2) = 4 - 4 + 1 = 1.$$

■

Proposition 8.1 *A function is invertible if and only if it is one-to-one.*

Example 8.3 Which of the functions in Example 2 have an inverse?

Solution Due to Proposition 8.1, a function is invertible if and only if it is one-to-one. It follows that the function f defined by $f(x) = \sin x$ is not invertible because it is not one-to-one, the function g defined by $g(x) = 2x - 3$ is invertible because it is one-to-one, and the function h defined by $h(x) = x^2 - 2x + 1$ is not invertible because it is not one-to-one. ■

Pictorially, the one-to-one criterion for invertibility of a function is satisfied if there is no horizontal line that intersects the graph of the function at more than one point. In other words, we have the following rule.

Horizontal Line Test

- If a horizontal line intersects the graph of a function at more than one point, then the function is *not* invertible.

- If no horizontal intersects the graph of a function at more than one point, then the function is invertible.

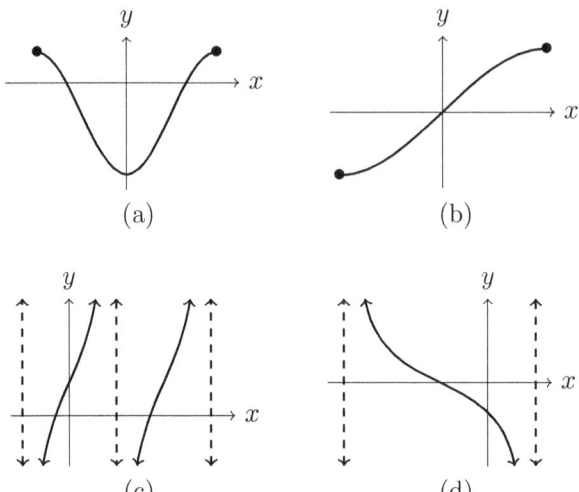

Example 8.4 Use the horizontal line test to determine which graphs correspond to invertible functions. Assume each function's domain is contained within the x-axis shown.

Solution

(a) Since we can draw a horizontal line which intersects the graph at more than one point, the corresponding function is not invertible.

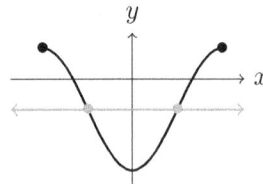

(b) This graph corresponds to an invertible function. No matter where a horizontal line is drawn, it intersects the graph no more than once.

(c) This function is not invertible. Any horizontal line intersects the graph at two points.

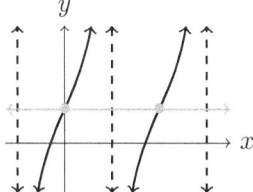

(d) No horizontal line intersects the graph at more than one point. Therefore, the corresponding function is invertible. ∎

8.1.1 Restricting the Domain

Though many functions are not invertible, we can restrict their domains to intervals on which they are one-to-one. The functions obtained from the restrictions are invertible. Sometimes we refer to the inverse of a function, when it is restricted to a particular domain, as "the inverse" of the function. This is not technically correct as the original function is not invertible, but it is a common practice which will be adopted within this text to simplify sentences.

Consider $f(x) = x^2$. It is not one-to-one, so it has no general inverse. However, $f(x) = x^2$ is one-to-one for $x \geq 0$. Within this domain its inverse is $g(x) = \sqrt{x}$.

No horizontal line intersects the graph of $f(x) = x^2$ at more than one point if we assume $x \geq 0$.

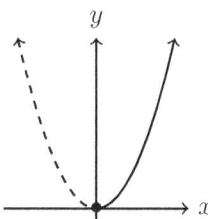

Another way to make $f(x) = x^2$ one-to-one is to restrict its domain

217

to $x \leq 0$. On this domain, its inverse is $g(x) = -\sqrt{x}$.

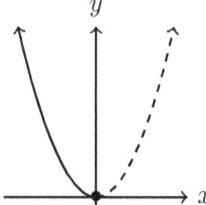

If we assume $x \leq 0$, the graph of $f(x) = x^2$ will intersect any horizontal line at most once.

Much like $f(x) = x^2$, many functions have multiple ways to obtain invertibility via a restriction of the domain. Usually the restriction is either based on a convention or the needs of the mathematician at the particular moment.

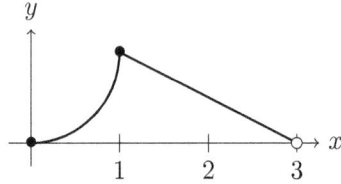

Example 8.5 Consider the graph above. (a) Determine the domain of the function. (b) Find two ways of restricting the domain to obtain invertibility.

Solution

(a) The domain of the function depicted in the graph is $0 \leq x < 3$.

(b) No horizontal line intersects the graph more than once if we assume either $0 \leq x \leq 1$ or $1 \leq x < 3$. Hence, the corresponding function is invertible if we restrict its domain to either $0 \leq x \leq 1$ or $1 \leq x < 3$.

8.2 Arc Sine

Consider $f(x) = \sin x$. As Example 2 (a) showed, sine is not one-to-one. As a result, it does not have an inverse on its entire domain. However, as can be seen from the graph below, f is invertible if we restrict the domain to $-\pi/2 \leq x \leq \pi/2$.

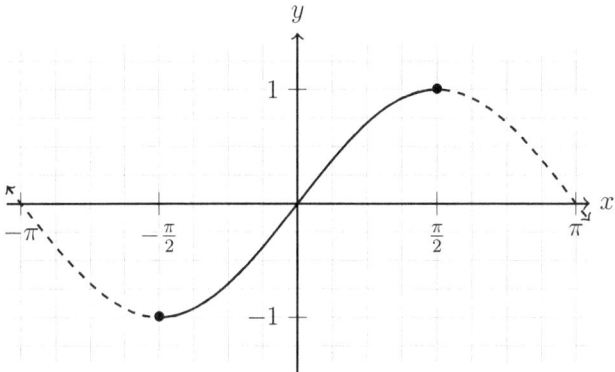

Definition 8.3 The **arc sine** of x, denoted $\arcsin x$, is the function defined by the relationship

$$y = \arcsin x \quad \text{if} \quad \sin y = x$$

for $-1 \leq x \leq 1$ and $-\pi/2 \leq y \leq \pi/2$.

Alternative notation for $\arcsin x$ is $\sin^{-1} x$. Usually when this notation is used, we say "inverse sine of x". Both forms will be utilized in examples and exercises. However, $\arcsin x$ is our preferred notation because $\sin^{-1} x$ is easily confused with $\csc x$.

Example 8.6 Evaluate without a calculator when possible.

(a) $\sin^{-1} 1$

(b) $\arcsin \dfrac{1}{2}$

(c) $\sin^{-1}\left(-\dfrac{1}{2}\right)$

(d) $\arcsin 11$

Solution

(a) We know
$$\sin^{-1} 1 = \frac{\pi}{2} \quad \text{because} \quad \sin \frac{\pi}{2} = 1$$
and $\pi/2$ is in the interval $[-\pi/2, \pi/2]$.

(b) Since $\sin(\pi/6) = 1/2$ and $\pi/6$ is in the interval $[-\pi/2, \pi/2]$,
$$\arcsin \frac{1}{2} = \frac{\pi}{6}.$$

(c) Result (b) tells us the reference angle of $\sin^{-1}(-1/2)$ is $\pi/6$. Since the range of inverse sine is the interval $[-\pi/2, \pi/2]$, the answer to (c) must be in quadrant IV and be a negative radian measure. Ergo,
$$\sin^{-1}\left(-\frac{1}{2}\right) = -\frac{\pi}{6}.$$

(d) The value 11 is not within the domain of arc sine. Hence, $\arcsin 11$ is undefined.

∎

Let us introduce a proposition which some students use to evaluate problems like Example 6 (c).

Proposition 8.2 *If $-1 \leq x \leq 1$, then*
$$\arcsin(-x) = -\arcsin x.$$
In other words, $f(x) = \arcsin x$ is an odd function.

Proof Suppose $-1 \le x \le 1$ and consider
$$y = \arcsin(-x).$$
Then
$$\begin{aligned} \Rightarrow \quad \sin y &= -x \\ \Rightarrow \quad -\sin y &= x. \end{aligned}$$
Proposition 5.4 (i) tells us $-\sin y = \sin(-y)$. So,
$$\begin{aligned} \sin(-y) &= x \\ \Rightarrow \quad -y &= \arcsin x \\ \Rightarrow \quad y &= -\arcsin x \\ \Rightarrow \quad \arcsin(-x) &= -\arcsin x. \end{aligned}$$

∎

Example 8.7 Use Proposition 8.2 to evaluate
$$\sin^{-1}\left(-\frac{\sqrt{2}}{2}\right).$$

Solution Since $\sin(\pi/4) = \sqrt{2}/2$, we have
$$\sin^{-1}\left(-\frac{\sqrt{2}}{2}\right) = -\sin^{-1}\frac{\sqrt{2}}{2} = -\frac{\pi}{4}.$$

∎

Proposition 8.3

(i) *Assume* $-1 \le x \le 1$. *Then*
$$\sin(\arcsin x) = x.$$

(ii) *Suppose* $-\pi/2 \le \theta \le \pi/2$. *Then*
$$\arcsin(\sin \theta) = \theta.$$

Part (i) of Proposition 8.3 does not merit careful thought in calculations. We use arc sine as the inverse of sine, and the range of sine is the interval $[-1, 1]$. As such, there is never a need to evaluate arc sine for values outside of the interval.

However, part (ii) of Proposition 8.3 requires careful consideration in computations. For example,

$$\sin\frac{3\pi}{2} = -1 \quad \text{but} \quad \arcsin\left(\sin\frac{3\pi}{2}\right) = -\frac{\pi}{2}.$$

Arc sine and sine are *not* general inverses of each other.

Example 8.8 Evaluate without a calculator.
(a) $\sin(\arcsin 0.8)$, (b) $\arcsin(\sin 60°)$, and (c) $\arcsin(\sin 4)$.

Solution

(a) Since $-1 \leq 0.8 \leq 1$, we immediately know

$$\sin(\arcsin 0.8) = 0.8.$$

(b) We have
$$\arcsin(\sin 60°) = 60°,$$
due to the fact that $-90° \leq 60° \leq 90°$.

(c) Because $\pi \approx 3.14 < 4 < 3\pi/2 \approx 4.71$, the terminal side of 4 lies in quadrant III. So, the reference angle of 4 is $4 - \pi$.

Furthermore, sine is negative is quadrant III. This means that $\arcsin(\sin 4)$ is between $-\pi/2$ and 0, because arc sine sends negative values in its domain to radian measures between $-\pi/2$ and 0.

Thus,
$$\arcsin(\sin 4) = -(4 - \pi) = \pi - 4.$$

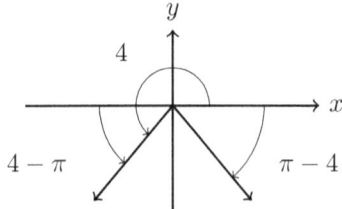

Example 8.9 Suppose
$$6\sin^2\theta + \sin\theta - 2 = 0.$$
Solve for θ when (a) $-90° \le \theta \le 90°$ and (b) $-180° < \theta \le 180°$.

Solution The first step is to factor and use the zero-product property:
$$6\sin^2\theta + \sin\theta - 2 = 0 \quad \text{implies} \quad (2\sin\theta - 1)(3\sin\theta + 2) = 0.$$
It follows that
$$2\sin\theta - 1 = 0 \quad \text{or} \quad 3\sin\theta + 2 = 0$$
Solving this equations for $\sin\theta$ yields
$$\sin\theta = \frac{1}{2} \quad \text{or} \quad \sin\theta = -\frac{2}{3}.$$
We are ready to answer the questions.

(a) We know
$$\sin\theta = \frac{1}{2} \quad \text{implies} \quad \theta = \arcsin\frac{1}{2} = 30°,$$
and
$$\sin\theta = -\frac{2}{3} \quad \text{implies} \quad \theta = \arcsin\left(-\frac{2}{3}\right) \approx -41.810°.$$
Since $-90° \le \theta \le 90°$ is the range of arc sine, we have not missed any angle measures.

(b) Now suppose $-180° < \theta \le 180°$. Consider
$$\sin\theta = \frac{1}{2}.$$
It is still the case that $\theta = 30°$ is a solution. But there is another solution θ whose terminal side lies in quadrant II. In particular,
$$\theta = 180° - 30° = 150°$$

is a solution.

Consider
$$\sin\theta = -\frac{2}{3}.$$
The degree measure $\arcsin(-2/3) \approx -41.810°$ is still a solution. But there is another solution θ whose terminal side lies in quadrant III. The reference angle is $\arcsin(2/3)$. Thus, this solution is
$$\theta = -180° + \arcsin\frac{2}{3} \approx -138.190°.$$

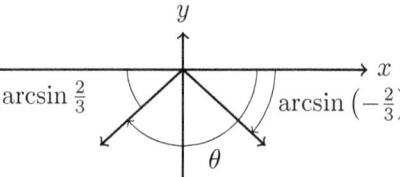

To summarize, $\theta =$

$-180° + \arcsin\dfrac{2}{3} \approx -138.190°$, $\arcsin\left(-\dfrac{2}{3}\right) \approx -41.810°$, $30°$,

or $150°$. ∎

8.3 Arc Cosine

The function $f(x) = \cos x$ is not one-to-one, so no general inverse exists. However, if we suppose $0 \leq x \leq \pi$, then no horizontal line intersects the graph more than once. This implies f is invertible within the domain $[0, \pi]$.

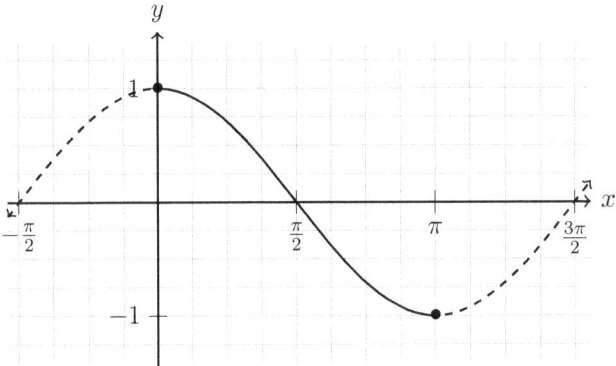

Definition 8.4 The **arc cosine** of x, denoted $\arccos x$, is the function defined by the relationship

$$y = \arccos x \quad \text{if} \quad \cos y = x$$

for $-1 \leq x \leq 1$ and $0 \leq y \leq \pi$.

An alternative notation for the arc cosine is $\cos^{-1} x$. When this notation is used, it is usually read as "cosine inverse of x" instead of "arc cosine of x".

Example 8.10 Compute without a calculator.

(a) $\cos^{-1}(-1)$

(b) $\arccos \dfrac{\sqrt{2}}{2}$

(c) $\cos^{-1}\left(-\dfrac{\sqrt{2}}{2}\right)$

(d) $\arccos 142$

Solution

(a) We conclude
$$\cos^{-1}(-1) = \pi \quad \text{because} \quad \cos \pi = -1$$
and π is an element of the closed interval $[0, \pi]$.

(b) Due to the fact $\cos(\pi/4) = \sqrt{2}/2$ and $\pi/4$ is in the interval $[0, \pi]$,
$$\arccos \dfrac{\sqrt{2}}{2} = \dfrac{\pi}{4}.$$

(c) From (b), we conclude that reference angle is $\pi/4$. Inverse cosine sends negative values to numbers between $\pi/2$ and π, so
$$\cos^{-1}\left(-\dfrac{\sqrt{2}}{2}\right) = \pi - \dfrac{\pi}{4} = \dfrac{3\pi}{4}.$$

(d) The value 142 is not within the domain of arc cosine, because 142 is not within the range of cosine. Thus, $\arccos 142$ is undefined.

∎

Proposition 8.4

(i) Suppose $-1 \leq x \leq 1$. Then
$$\cos(\arccos x) = x.$$

(ii) For $0 \leq \theta \leq \pi$, we have
$$\arccos(\cos \theta) = \theta.$$

Example 8.11 Evaluate without a calculator.
(a) $\cos\left(\arccos \dfrac{15}{49}\right)$ and (b) $\cos^{-1}(\cos 299°)$.

Solution

(a) Since $-1 \leq 15/49 \leq 1$,
$$\cos\left(\arccos \dfrac{15}{49}\right) = \dfrac{15}{49}.$$

(b) Because $270° < 299° < 360°$, the terminal side of $299°$ lies in quadrant IV. As a result, arc cosine does not "undo" cosine.

Since cosine is positive in quadrant IV, we know that
$$0 \leq \cos^{-1}(\cos 299°) \leq 90°$$
because inverse cosines sends positive values to numbers in the first quadrant. Furthermore, the reference angle will be the same as that of $299°$, so the reference angle of the solution is $360° - 299° = 61°$. When the original angle lies in the first quadrant, the reference angle has the same measure as the original angle. Therefore,
$$\cos^{-1}(\cos 299°) = 61°.$$

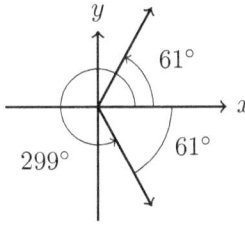

Example 8.12 Simplify. Assume $-1 \le x \le 1$.
(a) $\sin\left(\cos^{-1} x\right)$, (b) $\tan\left(\arccos x\right)$, and (c) $\sin\left(2\cos^{-1} x\right)$.

Solution Suppose $\arccos x$ is a standard position angle. To avoid the appearance of an unmerited correspondence, to will suppose this unit circle lies on the uv-plane and has equation $u^2 + v^2 = 1$. Because $\cos\left(\arccos x\right) = x$, the definition of cosine tells us that $\arccos x$ intersects the unit circle at a point with a u-coordinate of x. Building a triangle and using the Pythagorean Theorem, leads us to conclude that the v-coordinate is $\sqrt{1 - x^2}$.

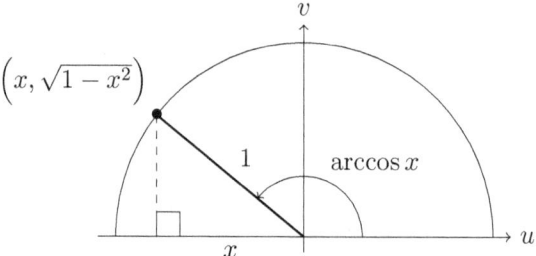

We are ready to answer our questions.

(a) Due to the definition of sine, we have
$$\sin\left(\cos^{-1} x\right) = \sqrt{1 - x^2}.$$

(b) Similarly, the definition of tangent tells us
$$\tan\left(\arccos x\right) = \frac{\sqrt{1 - x^2}}{x}.$$

(c) Because of Proposition 7.2 (i),
$$\sin\left(2\cos^{-1} x\right) = 2\sin\left(\cos^{-1} x\right)\cos\left(\cos^{-1} x\right)$$
$$= 2\left(\sqrt{1 - x^2}\right)x$$
$$= 2x\sqrt{1 - x^2}.$$

■

Let us examine arc cosine's version of Proposition 8.2.

Proposition 8.5 *For* $-1 \leq x \leq 1$,
$$\arccos(-x) = \pi - \arccos x.$$

Proof Theorem 7.1 (ii) tells us
$$\cos(\pi - \arccos x) = \cos \pi \cos(\arccos x) + \sin \pi \sin(\arccos x)$$
$$= -1 \cdot x + 0$$
$$= -x.$$

This implies
$$\arccos(-x) = \arccos\Big(\cos(\pi - \arccos x)\Big).$$

Proposition 8.4 (ii) tells us
$$\arccos\Big(\cos(\pi - \arccos x)\Big) = \pi - \arccos x$$

as long as $0 \leq \pi - \arccos x \leq \pi$.

Let us prove the inequality does, in fact, hold. We note
$$0 \leq \arccos x \leq \pi$$
$$\Rightarrow \quad 0 \geq -\arccos x \geq -\pi$$
$$\Rightarrow \quad \pi \geq \pi - \arccos x \geq 0$$

Hence,
$$\arccos(-x) = \pi - \arccos x.$$

∎

Example 8.13 Use Proposition 8.5 to compute
$$\arccos\left(-\frac{\sqrt{3}}{2}\right).$$

Solution Since $\cos(\pi/6) = \sqrt{3}/2$, Proposition 8.5 tells us
$$\arccos\left(-\frac{\sqrt{3}}{2}\right) = \pi - \arccos\frac{\sqrt{3}}{2}$$
$$= \pi - \frac{\pi}{6}$$
$$= \frac{5\pi}{6}.$$
∎

Example 8.14 Assume
$$8\cos\varphi + 8 = 5\sin^2\varphi.$$
Find φ when (a) $0 \leq \varphi \leq \pi$ and (b) $0 \leq \varphi < 2\pi$.

Solution We will use Pythagorean Identity (i) to convert the equation into one of only $\cos\varphi$. Then we can move all the terms to one side, factor, and utilize the zero-product property to solve the equation for $\cos\varphi$:

$$\begin{aligned}
& & 8\cos\varphi + 8 &= 5\sin^2\varphi \\
\Rightarrow & & 8\cos\varphi + 8 &= 5 - 5\cos^2\varphi \\
\Rightarrow & & 5\cos^2\varphi + 8\cos\varphi + 3 &= 0 \\
\Rightarrow & & (5\cos\varphi + 3)(\cos\varphi + 1) &= 0.
\end{aligned}$$

It follows that
$$\cos\varphi = -\frac{3}{5} \quad \text{or} \quad \cos\varphi = -1.$$
We are ready to answer our questions.

(a) Since $0 \leq \varphi \leq \pi$, we can take the arc cosine of both sides of each equation. Hence,
$$\varphi = \arccos\left(-\frac{3}{5}\right) \approx 2.214 \quad \text{or} \quad \varphi = \arccos(-1) = \pi \approx 3.142.$$

(b) Suppose $0 \leq \varphi < 2\pi$. This adds no new solutions to

$$\cos \varphi = -1.$$

However,

$$\cos \varphi = -3/5$$

has another solution in quadrant III. The reference angle of this solution is the same as that of $\arccos(-3/5)$, so its reference angle is $\arccos(3/5)$. Hence,

$$\varphi = \pi + \arccos \frac{3}{5} \approx 4.069$$

is also a solution.

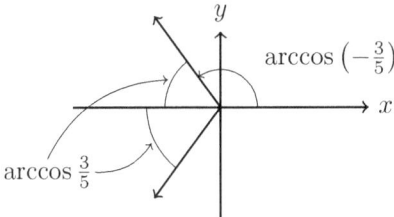

In summary, $\varphi = \arccos(-3/5) \approx 2.214$, $\pi + \arccos(3/5) \approx 4.069$, or $\pi \approx 3.142$. ∎

8.4 Arc Tangent

Let us examine an inverse for $f(x) = \tan x$. Tangent is not one-to-one everywhere. However, after a quick examination of its graph, we conclude that tangent is one-to-one, and therefore invertible, if we restrict its to domain $-\pi/2 < x < \pi/2$.

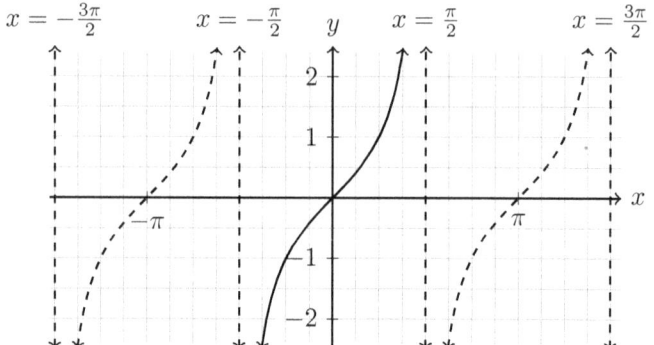

Definition 8.5 The **arc tangent** of x, denoted $\arctan x$, is the function defined by the relationship

$$y = \arctan x \quad \text{if} \quad \tan y = x$$

where x is any real number and $-\pi/2 < y < \pi/2$

Alternative notion for arc tangent is $\tan^{-1} x$. Usually, when this notation is used we say "inverse tangent of x".

Example 8.15 Find the exact values without a calculator.
(a) $\tan^{-1} 0$, (b) $\arctan \sqrt{3}$, and (c) $\tan^{-1}\left(-\sqrt{3}\right)$.

Solution

(a) We know

$$\tan^{-1} 0 = 0 \quad \text{because} \quad \tan 0 = 0$$

and $-\pi/2 < 0 < \pi/2$,

(b) Since $\tan(\pi/3) = \sqrt{3}$ and $\pi/3$ is in the interval $(-\pi/2, \pi/2)$,

$$\arctan \sqrt{3} = \frac{\pi}{3}.$$

(c) Using the result of (b), the reference angle has radian measure $\pi/3$. Because arc tangent sends negative values to numbers in the interval $(-\pi/2, 0)$, we know $-\pi/2 < \tan^{-1}\left(-\sqrt{3}\right) < 0$. Therefore,
$$\tan^{-1}\left(-\sqrt{3}\right) = -\frac{\pi}{3}.$$

∎

Proposition 8.6 *For x any real number,*
$$\arctan(-x) = -\arctan x.$$
In other words, $f(x) = \arctan x$ is an odd function.

Proof Consider
$$y = \arctan(-x).$$
Then
$$\tan y = -x \quad \text{implies} \quad -\tan y = x.$$
Proposition 5.4 (iii) tells us $-\tan y = \tan(-y)$. So,
$$\begin{aligned}
\tan(-y) &= x \\
\Rightarrow \quad -y &= \arctan x \\
\Rightarrow \quad y &= -\arctan x \\
\Rightarrow \quad \arctan(-x) &= -\arctan x.
\end{aligned}$$

∎

Example 8.16 Use Proposition 8.6 to find
$$\arctan(-1).$$

Solution Because $\tan(\pi/4) = 1$,
$$\arctan(-1) = -\arctan 1 = -\frac{\pi}{4}.$$

∎

Proposition 8.7

(i) For x a real number,
$$\tan(\arctan x) = x.$$

(ii) For $-\pi/2 < \theta < \pi/2$,
$$\arctan(\tan \theta) = \theta.$$

Example 8.17 Evaluate without a calculator.

(a) $\tan\left(\arctan\dfrac{5 - 2^{9\pi}}{2\pi}\right)$ and (b) $\tan^{-1}(\tan 2.5)$.

Solution

(a) Due to Proposition 8.7,
$$\tan\left(\arctan\dfrac{5 - 2^{9\pi}}{2\pi}\right) = \dfrac{5 - 2^{9\pi}}{2\pi}.$$

(b) Because $\pi/2 < 2.5 < \pi$, the terminal side of 2.5 lies in quadrant II. As a result, inverse tangent does not "undo" tangent. Since tangent is negative in quadrant II,
$$-\dfrac{\pi}{2} < \tan^{-1}(\tan 2.5) < 0.$$
The reference angle of 2.5 is $\pi - 2.5$. Hence,
$$\tan^{-1}(\tan 2.5) = -(\pi - 2.5) = 2.5 - \pi.$$

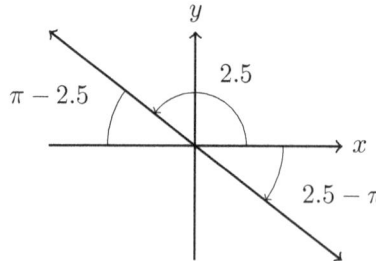

■

Example 8.18 Simplify.

(a) $\csc\left(\tan^{-1} x\right)$

(b) $\tan\left(2\arctan x\right)$

(c) $\cos\left(2\tan^{-1} x\right)$

(d) $\sin\left(\dfrac{\arctan x}{2}\right)$, $x < 0$

Solution Suppose $\arctan x = \tan^{-1} x$ is a standard position angle. Much like in Example 12, we will suppose the unit circle lies in the uv-plane and has equation $u^2 + v^2 = 1$. We do this because the x-value in our example does not correspond to a value on the horizontal axis. Assume the terminal side of $\arctan x$ intersects a circle with center the origin and radius r. Because

$$\tan\left(\arctan x\right) = x = \frac{x}{1},$$

we can suppose the terminal side of $\arctan x$ intersects the circle at the point $(1, x)$. Then building a triangle and utilizing the Pythagorean Theorem leads us to conclude that the radius of the circle is $r = \sqrt{1 + x^2}$.

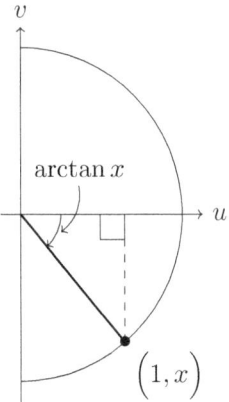

We are ready to simplify the expressions.

(a) Using Theorem 5.1,

$$\csc\left(\tan^{-1} x\right) = \frac{\sqrt{1+x^2}}{x}.$$

(b) The Double Angle Identity (iii) gives

$$\tan\left(2\arctan x\right) = \frac{2\tan\left(\arctan x\right)}{1 - \tan^2\left(\arctan x\right)}.$$

Then, because $\tan\left(\arctan x\right) = x$, we conclude

$$\tan\left(2\arctan x\right) = \frac{2x}{1 - x^2}.$$

(c) The Double Angle Identity (ii) tells us

$$\cos\left(2\tan^{-1} x\right) = 2\cos^2\left(\tan^{-1} x\right) - 1.$$

Because of Theorem 5.1, we know

$$\cos\left(\tan^{-1} x\right) = \frac{1}{\sqrt{1 + x^2}}.$$

Substituting yields

$$\cos\left(2\tan^{-1} x\right) = 2\left(\frac{1}{\sqrt{1 + x^2}}\right)^2 - 1$$

$$= 2\left(\frac{1}{1 + x^2}\right) - 1$$

$$= \frac{2}{1 + x^2} - \frac{1 + x^2}{1 + x^2}$$

$$= \frac{1 - x^2}{1 + x^2}.$$

(d) Using the Half Angle Identity (i),

$$\sin\left(\frac{\arctan x}{2}\right) = \pm\sqrt{\frac{1 - \cos\left(\arctan x\right)}{2}}.$$

Then Theorem 5.1, once again, leads us to the equation

$$\cos\left(\tan^{-1} x\right) = \frac{1}{\sqrt{1 + x^2}}.$$

So,

$$\sin\left(\frac{\arctan x}{2}\right) = \pm\sqrt{\frac{1 - 1/\sqrt{1+x^2}}{2}}$$

$$= \pm\sqrt{\frac{\sqrt{1+x^2} - 1}{2\sqrt{1+x^2}}}$$

$$= \pm\sqrt{\frac{1 + x^2 - \sqrt{1+x^2}}{2(1+x^2)}}.$$

Because $x < 0$ we know $-90° < \arctan x < 0$. This implies

$$-45° < \frac{\arctan x}{2} < 0.$$

Since $\arctan(x)/2$ is in quadrant IV, sine is negative. Thus,

$$\sin\left(\frac{\arctan x}{2}\right) = -\sqrt{\frac{1 + x^2 - \sqrt{1+x^2}}{2(1+x^2)}}.$$

■

Example 8.19 Find the exact value of

$$\arctan\left(\frac{5}{3}\right) + \arctan\left(-\frac{1}{4}\right).$$

Solution Suppose that $\alpha = \arctan(5/3)$ and $\beta = \arctan(-1/4)$. Then Theorem 7.1 (iii) says

$$\tan(\alpha + \beta) = \frac{\tan\alpha + \tan\beta}{1 - \tan\alpha\tan\beta}$$

$$= \frac{5/3 - 1/4}{1 - (5/3)(-1/4)}$$

$$= \frac{17/12}{17/12}$$

$$= 1.$$

We want to take arc tangent of both sides to "undo" tangent. But Proposition 8.7 tells us

$$\arctan\Big(\tan(\alpha+\beta)\Big) = \alpha+\beta,$$

only if $-\pi/2 < \alpha+\beta < \pi/2$.

Let us prove this inequality holds. We know $0 < \alpha < \pi/2$ and $-\pi/2 < \beta < 0$, because arc tangent sends positive numbers to values in the interval $(0, \pi/2)$ and negative numbers to values in the interval $(-\pi/2, 0)$. It follows that

$$-\frac{\pi}{2} < \alpha+\beta < \frac{\pi}{2}.$$

Therefore,

$$\tan(\alpha+\beta) = 1 \quad \text{implies} \quad \alpha+\beta = \arctan 1 = \frac{\pi}{4}.$$

∎

8.5 Other Inverse Trigonometric Functions

In this section, we explore the remaining three inverses: an inverse for secant, cosecant, and cotangent. These inverses are a bit more obscure and will, therefore, receive less treatment. The functions secant, cosecant, and cotangent are not invertible on their entire domains. As a result, we once again restrict their domains to obtain invertibility.

Definition 8.6

- The **arc secant** of x, denoted $\operatorname{arcsec} x$, is the function defined by the relationship

$$y = \operatorname{arcsec} x \quad \text{if} \quad \sec y = x$$

for $x \leq -1$ or $1 \leq x$ and $0 \leq y < \pi/2$ or $\pi/2 < y \leq \pi$.

- The **arc cosecant** of x, denoted $\operatorname{arccsc} x$, is the function defined by the relationship

$$y = \operatorname{arccsc} x \quad \text{if} \quad \csc y = x$$

for $x \leq -1$ or $1 \leq x$ and $-\pi/2 \leq y < 0$ or $0 < y \leq \pi/2$.

- The **arc cotangent** of x, denoted $\operatorname{arccot} x$, is the function defined by the relationship

$$y = \operatorname{arccot} x \quad \text{if} \quad \cot y = x$$

for x any real number and $-\pi/2 < y < 0$ or $0 < y \leq \pi/2$.

The notation $\sec^{-1} x$, $\csc^{-1} x$, and $\cot^{-1} x$ is also common. Usually when this notation is used, we refer to it as the "inverse" of the corresponding trigonometric function, e.g. we would read $\csc^{-1} x$ as "inverse cosecant of x".

To aid in the evaluation of the inverse trigonometric functions, we include the following table.

θ	0	$\dfrac{\pi}{6}$	$\dfrac{\pi}{4}$	$\dfrac{\pi}{3}$	$\dfrac{\pi}{2}$
$\sec \theta$	1	$\dfrac{2\sqrt{3}}{3}$	$\sqrt{2}$	2	undefined
$\csc \theta$	undefined	2	$\sqrt{2}$	$\dfrac{2\sqrt{3}}{3}$	1
$\cot \theta$	undefined	$\sqrt{3}$	1	$\dfrac{\sqrt{3}}{3}$	0

Example 8.20 Find the exact value of (a) arcsec $\sqrt{2}$ and (b) $\sec^{-1}(-\sqrt{2})$.

Solution

(a) The conclusion
$$\text{arcsec}\,\sqrt{2} = 45°,$$
follows from the fact that $\sec 45° = \sqrt{2}$ and $45°$ is in the set $[0, 90°) \cup (90°, 180°]$,

(b) Inverse secant produces values in the interval $(90°, 180°]$ when evaluated at a negative number. Furthermore, due to (a), we know the reference angle is $45°$. Hence,
$$\sec^{-1}\left(-\sqrt{2}\right) = 180° - 45° = 135°.$$

■

Example 8.21 Evaluate (a) arccsc 2 and (b) $\csc^{-1}(-2)$ without a calculator.

Solution

(a) Since $\csc(\pi/6) = 2$ and $\pi/6$ is in the set $[-\pi/2, 0) \cup (0, \pi/2]$,
$$\text{arccsc}\,2 = \frac{\pi}{6}.$$

(b) Cosecant is negative when evaluated at angle measures in the interval $[-\pi/2, 0)$. From (a), we know the reference angle is $\pi/6$. Hence,
$$\csc^{-1}(-2) = -\frac{\pi}{6}.$$

■

Example 8.22 What are the exact numeric values of (a) arccot $\sqrt{3}$ and (b) $\cot^{-1}(-\sqrt{3})$?

Solution

(a) We have
$$\operatorname{arccot} \sqrt{3} = 30° \quad \text{because} \quad \cot 30° = \sqrt{3}$$
and $30°$ is in the set $(-90°, 0) \cup (0, 90°]$,

(b) When the input of inverse cotangent is negative, its output is in the interval $(-90°, 0)$. Furthermore, the result of (a) tells us that the reference angle is $30°$. It follows that
$$\cot^{-1}\left(-\sqrt{3}\right) = -30°.$$

■

Example 8.23 Find (a) $\operatorname{arcsec}\left(\sec \dfrac{9\pi}{8}\right)$ and (b) $\cot^{-1}(\cot 271°)$.

Solution

(a) The terminal side of $9\pi/8$ lies in quadrant III, so arc secant does not "undo" secant. Since secant is negative in quadrant III,
$$\frac{\pi}{2} < \operatorname{arcsec}\left(\sec \frac{9\pi}{8}\right) \leq \pi.$$
The reference angle of $9\pi/8$ is $9\pi/8 - \pi = \pi/8$. Thus,
$$\operatorname{arcsec}\left(\sec \frac{9\pi}{8}\right) = \pi - \frac{\pi}{8} = \frac{7\pi}{8}.$$

(b) The terminal side of $271°$ lies in quadrant IV, so inverse cotangent does not "undo" cotangent. Since cotangent is negative in quadrant IV,
$$-90° \leq \cot^{-1}(\cot 271°) < 0.$$
The reference angle of $271°$ is $360° - 271° = 89°$. Therefore,
$$\cot^{-1}(\cot 271°) = -89°.$$

■

In the remaining portion of this section, we will introduce a few tricks to avoid arc secant, arc cosecant, and arc cotangent.

Proposition 8.8

(i) For $x \leq -1$ or $1 \leq x$,
$$\operatorname{arcsec} x = \arccos \frac{1}{x}.$$

(ii) For $x \leq -1$ or $1 \leq x$,
$$\operatorname{arccsc} x = \arcsin \frac{1}{x}.$$

(iii) For all real numbers $x \neq 0$,
$$\operatorname{arccot} x = \arctan \frac{1}{x}.$$

Proof We will prove (i) and leave the rest as exercises. Let $\operatorname{arcsec} x = y$. Then
$$\sec y = x \quad \text{implies} \quad \cos y = \frac{1}{x}.$$
It follows that $y = \arccos(1/x)$. Thus,
$$\operatorname{arcsec} x = \arccos \frac{1}{x}.$$
∎

Example 8.24 Evaluate using Proposition 8.8.
(a) $\operatorname{arccsc} \dfrac{2}{\sqrt{3}}$ and (b) $\operatorname{arccot}(-1)$.

Solution

(a) $\operatorname{arcsec} \dfrac{2}{\sqrt{3}} = \arccos \dfrac{\sqrt{3}}{2} = \dfrac{\pi}{6}.$

(b) $\operatorname{arccot}(-1) = \arctan\left(\dfrac{1}{-1}\right) = \arctan(-1) = -\dfrac{\pi}{4}.$ ∎

8.6 Graphing Arc Sine, Arc Cosine, and Arc Tangent

In this section, we will graph expressions containing arc sine, arc cosine, and arc tangent. We will rely on knowledge of transformations. For information about how transformations affect graphs, see Appendix C.

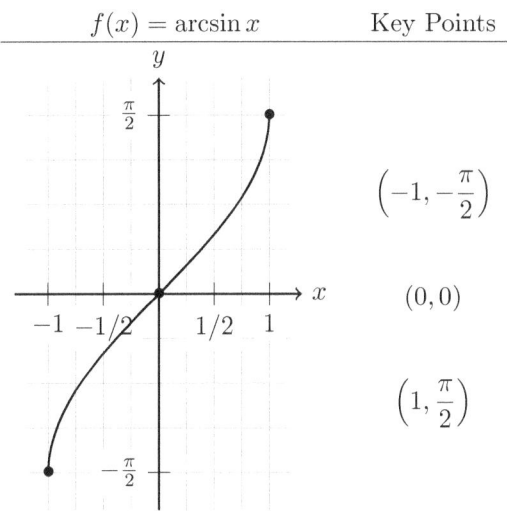

$f(x) = \arcsin x$ | Key Points
$\left(-1, -\dfrac{\pi}{2}\right)$
$(0, 0)$
$\left(1, \dfrac{\pi}{2}\right)$

Example 8.25 Graph

$$y = \arcsin(x-2) + \frac{\pi}{4}.$$

Solution This is the graph of $y = \arcsin x$ shifted right 2 units and up $\pi/4$ units. Since the graph is shifted right 2 units, we increase the x-values by 2. Similarly, since the graph is shifted up $\pi/4$ units, we increase the y-values by $\pi/4$.

x	$y = \arcsin x$		x	$y = \arcsin(x-2) + \frac{\pi}{4}$
-1	$-\frac{\pi}{2}$		$-1+2=1$	$-\frac{\pi}{2} + \frac{\pi}{4} = -\frac{\pi}{4}$
0	0	$\longrightarrow$	$0+2=2$	$0 + \frac{\pi}{4} = \frac{\pi}{4}$
1	$\frac{\pi}{2}$		$1+2=3$	$\frac{\pi}{2} + \frac{\pi}{4} = \frac{3\pi}{4}$

Hence, we have the following graph.

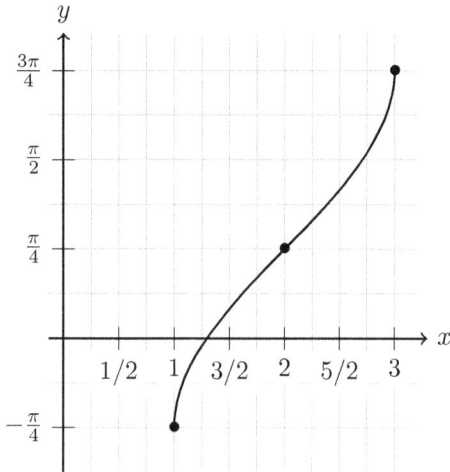

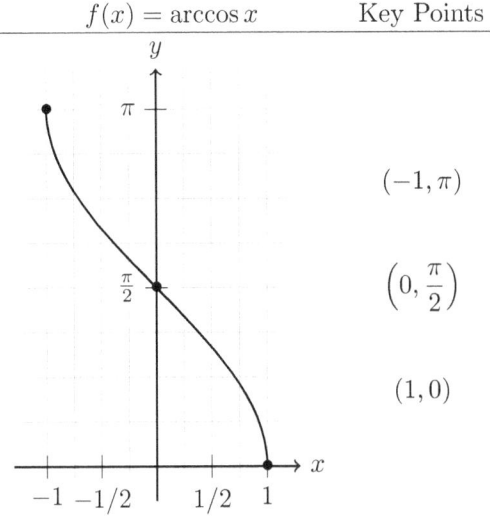

Example 8.26 Graph

$$y = -\frac{3}{2}\cos^{-1} x.$$

Solution This is the graph of $y = \cos^{-1} x$ stretched vertically by a factor of 3/2 and reflected about the x-axis. As a result, we multiply the y-values by $-3/2$, and leave the x-values unchanged.

x	$y = \cos^{-1} x$
-1	π
0	$\dfrac{\pi}{2}$
1	0

$\longrightarrow$

x	$y = -\frac{3}{2}\cos^{-1} x$
-1	$-\dfrac{3\pi}{2}$
0	$-\dfrac{3}{2}\left(\dfrac{\pi}{2}\right) = -\dfrac{3\pi}{4}$
1	$-\dfrac{3}{2}(0) = 0$

Thus, the graph of $y = (-3/2)\cos^{-1} x$ must be as shown below.

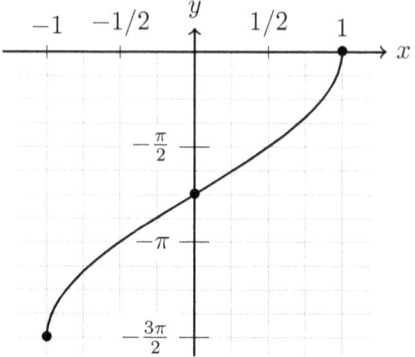

∎

	$f(x) = \arctan x$	Key Points	Horizontal Asymptotes
	[graph of arctan x with $y = \pi/2$ and $y = -\pi/2$ as dashed horizontal asymptotes; key points marked at $(-1, -\pi/4)$, $(0,0)$, $(1, \pi/4)$]	$\left(-1, -\dfrac{\pi}{4}\right)$ $(0, 0)$ $\left(1, \dfrac{\pi}{4}\right)$	$y = \dfrac{\pi}{2}$ $y = -\dfrac{\pi}{2}$

Example 8.27 Graph

$$y = \arctan \frac{x}{2}.$$

Solution This is the graph of $y = \arctan x$ stretched horizontally by a factor of 2. So, we double all of the x-values, and leave the y-values unchanged.

x	$y = \arctan x$		x	$y = \arctan \frac{x}{2}$
-1	$-\frac{\pi}{4}$		$2(-1) = -2$	$-\frac{\pi}{4}$
0	0	$\longrightarrow$	$2(0) = 0$	0
1	$\frac{\pi}{4}$		$2(1) = 2$	$\frac{\pi}{4}$

Horizontal stretches do not affect horizontal asymptotes. It follows that the horizontal asymptotes are still

$$y = -\frac{\pi}{2} \quad \text{and} \quad y = \frac{\pi}{2}.$$

Therefore, the graph of $y = \arctan(x/2)$ is as shown below.

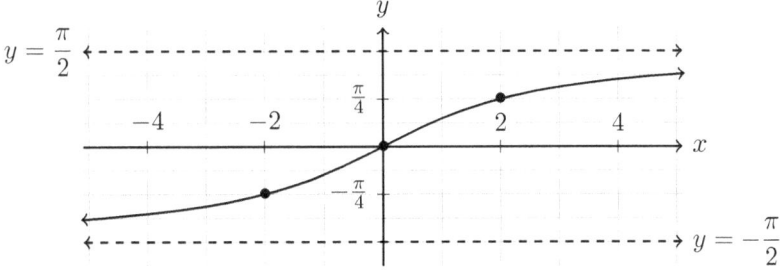

∎

Example 8.28 Graph
$$y = \tan^{-1}(1-x) - \pi.$$

Solution Let us rewrite this into a more tractable form:
$$y = \tan^{-1}\left(-(x-1)\right) - \pi.$$
The graph is $y = \arctan x$ reflected about the y-axis, then shifted right 1 unit. It is also shifted π units down. So, negate the x-values, add 1 to them, and subtract π from the y-values.

x	$y = \arctan x$		x	$y = \tan^{-1}(1-x) - \pi$
-1	$-\dfrac{\pi}{4}$		$-(-1)+1 = 2$	$-\dfrac{\pi}{4} - \pi = -\dfrac{5\pi}{4}$
0	0	$\longrightarrow$	$-0+1 = 1$	$0 - \pi = -\pi$
1	$\dfrac{\pi}{4}$		$-1+1 = 0$	$\dfrac{\pi}{4} - \pi = -\dfrac{3\pi}{4}$

Furthermore, because the graph of $y = \arctan x$ is shifted π units down, the horizontal asymptotes are
$$y = -\frac{\pi}{2} - \pi = -\frac{3\pi}{2} \quad \text{and} \quad y = \frac{\pi}{2} - \pi = -\frac{\pi}{2}.$$
This yields the graph shown.

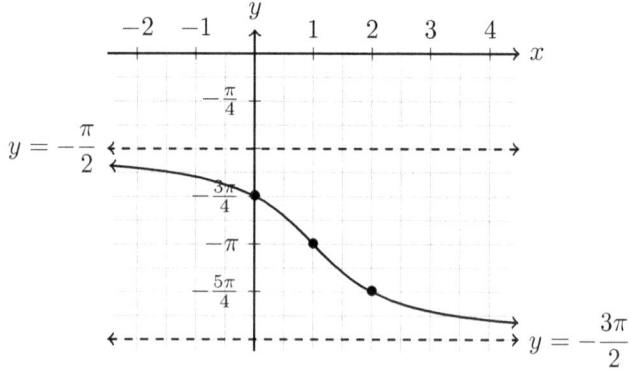

■

8.7 Exercises

x	$f(x)$
-5	1
-3	-3
-1	-7
0	5
1	2
3	-5
5	-1

** Exercise 1

Consider the table above. Suppose the inverse of f is the function g. Find each of the following.

(a) $g(-7)$ (e) $f\big(g(2)\big)$

(b) $g(1)$ (f) $f\big(g(f(0))\big)$

(c) $g(2)$ (g) $g\big(g(-3)\big)$

(d) $g\big(f(-5)\big)$ (h) $g\big(g(-1)\big)$

** Exercise 2

Use the horizontal line test to determine the invertibility of the functions corresponding to the graphs on page 254. Suppose each functions' domain is contained in the x-axis shown.

** Exercise 3

Repeat Exercise 2, but use the graphs on pages 180 and 181.

** Exercise 4

For each graph on page 254 corresponding to a non-invertible function, find two connected intervals on which the function can be restricted to make it invertible.

** Exercise 5

Prove f is *not* one-to-one by finding u and v such that

$$u \neq v \quad \text{and} \quad f(u) = f(v),$$

where $f(x) = \ldots$

(a) $x^2 - 6x + 5$

(b) $\cos 2x$

(c) $|x|$

(d) $2 - \tan x$

** Exercise 6

Determine which functions are invertible.

(a) $f(x) = 2x - 7$

(b) $g(x) = 1 - \sin \pi x$

(c) $h(x) = x^2 + 4x + 4$

(d) $i(x) = -|x - 2|$

249

* Exercise 7

Find the degree measure without a calculator.

(a) arctan 1

(b) $\sin^{-1} \frac{\sqrt{3}}{2}$

(c) arccos 1

(d) $\cos^{-1} 0$

(e) $\arcsin \frac{1}{2}$

(f) $\tan^{-1} \frac{\sqrt{3}}{3}$

* Exercise 8

Compute the radian measure without a calculator.

(a) arcsin 0

(b) $\cos^{-1} \frac{\sqrt{2}}{2}$

(c) arcsin 1

(d) $\tan^{-1} \sqrt{3}$

(e) arctan 0

(f) $\cos^{-1} \frac{1}{2}$

** Exercise 9

Determine the degree measure without a calculator. Some values may be undefined.

(a) arccot 0

(b) $\tan^{-1}\left(-\frac{3}{\sqrt{3}}\right)$

(c) $\arcsin\left(-\frac{\sqrt{3}}{2}\right)$

(d) $\sin^{-1}(-1)$

(e) $\arccos\left(-\frac{1}{2}\right)$

(f) $\sec^{-1} \frac{2\sqrt{3}}{3}$

** Exercise 10

Find the radian measure without a calculator. Some values may be undefined.

(a) $\operatorname{arccsc}\left(-\sqrt{2}\right)$

(b) $\cot^{-1} 1$

(c) $\arcsin\left(-\frac{3}{\sqrt{12}}\right)$

(d) $\cos^{-1} \frac{\sqrt{5}}{2}$

(e) $\operatorname{arccot}\left(-\frac{3}{\sqrt{3}}\right)$

(f) $\sec^{-1} \frac{1}{2}$

* Exercise 11

Suppose $x = 2/5$. Evaluate.

(a) $\sin(\arcsin x)$

(b) $\cos\left(\cos^{-1} x\right)$

(c) $\tan(\arctan x)$

(d) $\csc(\arcsin x)$

(e) $\cos\left(\sec^{-1}\frac{1}{x}\right)$

(f) $\cot(\arctan x)$

* Exercise 12

Let $\theta = 25°$. Evaluate.

(a) $\arccos(\cos\theta)$

(b) $\sin^{-1}(\sin\theta)$

(c) $\arctan(\cot\theta)$

(d) $\cos^{-1}(\sin\theta)$

** Exercise 13

Repeat Exercise 12 but with $\theta = 155°$.

** Exercise 14

Assume $\varphi = 6\pi/5$. Evaluate.

(a) $\arccos(\cos\varphi)$

(b) $\tan^{-1}(\tan\varphi)$

(c) $\arcsec(\sec\varphi)$

(d) $\csc^{-1}(\sec\varphi)$

(e) $\arccot(\tan\varphi)$

(f) $\sec^{-1}(\csc\varphi)$

** Exercise 15

Repeat Exercise 14 but with $\varphi = 9\pi/5$.

** Exercise 16

Find all θ in the interval $[0, 360°)$.

(a) $3 - 5\sin\theta = 4$

(b) $7\cos\theta + 4 = 9$

(c) $\tan\theta + 11 = 5$

(d) $3\cos\theta + 5 = 1$

(e) $9\sin^2\theta = 1$

** Exercise 17

Solve for φ. Suppose φ is in the interval $[-\pi/2, \pi/2]$.

(a) $5\csc\varphi + 2\cot^2\varphi = 1$

(b) $2\sec^2\varphi = 3\tan\varphi + 1$

(c) $\csc^2\varphi + 8\cot\varphi = -6$

(d) $28\sin\varphi + 5\cos 2\varphi = -1$

** Exercise 18

Repeat Exercise 17 but with φ in the interval $[0, 2\pi)$.

** Exercise 19

Find all θ in the interval $[0, 180°]$.

(a) $6\sin^2\theta = 13\cos\theta + 1$

(b) $2\tan^2\theta + 13\sec\theta = 5$

(c) $\cos 2\theta + 2\cos\theta = 11$

(d) $\tan^2\theta - 3\sec\theta = 53$

** Exercise 20

Repeat Exercise 19 but with θ in the interval $(-180°, 180°]$

** Exercise 21

Write each of the following in the form

$$y = A\sin(Bx + C).$$

Use Example 7.4 as a templet.

(a) $y = 3\sin 2x + 4\cos 2x$

(b) $y = 12\cos x - 5\sin x$

(c) $y = 12\sin\dfrac{x}{2} - 9\cos\dfrac{x}{2}$

(d) $y = -7\sin x - 24\cos x$

** Exercise 22

Suppose $\alpha = \arcsin x$. Find the values of the six trigonometric functions evaluated at α.

** Exercise 23

Let $\beta = \cot^{-1} 2x$. What are the values of the six trigonometric functions evaluated at β?

** Exercise 24

Assume $\theta = \arccos(x/3)$. Simplify each of the following.

(a) $\sin 2\theta$ (c) $\tan 2\theta$

(b) $\cos 2\theta$ (d) $\csc 2\theta$

*** Exercise 25

Suppose $\varphi = \operatorname{arccsc} x$ and $x > 0$. Compute each of the following.

(a) $\sin\dfrac{\varphi}{2}$ (c) $\tan\dfrac{\varphi}{2}$

(b) $\cos\dfrac{\varphi}{2}$ (d) $\cot\dfrac{\varphi}{2}$

*** Exercise 26

Let $\theta = \operatorname{arcsec} 2x$ and $x < 0$. Simplify each of the following.

(a) $\sin\dfrac{\theta}{2}$ (c) $\tan\dfrac{\theta}{2}$

(b) $\cos\dfrac{\theta}{2}$ (d) $\cot\dfrac{\theta}{2}$

** Exercise 27

Assume $\alpha = \arcsin 4x$ and $\beta = \arccos 5x$. Compute each of the following.

(a) $\sin(\alpha + \beta)$

(b) $\cos(\alpha - \beta)$

(c) $\tan(\alpha + \beta)$

*** Exercise 28

Find the exact value of $\alpha - \beta$ for each pair of α and β.

(a) Suppose

$$\alpha = \arctan\dfrac{2\sqrt{3}}{3}$$

$$\beta = \arctan\dfrac{\sqrt{3}}{5}$$

(b) Suppose

$$\alpha = \arctan \frac{1}{7}$$
$$\beta = \arctan \frac{4}{3}$$

*** Exercise 29

Prove the following are identities.

(a) $\sin^{-1} x + \cos^{-1} x = \frac{\pi}{2}$

(b) $\arctan x + \text{arccot}\, x = \frac{\pi}{2}$

(c) $\sec^{-1} x + \csc^{-1} x = \frac{\pi}{2}$

** Exercise 30

Prove Proposition 8.8 (ii) and (iii).

** Exercise 31

Graph.

(a) $y = \pi + \arcsin\left(x - \frac{1}{2}\right)$

(b) $y = -\cos^{-1}(x+1)$

(c) $y = 2\arcsin(-x)$

(d) $y = 1 - \arccos\left(\frac{x}{3}\right)$

(e) $y = \frac{1}{2}\sin^{-1}\left(\frac{x-4}{7}\right)$

(f) $y = -2\arccos(2x+4)$

** Exercise 32

Graph. Include asymptotes.

(a) $y = \frac{3}{4}\arctan(-x)$

(b) $y = 2 - \arctan\left(\frac{2x}{5}\right)$

(c) $y = \frac{\pi}{4} + \arctan\left(x + \frac{2}{3}\right)$

(d) $y = -\tan^{-1}(x-2)$

** Exercise 33

Write corresponding functions.

(a) The graph of arc sine stretched horizontally by a factor of 2 and shifted up $\pi/2$ units.

(b) The graph of arc cosine reflected about the y-axis and stretched vertically by a factor of 7.

(c) The graph of arc tangent reflected about the x-axis shifted left 3 units.

** Exercise 34

Write equations for the graphs shown on page 255.

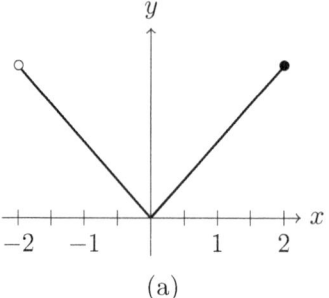
(a)

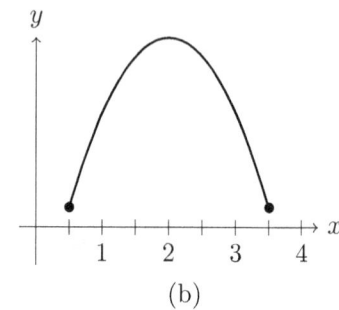
(b)

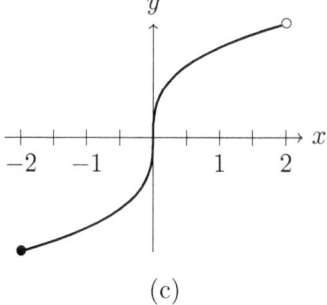
(c)

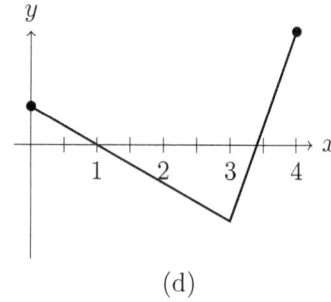
(d)

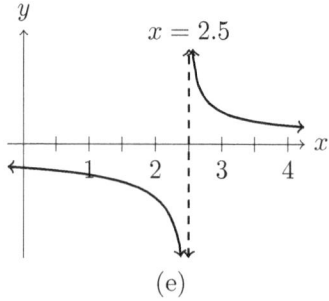
(e)

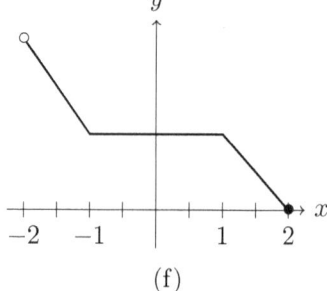
(f)

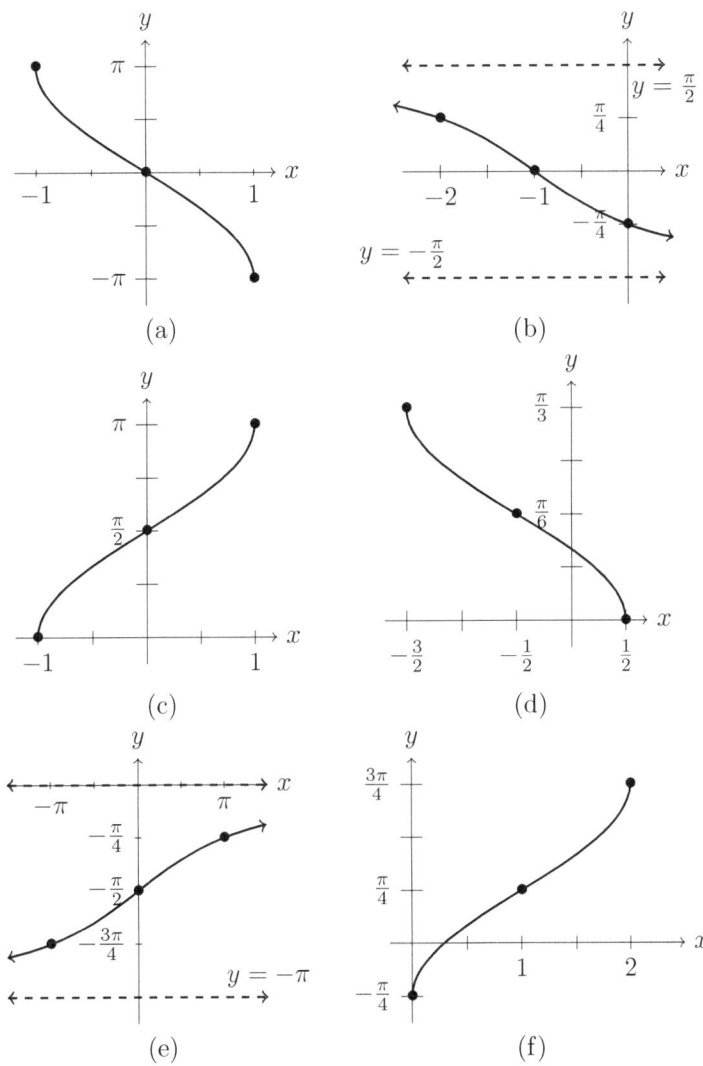

Chapter 9

Oblique Triangles

Definition 9.1

- A triangle is **oblique** if contains no right angles. In other words, an oblique triangle is a triangle that is either acute or obtuse.

- To **solve a triangle** is to find all of its side lengths and angle measures.

We dedicate this chapter to the study of solving oblique triangles. Readers are assumed to have knowledge of Chapters 1, 2, 5, and 8. In particular, the reader needs a solid command of the Isosceles Triangles Theorem (Theorem 1.2), knowledge of the triangle congruence postulates and theorems (Section 2.2), and an understanding of arc sine and arc cosine (Sections 8.2 and 8.3).

The congruence postulates and theorems provide a criteria for uniqueness of a triangles. When a triangle is uniquely determined by the given information, we can find its side lengths and angle measures. Sometimes this will require inverse trigonometric functions.

It is imperative you have a scientific calculator and you are knowledgeable of its functionality with regard to trigonometry. It will be used extensively in the examples and exercises. We will use degree measures exclusively, so we recommend changing your calculator settings to degree mode.

Before we begin solving oblique triangles, let us introduce a convention to simplify our notation. Some readers may note that this is the same convention we used in Chapter 4 for right triangles.

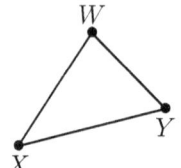

When referencing triangles, uppercase letters represent vertices, and their corresponding lowercase letters represent the sides opposite.

For example, in $\triangle WXY$

$$w = XY, \quad x = WY, \quad \text{and} \quad y = WX.$$

9.1 Law of Cosines

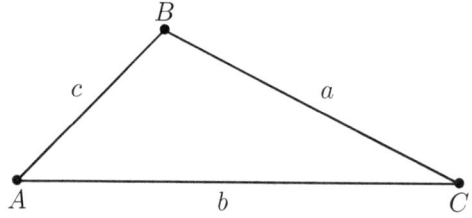

Theorem 9.1 (Law of Cosines) *Consider $\triangle ABC$.*

$$a^2 = b^2 + c^2 - 2bc\cos A$$
$$b^2 = a^2 + c^2 - 2ac\cos B$$
$$c^2 = a^2 + b^2 - 2ab\cos C$$

We will use the Law of Cosines to solve triangles that are unique due to the SAS or SSS congruence postulate.

9.1.1 Unique Triangle Due to SAS

Example 9.1 In $\triangle TUV$, $u = 3$, $v = 4$, and $m\angle T = 140°$. Solve $\triangle TUV$.

Solution

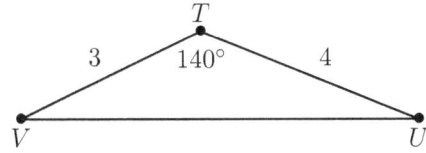

The Law of Cosines tells us

$$\begin{aligned} t^2 &= u^2 + v^2 - 2uv\cos T \\ &= 3^2 + 4^2 - 2(3)(4)\cos 140° \\ &= 25 - 24\cos 140° \\ &\approx 43.385 \\ \Rightarrow \quad t &\approx \sqrt{43.385} \\ &\approx 6.587. \end{aligned}$$

Let us find $m\angle U$ next. Using the Law of Cosines, we have

$$\begin{aligned} t^2 + v^2 - 2tv\cos U &= u^2 \\ \Rightarrow \quad 6.587^2 + 4^2 - 2(6.587)(4)\cos U &\approx 3^2 \\ \Rightarrow \quad 59.385 - 52.696\cos U &\approx 9 \\ \Rightarrow \quad -52.696\cos U &\approx -50.385 \\ \Rightarrow \quad \cos U &\approx 0.956 \end{aligned}$$

Therefore,
$$m\angle U \approx \arccos(0.956) \approx 17.031°.$$

Because the Triangle Sum Theorem states that the sum of the interior angle measures of a triangle is 180°,

$$\begin{aligned} m\angle T + m\angle U + m\angle V &= 180° \\ \Rightarrow \quad 140° + 17.031° + m\angle V &\approx 180° \\ \Rightarrow \quad m\angle V &\approx 22.969°. \end{aligned}$$

■

9.1.2 Unique Triangle Due to SSS

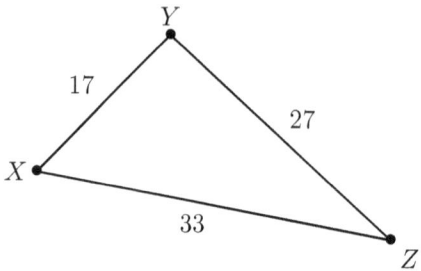

Example 9.2 Solve $\triangle XYZ$.

Solution Let us find $m\angle X$. The Law of Cosines says

$$
\begin{aligned}
y^2 + z^2 - 2yz \cos X &= x^2 \\
\Rightarrow \quad 33^2 + 17^2 - 2(33)(17) \cos X &= 27^2 \\
\Rightarrow \quad 1378 - 1122 \cos X &= 729 \\
\Rightarrow \quad -1122 \cos X &= -649 \\
\Rightarrow \quad \cos X &= \frac{59}{102}.
\end{aligned}
$$

It follows that

$$m\angle X = \arccos \frac{59}{102} \approx 54.660°.$$

To find $m\angle Y$, we again use the Law of Cosines:

$$
\begin{aligned}
x^2 + z^2 - 2xz \cos Y &= y^2 \\
\Rightarrow \quad 27^2 + 17^2 - 2(27)(17) \cos Y &= 33^2 \\
\Rightarrow \quad 1018 - 918 \cos Y &= 1089 \\
\Rightarrow \quad -918 \cos Y &= 71 \\
\Rightarrow \quad \cos Y &= -\frac{71}{918}.
\end{aligned}
$$

So,

$$m\angle Y = \arccos\left(-\frac{71}{918}\right) \approx 94.436°.$$

All that is left is $m\angle Z$. Since the sum of the interior angle measures of a triangle is 180°,

$$\begin{aligned} m\angle X + m\angle Y + m\angle Z &= 180° \\ \Rightarrow \quad 54.660° + 94.436° + m\angle Z &\approx 180° \\ \Rightarrow \quad m\angle Z &\approx 30.904°. \end{aligned}$$

∎

9.2 Law of Sines in Unambiguous Cases

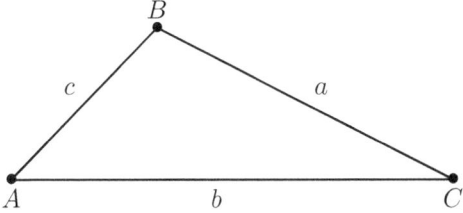

Theorem 9.2 (Law of Sines) *Consider $\triangle ABC$.*

$$\frac{\sin A}{a} = \frac{\sin B}{b} = \frac{\sin C}{c}.$$

We can use the Law of Sines when the given information falls into either the AAS, ASA, or SSA category. The AAS and ASA congruence postulates guarantee a unique triangle, so solving triangles when the givens are in the AAS or ASA category is more straightforward. In contrast, SSA does not always guarantee a unique triangle, so it is harder to use the Law of Sines to solve this type of problem. We therefore delay the analysis of the SSA category until Section 9.3. Because the Law of Sines is computationally easier than the Law of Cosines, we will sometimes use it to simplify calculations within the SAS and SSS categories.

9.2.1 Unique Triangle Due to AAS

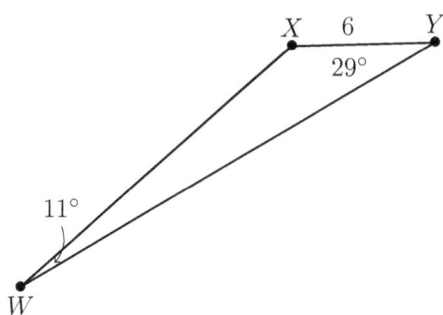

Example 9.3 Solve $\triangle WXY$.

Solution Let us find y. Using the Law of Sines, we have

$$\frac{\sin W}{w} = \frac{\sin Y}{y} \quad \text{implies} \quad \frac{\sin 11°}{6} = \frac{\sin 29°}{y}.$$

It follows that
$$y = \frac{6 \sin 29°}{\sin 11°} \approx 15.245.$$

Let us find $m\angle X$. Due to the Triangle Sum Theorem,

$$m\angle W + m\angle X + m\angle Y = 180° \quad \text{implies} \quad m\angle X = 140°.$$

Lastly, we obtain x via the Law of Sines, which gives

$$\frac{\sin W}{w} = \frac{\sin X}{x} \quad \text{implies} \quad \frac{\sin 11°}{6} = \frac{\sin 140°}{x}.$$

A bit of algebra shows
$$x = \frac{6 \sin 140°}{\sin 11°} \approx 20.212.$$

■

9.2.2 Unique Triangle Due to ASA

When the uniqueness of a triangle is obtained by ASA, the ideas needed to solve the triangle are almost identical to those in the AAS category. Simply use the Triangle Sum Theorem to determine the angle measure opposite the given side length and then continue as previously demonstrated. Despite ASA's computational similarity to AAS, we will include an example for easy reference and for completeness.

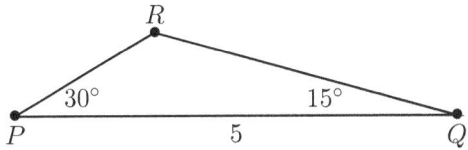

Example 9.4 Solve $\triangle PQR$.

Solution We need an angle opposite a given side, so let us find $m\angle R$. Due to the Triangle Sum Theorem,

$$m\angle P + m\angle Q + m\angle R = 180° \quad \text{implies} \quad m\angle R = 135°.$$

To find p and q, we will use the Law of Sines. We have

$$\frac{\sin P}{p} = \frac{\sin R}{r} \quad \text{implies} \quad \frac{\sin 30°}{p} = \frac{\sin 135°}{5}.$$

It follows that
$$p = \frac{5\sin 30°}{\sin 135°} \approx 3.536.$$

Furthermore,

$$\frac{\sin Q}{q} = \frac{\sin R}{r} \quad \text{implies} \quad \frac{\sin 15°}{q} = \frac{\sin 135°}{5}.$$

So,
$$q = \frac{5\sin 15°}{\sin 135°} \approx 1.830.$$

■

9.2.3 Law of Sines to Simplify Calculations

We can use the Law of Sines to find the angle measures of a triangle. This is helpful because the Law of Sines is computationally easier than the Law of Cosines. However, be careful because the range of arc sine does not include angle measures greater than $90°$. Due to this quirk, the Law of Sines is not well suited for finding obtuse angle measures.

We recommend avoiding the Law of Sines when the angle could be obtuse. The Isosceles Triangle Theorem tells us that the angle of largest measure is opposite the longest side. Since there can only be one obtuse angle in a triangle, it would have to be opposite the longest side. So, let us adopt the following convention:

> Do not use the Law of Sines to find the angle measure opposite the longest side. Either find a different angle measure or use the Law of Cosines.

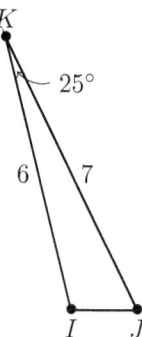

Example 9.5 Solve $\triangle IJK$.

Solution We need to have the side length opposite an angle to use the Law of Sines. So, we will use the Law of Cosines first to find

k:

$$\begin{aligned} k^2 &= i^2 + j^2 - 2ij\cos K \\ &= 7^2 + 6^2 - 2(7)(6)\cos 25° \\ &= 85 - 84\cos 25° \\ &\approx 8.870 \end{aligned}$$
$$\Rightarrow \quad k \approx \sqrt{8.870}$$
$$\approx 2.978.$$

We now have the Law of Sines at our disposal. Since $\angle I$ could be obtuse, we will avoid using the Law of Sines to find its measure.

Instead, we will use the Law of Sines to find $m\angle J$. The Law of Sines says

$$\frac{\sin J}{j} = \frac{\sin K}{k} \quad \text{implies} \quad \frac{\sin J}{6} \approx \frac{\sin 25°}{2.978}.$$

Multiplying by six yields

$$\sin J \approx \frac{6\sin 25°}{2.978}.$$

It follows that

$$m\angle J \approx \arcsin\left(\frac{6\sin 25°}{2.978}\right) \approx 58.364°.$$

We will find $m\angle I$ using the Triangle Sum Theorem:

$$\begin{aligned} m\angle I + m\angle J + m\angle K &= 180° \\ \Rightarrow \quad m\angle I + 58.364° + 25° &\approx 180° \\ \Rightarrow \quad m\angle I &\approx 96.636°. \end{aligned}$$

∎

To demonstrate why we do not use the Law of Sines to find the angle measure opposite the longest side, let us try to find $m\angle I$ in Example 5 using the Law of Sines:

$$\frac{\sin I}{i} = \frac{\sin K}{k} \quad \text{implies} \quad \frac{\sin I}{7} \approx \frac{\sin 25°}{2.978}.$$

Then multiplying by 7 gives us

$$\sin I \approx \frac{7\sin 25°}{2.978} \approx 0.993.$$

This statement is true. However,
$$\arcsin(\sin I) \neq m\angle I,$$
because $\angle I$ is obtuse, so we cannot use arc sine to solve for $m\angle I$. Indeed, Example 5 showed
$$m\angle I \approx 96.636° \quad \text{but} \quad \arcsin(0.993) \approx 83.411°.$$

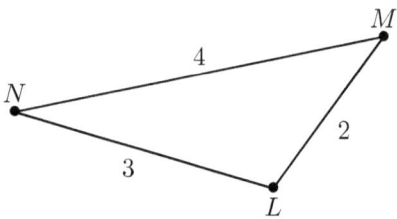

Example 9.6 Solve $\triangle LMN$.

Solution We need to use the Law of Cosines to find the first angle measure, because the Law of Sines only works when an angle measure and its opposite side length are known. Let us find $m\angle L$ first, because it is opposite the longest side, so we would not the Law of Sines to find it anyway.

$$\begin{aligned} m^2 + n^2 - 2mn\cos L &= \ell^2 \\ \Rightarrow \quad 3^2 + 2^2 - 2(3)(2)\cos L &= 4^2 \\ \Rightarrow \quad 13 - 12\cos L &= 16 \\ \Rightarrow \quad \cos L &= -\frac{1}{4}. \end{aligned}$$

Therefore,
$$m\angle L = \arccos\left(-\frac{1}{4}\right) \approx 104.478°.$$

We can use the Law of Sines to find either of the other two angles. Let us find $m\angle M$. The Law of Sines says
$$\frac{\sin M}{m} = \frac{\sin L}{\ell} \quad \text{implies} \quad \frac{\sin M}{3} = \frac{\sin(104.478°)}{4}.$$

Multiplying by three yields

$$\sin M = \frac{3\sin(104.478°)}{4}.$$

This implies

$$m\angle M = \arcsin\left(\frac{3\sin 104.478°}{4}\right) \approx 46.567°.$$

Then the Triangle Sum Theorem tells us:

$$\begin{aligned} m\angle L + m\angle M + m\angle N &= 180° \\ \Rightarrow \quad 104.478° + 46.567° + m\angle N &\approx 180° \\ \Rightarrow \quad m\angle N &\approx 28.955°. \end{aligned}$$

∎

9.3 Law of Sines and SSA

We will now consider triangles where the given information falls into the SSA category. When two side lengths and the non-included angle measure of an oblique triangle are given, we will use the Law of Sines to solve when possible. However, great care must be taken, because SSA is not sufficient to determine the existence of a unique triangle.

Indeed, if the given information falls into the SSA category, there are three potential outcomes: either no triangle, one triangle, or two triangles can be formed. Our goal is to develop a protocol to determine which of these outcomes occurs for each particular set of givens.

Consider $\triangle ABC$. Assume that the givens are $m\angle A$, length a, and length c. Our analysis breaks down into three subsections:

- $\angle A$ obtuse,
- $\angle A$ right, and
- $\angle A$ acute.

9.3.1 Given Angle is Obtuse

When $m\angle A > 90°$, either one or no triangle can be formed with the given information.

(i) If $a > c$, then $\triangle ABC$ is unique.

(ii) If $a \leq c$, then no triangle can be formed with the given information.

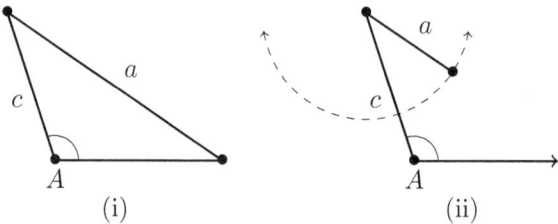

If the given information falls into case (i), use the Law of Sines to determine all the lengths and unknown angle measures.

On the other hand, if the givens put us into case (ii), no triangle can be formed. As a result, we are done because it is impossible to solve a triangle that does not exist.

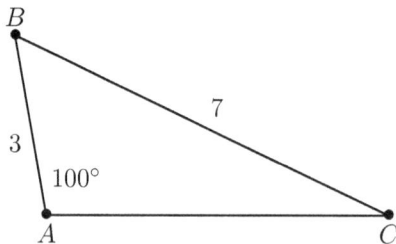

Example 9.7 Suppose $m\angle A = 100°$, $a = 7$, and $c = 3$. Solve $\triangle ABC$.

Solution Because $\angle C$ is obtuse and $a > c$, we are in case (i). Let us use the Law of Sines.

We will find $m\angle C$ first:

$$\frac{\sin C}{c} = \frac{\sin A}{a} \quad \text{implies} \quad \frac{\sin C}{3} = \frac{\sin 100°}{7}.$$

It follows that

$$\sin C = \frac{3\sin 100°}{7}.$$

Ergo,

$$m\angle C = \arcsin\left(\frac{3\sin 100°}{7}\right) \approx 24.965°.$$

To find $m\angle B$, we will utilize the Triangle Sum Theorem. It tells us

$$m\angle A + m\angle B + m\angle C = 180° \quad \text{implies} \quad m\angle B \approx 55.035°.$$

Let us find b via the Law of Sines. We have

$$\frac{\sin B}{b} = \frac{\sin A}{a} \quad \text{implies} \quad \frac{\sin 55.035°}{b} \approx \frac{\sin 100°}{7}.$$

Therefore,

$$b \approx \frac{7\sin 55.035°}{\sin 100°} \approx 5.825.$$

■

Example 9.8 Use a theorem from the text to explain why no triangle can be formed if $\angle A$ is obtuse and $a \leq c$.

Solution The Isosceles Triangles Theorem (Theorem 1.2) says that the angle of largest measure is opposite the longest side. A triangle can have at most one obtuse angle, which means that the obtuse $\angle A$ must have the largest measure. Hence, a must have the longest length. This would contradict the assumption $a \leq c$. Therefore, no such triangle can be formed. ■

9.3.2 Given Angle is Right

When $\angle A$ is right, either one or no triangle can be formed with the given information.

(i) If $a > c$, then $\triangle ABC$ is unique.

(ii) If $a \leq c$, then no triangle can be formed with the given information.

If $\angle A$ is a right angle and the length of hypotenuse a is longer than the length of leg c, then $\triangle ABC$ is a right triangle. As a result, there is no need for the Law of Cosines or Sines. We can use the techniques studied in Chapter 4 to solve the triangle.

In contrast, if $\angle A$ is right and a is not longest, then no triangle can be formed. The hypotenuse is always the longest side of a triangle. Supposing otherwise is a contradiction of the Isosceles Triangle Theorem (Theorem 1.2).

Example 9.9 Solve $\triangle ABC$ or explain why no triangle can be formed with the given information.

(a) $m\angle A = 90°$, $a = 10$, and $c = 7$.

(b) $m\angle A = 90°$, $a = 10$, and $c = 12$.

Solution

(a) Since $a > c$, a triangle can be formed. Using the Pythagorean Theorem,

$$10^2 = 7^2 + b^2 \quad \text{implies} \quad b = \sqrt{51} \approx 7.141.$$

Furthermore,

$$\cos B = \frac{7}{10}$$

implies

$$m\angle B = \arccos \frac{7}{10} \approx 46.573°.$$

To find $m\angle C$, we note

$$m\angle A + m\angle B + m\angle C = 180° \quad \text{implies} \quad m\angle C \approx 44.427°.$$

(b) The hypotenuse a is always the longest side of a right triangle, but $a < c$. This means the given information does not form a triangle. ∎

9.3.3 Given Angle is Acute

When $\angle A$ is acute, either zero, one, or two triangles can be formed with the given information.

(i) If $a < c\sin A$, then no triangle can be formed with the given information.

(ii) If $a = c\sin A$, then $\triangle ABC$ is unique.

(iii) If $c\sin A < a < c$, then two triangles can be formed with the given information.

(iv) If $a \geq c$, then $\triangle ABC$ is unique.

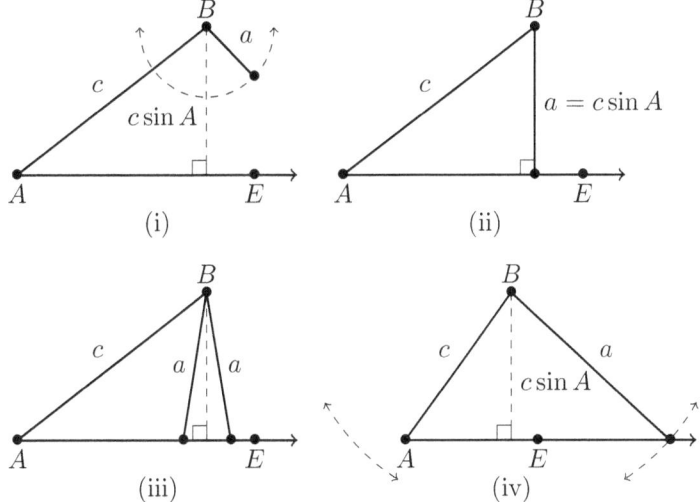

The shortest segment between B and ray $\overrightarrow{AE}$ is perpendicular to the ray. This implies its length is $c\sin A$.

In case (i), we assumed $a < c\sin A$. Because the shortest distance between B and $\overrightarrow{AE}$ is $c\sin A$, it is impossible for the segment to reach the ray. Hence, no triangle can be formed.

In case (ii), we supposed $a = c\sin A$. Since the shortest distance that can reach $\overrightarrow{AE}$ is $a = BC$, the segment corresponding to a is perpendicular to $\overrightarrow{AE}$. Uniqueness follows from the HL Theorem (Theorem 2.2).

In case (iii), we hypothesized that $c\sin A < a < c$. As you can see from the diagram, the segment corresponding to a could intersect $\overrightarrow{AE}$ at either of two points. This implies that two triangles can be formed under this scenario. Use the Law of Sines to determine the side lengths and angle measures of both triangles. A helpful observation is that $\angle C$ is acute in one of the triangles and obtuse in the other.

In case (iv), we presumed that $a \geq c$. The segment corresponding to a only intersects $\overrightarrow{AE}$ at one point. There is no intersection due to a leftward swing because it would contradict the assumption that $\angle A$ is an interior angle. Use the Law of Sines to solve the one triangle. Note that $\angle C$ is acute in this triangle.

Example 9.10 Suppose $m\angle A = 35°$, $a = 1$, and $c = 5$. Solve $\triangle ABC$ or state that no such triangle exists.

Solution

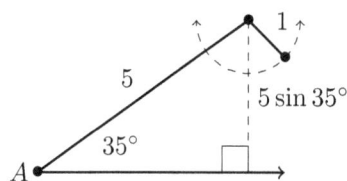

Because
$$c\sin A = 5\sin 35° \approx 2.867$$
and $a = 1$, we have
$$a < c\sin A.$$

This is case (i). Thus, no triangle can be formed with the given information. The side a is too short to reach the ray. ∎

Example 9.11 Consider $\triangle ABC$, where $m\angle A = 45°$, $a = 9$, and $c = 12$. Solve all triangles $\triangle ABC$ which satisfy the given criteria.

Solution We have
$$c \sin A = 12 \sin 45° \approx 8.485.$$
Since $c \sin A < a < c$, this is case (iii) and two triangles can be formed. In one triangle $\angle C$ is acute, and $\angle C$ is obtuse in the other.

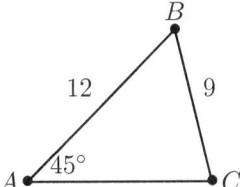

$\angle C$ acute: Let us find $m\angle C$. The Law of Sines says
$$\frac{\sin C}{c} = \frac{\sin A}{a} \quad \text{implies} \quad \frac{\sin C}{12} = \frac{\sin 45°}{9}.$$
It follows that
$$\sin C = \frac{4 \sin 45°}{3}.$$
Because we are assuming $\angle C$ is acute,
$$m\angle C = \arcsin\left(\frac{4 \sin 45°}{3}\right) \approx 70.529°.$$
To find $m\angle B$, note
$$m\angle A + m\angle B + m\angle C = 180° \quad \text{implies} \quad m\angle B \approx 64.471°.$$
Lastly, we need b, which we obtain using the Law of Sines:
$$\frac{\sin B}{b} = \frac{\sin A}{a} \quad \text{implies} \quad \frac{\sin 64.471°}{b} \approx \frac{\sin 45°}{9}.$$
Hence,
$$b \approx \frac{9 \sin 64.471°}{\sin 45°} \approx 11.485.$$

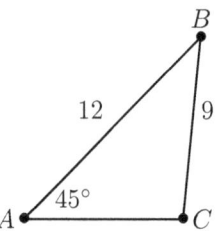

∠**C** **obtuse:** Due to the Law of Sines,

$$\frac{\sin C}{c} = \frac{\sin A}{a} \quad \text{implies} \quad \sin C = \frac{4\sin 45°}{3}.$$

Since ∠C is obtuse, its terminal side lies in quadrant II when placed in standard position. As a result, its reference angle is $\arcsin\left(4\sin(45°)/3\right)$. It follows that

$$m\angle C = 180° - \arcsin\left(\frac{4\sin 45°}{3}\right) \approx 109.471°.$$

Let us find $m\angle B$. We know

$$m\angle A + m\angle B + m\angle C = 180° \quad \text{implies} \quad m\angle B \approx 25.529°.$$

Lastly, we get the length b using the Law of Sines. We have

$$\frac{\sin B}{b} = \frac{\sin A}{a} \quad \text{implies} \quad \frac{\sin 25.529°}{b} \approx \frac{\sin 45°}{9}.$$

So,

$$b \approx \frac{9\sin 25.529°}{\sin 45°} \approx 5.485.$$

■

Example 9.12 Consider $\triangle ABC$, where $m\angle A = 45°$, $a = 13$, and $c = 12$. Solve all triangles $\triangle ABC$ which satisfy the given criteria.

Solution Since $a \geq c$, the criteria above uniquely determines $\triangle ABC$.

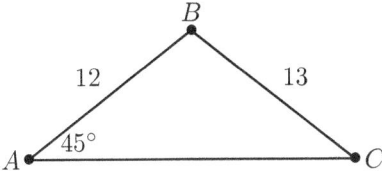

We obtain $m\angle C$ using the Law of Sines:
$$\frac{\sin C}{c} = \frac{\sin A}{a} \quad \text{implies} \quad \frac{\sin C}{12} = \frac{\sin 45°}{13}.$$

Multiplying by 12 gives
$$\sin C = \frac{12 \sin 45°}{13}.$$

Since $\angle C$ is acute,
$$m\angle C = \arcsin\left(\frac{12 \sin 45°}{13}\right) \approx 40.747°.$$

To find $m\angle B$, we utilize the Triangle Sum Theorem:
$$m\angle A + m\angle B + m\angle C = 180° \quad \text{implies} \quad m\angle B \approx 94.253°.$$

We can find b by observing
$$\frac{\sin B}{b} = \frac{\sin A}{a} \quad \text{implies} \quad \frac{\sin 94.253°}{b} \approx \frac{\sin 45°}{13}.$$

Thus,
$$b \approx \frac{13 \sin 94.253°}{\sin 45°} \approx 18.334.$$

∎

9.3.4 Summary

Consider $\triangle ABC$. Suppose we are given $m\angle A$, a, and c.

- $\angle A$ is obtuse.

  (i) If $a > c$, then $\triangle ABC$ is unique.

  (ii) If $a \leq c$, then no triangle can be formed with the given information.

- $\angle A$ is right.

  (i) If $a > c$, then $\triangle ABC$ is unique.

  (ii) If $a \leq c$, then no triangle can be formed with the given information.

- $\angle A$ is acute.

  (i) If $a < c \sin A$, then no triangle can be formed with the given information.

  (ii) If $a = c \sin A$, then $\triangle ABC$ is unique.

  (iii) If $c \sin A < a < c$, then two triangles can be formed with the given information.

  (iv) If $a \geq c$, then $\triangle ABC$ is unique.

Example 9.13 Determine whether zero, one, or two triangles satisfy the given information.

(a) $m\angle A = 25°$, $a = 10$, and $c = 10$.

(b) $m\angle A = 90°$, $a = 15$, and $c = 7$.

(c) $m\angle A = 115°$, $a = 3$, and $c = 11$.

(d) $m\angle A = 30°$, $a = 7$, and $c = 14$.

(e) $m\angle A = 45°$, $a = 25$, and $c = 30$.

(f) $m\angle A = 90°$, $a = 9$, and $c = 11$.

(g) $m\angle A = 170°$, $a = 100$, and $c = 5$.

(h) $m\angle A = 80°$, $a = 17$, and $c = 19$.

Solution

(a) Since $\angle A$ is acute and $a \geq c$, there is one triangle which satisfies the given information.

(b) Because $\angle A$ is right and $a > c$, only one triangle can be formed using the information.

(c) Since $\angle A$ is obtuse and $a < c$, no triangle meets the listed criteria.

(d) Due to the fact that $\angle A$ is acute and
$$c \sin A = 14 \sin 30° = 7 = a,$$
it follows that one triangle satisfies the given information.

(e) We were given that $\angle A$ is acute. Furthermore, because
$$c \sin A = 30 \sin 45° \approx 21.213,$$
we have $c \sin A < a < c$. Hence, there are two triangles.

(f) Because $\angle A$ is right and $a < c$, it is impossible to form a triangle using the given criteria.

(g) Since $\angle A$ is obtuse and $a > c$, one triangle can be constructed using the information provided.

(f) We have
$$c \sin A = 19 \sin 80° \approx 18.711.$$
So, $a < c \sin A$. It follows that no triangle can be formed with the listed criteria. ∎

9.4 Proofs of the Laws of Cosines and Sines

First, we will prove the Law of Cosines (Theorem 9.1) for $\triangle ABC$. In particular, we will prove
$$c^2 = a^2 + b^2 - 2ab \cos C.$$
The other two cases follow via relabeling.

Proof Place side b on the positive x-axis, and point C at the origin.

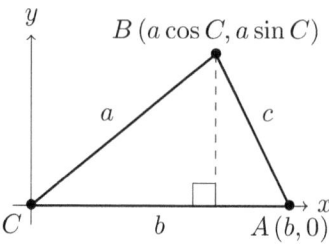

Point A has coordinates $(b, 0)$ and Theorem 5.1 tells us that B has coordinates $(a \cos C, a \sin C)$.

Using the distance formula,

$$c = \sqrt{(a \cos C - b)^2 + (a \sin C - 0)^2}.$$

It follows that

$$\begin{aligned} c^2 &= (a \cos C - b)^2 + (a \sin C)^2 \\ &= a^2 \cos^2 C - 2ab \cos C + b^2 + a^2 \sin^2 C \\ &= \underbrace{a^2 \sin C + a^2 \cos^2 C}_{a^2} + b^2 - 2ab \cos C \\ &= a^2 + b^2 - 2ab \cos C. \end{aligned}$$

∎

The following proposition will help us prove the Law of Sines.

Proposition 9.1 *Suppose that θ is an angle measure of a triangle.*

$$\sin(180° - \theta) = \sin \theta.$$

Proof Using Theorem 7.1 (i),

$$\begin{aligned} \sin(180° - \theta) &= \sin 180° \cos \theta - \cos 180° \sin \theta \\ &= 0 \cos \theta - (-1) \sin \theta \\ &= \sin \theta. \end{aligned}$$

It is time to prove the Law of Sines (Theorem 9.2). It says for $\triangle ABC$, we have
$$\frac{\sin A}{a} = \frac{\sin B}{b} = \frac{\sin C}{c}.$$

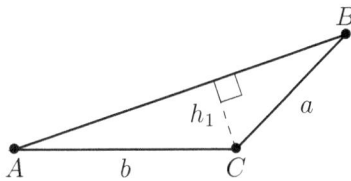

Proof Assume that $\angle C$ has the largest measure of the three angles. Our diagram shows $\angle C$ as obtuse, but this need not be the case.

We see
$$b \sin A = h_1 \quad \text{and} \quad a \sin B = h_1.$$

So,
$$b \sin A = a \sin B \quad \text{implies} \quad \frac{\sin A}{a} = \frac{\sin B}{b}.$$

To obtain the last equation, we will break $\angle C$ into two cases:

(i) $m\angle C < 90°$ and (ii) $m\angle C \geq 90°$.

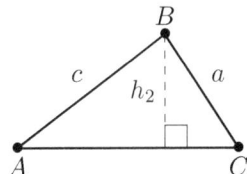

(i) We see
$$c \sin A = h_2 \quad \text{and} \quad a \sin C = h_2.$$

So,
$$c \sin A = a \sin C \quad \text{implies} \quad \frac{\sin A}{a} = \frac{\sin C}{c}.$$

We conclude
$$\frac{\sin A}{a} = \frac{\sin B}{b} = \frac{\sin C}{c}$$
for $m\angle C < 90°$.

(ii) Let $m\angle C = \gamma$.

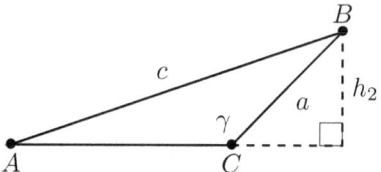

We see
$$c \sin A = h_2 \quad \text{and} \quad a \sin(180° - \gamma) = h_2.$$
Due to Proposition 9.1,
$$a \sin(180° - \gamma) = a \sin \gamma = h_2.$$
So, we have
$$c \sin A = a \sin \gamma \quad \text{implies} \quad \frac{\sin A}{a} = \frac{\sin \gamma}{c}.$$
Since $m\angle C = \gamma$,
$$\frac{\sin A}{a} = \frac{\sin B}{b} = \frac{\sin C}{c}.$$

9.5 Exercises

* Exercise 1

State whether the given information in $\triangle ABC$ places the triangle into the category of SSS, SAS, ASA, AAS, or SSA.

(a) $m\angle C = 37°$, $a = 4$, and $b = 7$.

(b) $m\angle A = 115°$, $m\angle C = 13°$, and $c = 15$.

(c) $a = 3$, $b = 9$, and $c = 11$.

(d) $m\angle C = 100°$, $b = 5$, and $c = 7$.

(e) $m\angle A = 25°$, $m\angle C = 30°$, and $b = 19$.

(f) $m\angle B = 12°$, $a = 10$, and $c = 14$.

(g) $m\angle A = 91°$, $c = 5$, and $a = 5$.

(h) $m\angle B = 30°$, $m\angle C = 111°$, and $c = 12$.

* Exercise 2

For each set of givens in Exercise 1, is the Law of Cosines or Sines needed to solve the triangle?

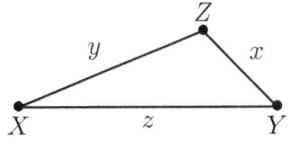

Figure 1

** Exercise 3

Consider Figure 1, which is not drawn to scale. Solve $\triangle XYZ$.

(a) $x = 5$, $y = 7$, and $z = 8$.

(b) $m\angle Z = 174°$, $x = 3$, and $y = 17$.

(c) $m\angle Y = 70°$, $x = 50$, and $z = 59$.

(d) $x = 4$, $y = 5$, and $z = 7$.

** Exercise 4

Suppose the vertices and sides are as shown in Figure 1. Solve $\triangle XYZ$. The triangle is not drawn to scale.

(a) $m\angle X = 24°$, $m\angle Z = 111°$, and $y = 3$.

(b) $m\angle Y = 54°$, $m\angle Z = 83°$, and $x = 7$.

(c) $m\angle Y = 71°$, $m\angle Z = 73°$, and $z = 2$

(d) $m\angle X = 10°$, $m\angle Z = 97°$, and $z = 5$.

** Exercise 5

Solve $\triangle TUV$ when possible, and say "no triangle" when it is not. Assume ...

(a) ... $m\angle T = 155°$, $t = 12$, and $u = 11$.

(b) ... $m\angle V = 130°$, $t = 21$, and $v = 12$.

(c) ... $m\angle T = 90°$, $t = 5$, and $u = 12$.

(d) ... $m\angle U = 108°$, $u = 39$, and $v = 37$.

(e) ... $m\angle V = 90°$, $t = 5$, and $v = 8$.

** Exercise 6

Determine whether there is a $\triangle WXY$ that satisfies the given criteria.

(a) $m\angle W = 30°$, $w = 5$, and $y = 10$.

(b) $m\angle X = 45°$, $x = 7$, and $y = 12$.

(c) $m\angle Y = \arccos \frac{3}{5}$, $x = 10$, and $y = 8$.

** Exercise 7

There are two triangles $\triangle DEF$ which satisfy the given criteria. Solve them.

(a) $m\angle D = 35°$, $d = 10$, and $f = 15$.

(b) $m\angle E = 65°$, $d = 25$, and $e = 23$.

(c) $m\angle F = 22°$, $e = 100$, and $f = 49$.

(d) $m\angle D = 59°$, $d = 17$, and $e = 18$.

** Exercise 8

Consider $\triangle GHI$. State whether the given information forms zero, one, or two triangles.

(a) $m\angle H = 115°$, $g = 12$, and $h = 17$.

(b) $m\angle I = 45°$, $g = 13$, and $i = 19$.

(c) $m\angle G = 90°$, $g = 90$, and $h = 100$.

(d) $m\angle H = 20°$, $h = 15$, and $i = 53$.

(e) $m\angle G = 42°$, $g = 23$, and $h = 32$.

(f) $m\angle G = 100°$, $g = 15$, and $h = 10$.

(g) $m\angle I = 30°$, $g = 6$, and $i = 3$.

(h) $m\angle H = 90°$, $g = 4$, and $h = 7$.

** Exercise 9

Use the given information of $\triangle JKL$ to determine whether it defines zero, one, or two triangles.

(a) $m\angle L = 113°$, $k = 20$, and $\ell = 22$.

(b) $m\angle K = 23°$, $j = 50$, and $k = 19$.

(c) $m\angle L = 90°$, $k = 35$, and $\ell = 21$.

(d) $m\angle K = 170°$, $j = 100$, and $k = 22$.

(e) $m\angle J = 20°$, $j = 11$, and $k = 10$.

(f) $m\angle J = 77°$, $j \approx 116.924$, and $\ell = 120$.

(g) $m\angle L = 61°$, $k = 50$, and $\ell = 44$.

(h) $m\angle J = 90°$, $j = 13$, and $k = 5$.

** Exercise 10

Information about $\triangle PQR$ is given. Use it to solve all possible triangles which satisfy the given information. When it is impossible to form a triangle using the given information, say so.

(a) $m\angle Q = 38°$, $q = 42$, and $r = 32$.

(b) $m\angle R = 95°$, $q = 8$, and $r = 9$.

(c) $m\angle P \approx 67.380°$, $p = 12$, and $r = 13$.

(d) $m\angle Q = 23°$, $q = 7$, and $r = 14$.

(e) $m\angle Q = 108°$, $q = 19$, and $r = 13$.

(f) $m\angle P = 45°$, $p = 20$, and $r = 14$.

(g) $m\angle R = 90°$, $q = 5$, and $r = 3$.

(h) $m\angle Q = 25°$, $p = 51$, and $q = 23$.

(i) $m\angle P = 102°$, $p = 10$, and $r = 17$.

** Exercise 11

Side lengths and an angle measure of $\triangle ABC$ are listed. Solve the triangle(s) which satisfy the given criteria.

(a) $m\angle C = 110°$, $b = 10$, and $c = 17$.

(b) $m\angle A = 100°$, $m\angle C = 25°$, and $c = 10$

(c) $m\angle A = 40°$, $a = 21$, and $b = 30$.

(d) $m\angle A = 90°$, $a = 25$, and $b = 15$.

(e) $m\angle B = 25°$, $m\angle C = 45°$, and $a = 40$.

(f) $m\angle B = 22°$, $a = 10$, and $c = 15$.

(g) $m\angle C = 31°$, $b = 37$, and $c = 40$.

(h) $a = 10$, $b = 12$, and $c = 17$

(i) $m\angle A = 60°$, $a = 5\sqrt{3}$, and $c = 10$.

Chapter 10

Area and Perimeter

In this chapter, we will use trigonometry to solve area and perimeter problems. A solid understanding of Section 1.5, Chapter 3, Chapter 5, and Chapter 9 is necessary. We will use scientific calculators in this chapter.

10.1 Triangles

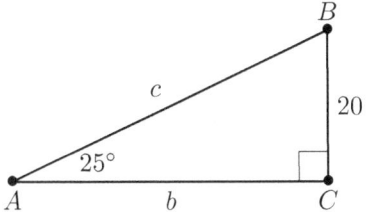

Example 10.1 Find (a) the area and (b) the perimeter of $\triangle ABC$.

Solution

(a) Recall that the area of a triangle is

$$\frac{1}{2}bh,$$

285

where b is the base and h is the height.

On $\triangle ABC$, if a is the height, then b is the base. As a result, we need to find b. Notice

$$\tan 25° = \frac{20}{b} \quad \text{implies} \quad b = \frac{20}{\tan 25°}.$$

It follows that the area of $\triangle ABC$ is

$$\frac{1}{2}\left(\frac{20}{\tan 25°}\right)(20) = \frac{200}{\tan 25°} \approx 428.901.$$

(b) The perimeter is the sum of the side lengths. We have a and b, and we need to find c. Using right triangle trigonometry, we have

$$\sin 25° = \frac{20}{c} \quad \text{implies} \quad c = \frac{20}{\sin 25°}.$$

Thus, the perimeter of $\triangle ABC$ is

$$a + b + c = 20 + \frac{20}{\tan 25°} + \frac{20}{\sin 25°} \approx 110.214.$$

∎

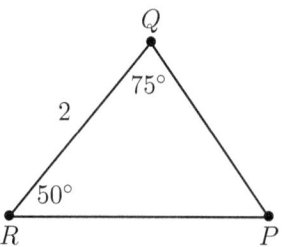

Example 10.2 Find the perimeter of $\triangle PQR$.

Solution Our goal is to use the Law of Sines (Theorem 9.2), which requires $m\angle P$. Since the sum of the interior angle measures is 180°,

$$m\angle P + m\angle Q + m\angle R = 180° \quad \text{implies} \quad m\angle P = 55°.$$

So,
$$\frac{\sin P}{p} = \frac{\sin Q}{q} \quad \text{implies} \quad q = \frac{2\sin 75°}{\sin 55°}$$
and
$$\frac{\sin P}{p} = \frac{\sin R}{r} \quad \text{implies} \quad r = \frac{2\sin 50°}{\sin 55°}.$$

Thus, the perimeter of $\triangle PQR$ is

$$p + q + r = 2 + \frac{2\sin 75°}{\sin 55°} + \frac{2\sin 50°}{\sin 55°} \approx 6.229.$$

∎

We could use basic trigonometry to find the area of oblique triangles, like the one in Example 2. However, the next two propositions provide easier routes.

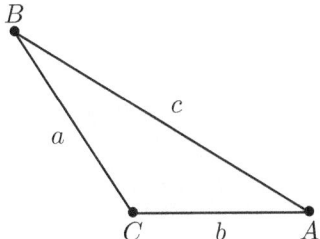

Proposition 10.1 *The area of $\triangle ABC$ is*

$$\frac{bc\sin A}{2} = \frac{ac\sin B}{2} = \frac{ab\sin C}{2}.$$

Proof We will prove that the area of $\triangle ABC$ is $ab\sin(C)/2$. The proofs for the other two are nearly identical.

The area is one-half base times height. Suppose b is the length of the base. All we need is the height. Place $\triangle ABC$ on the xy-coordinate plane such that C is on the origin, and $\overline{AC}$ lies on the positive x-axis.

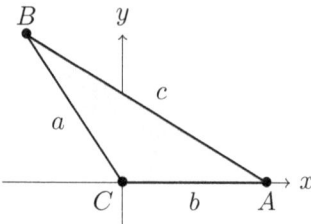

Theorem 5.1 tells us that B has coordinates $(a\cos C, a\sin C)$. The value $a\sin C$ is the height of triangle for $0 < m\angle C < 180°$. Thus, the area of $\triangle ABC$ is

$$\frac{1}{2}b(c\sin A) = \frac{bc\sin A}{2}.$$

■

Example 10.3 Suppose $m\angle Y = 51°$ and x, y, and z are 5, 7, and 9, respectively. What is the area of $\triangle XYZ$?

Solution We know the measure of $\angle Y$. The two adjacent sides, x and z, have lengths 5 and 9. Hence, the area of triangle $\triangle XYZ$ is

$$\frac{xz\sin Y}{2} = \frac{(5)(9)\sin 51°}{2} \approx 17.486.$$

■

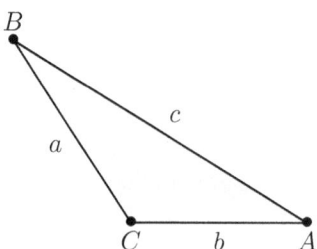

Theorem 10.1 (Horen) *Suppose* $s = \dfrac{a+b+c}{2}$. *Then the area of* $\triangle ABC$ *is*

$$A = \sqrt{s(s-a)(s-b)(s-c)}.$$

288

We omit the proof because it is very computationally intensive. It follows from Proposition 10.1 and the Law of Cosines (Theorem 9.1).

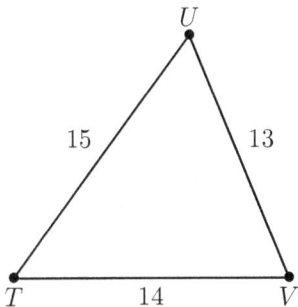

Example 10.4 Find the area of $\triangle TUV$.

Solution We have
$$s = \frac{t + u + v}{2}$$
$$= \frac{13 + 14 + 15}{2}$$
$$= 21.$$

Hence, the theorem of Horen tells us that the area of $\triangle TUV$ is
$$A = \sqrt{s(s-t)(s-u)(s-v)}$$
$$= \sqrt{21(8)(7)(6)}$$
$$= 84.$$

10.2 Regular Polygons

Definition 10.1 A **polygon** is a closed geometric object that is bounded by line segments.

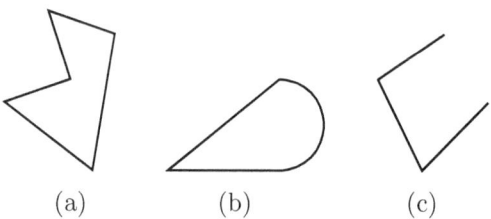

(a)   (b)   (c)

Image (a) is a polygon because it is closed and is bounded by segments. Image (b) is not a polygon because a portion of its boundary is an arc instead of a line segment. Image (c) is not a polygon because it is not closed.

Polygons are usually classified based on the number of sides they contain.

Sides	Name
3	triangle
4	quadrilateral
5	pentagon
6	hexagon
7	heptagon
8	octagon
9	nonagon
10	decagon
12	dodecagon

We refer to a polygon with n sides as an n-gon for $n > 12$ or $n = 11$. A polygon of an unknown number of sides is sometimes referred to as an n-gon as well.

Definition 10.2 A polygon is **regular** when all its interior angles are congruent and all its side lengths are equal.

Examples of regular polygons include equilateral triangles and squares.

Definition 10.3

- A **circumradius** is any of the line segments from the center of a regular polygon to a vertex.

- An **apothem** is the line segment from the center of a regular polygon to the midpoint of a side.

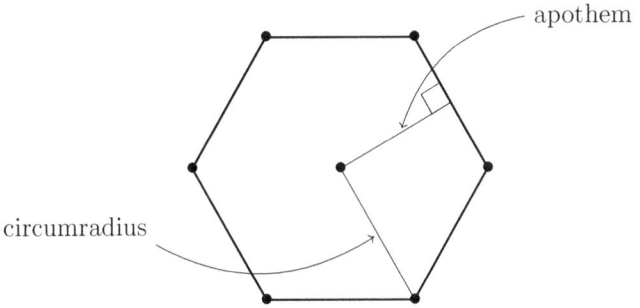

For any regular n-gon, all of the apothems have the same length, and all of the circumradii have the same length as well.

Proposition 10.2 *An apothem is perpendicular to the side it intersects.*

Proof Divide the regular polygon into isosceles triangles. One triangle is shown below. We have labeled its vertices for ease of reference; the segment $\overline{AC}$ is a side of the polygon and $\overline{BD}$ is its apothem.

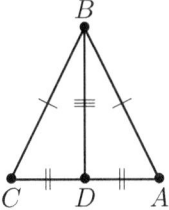

The length of apothem $\overline{BD}$ is equal to itself. The circumradii $\overline{AB}$ and $\overline{CB}$ have equal lengths as well. We have that $\overline{AD}$ and $\overline{CD}$ have the same length because the apothem bisects the side of the polygon. Hence, $\triangle ABD$ is congruent to $\triangle CBD$ by the SSS congruence postulate.

It follows that $\angle ADB$ is congruent to $\angle CDB$. Call $m\angle ADB = m\angle CDB = x$.

Since the two angles form a linear pair, they are supplementary. So,
$$m\angle ADB + m\angle CDB = 180° \quad \text{implies} \quad x = 90°.$$
We conclude that $\angle ADB$ and $\angle CDB$ are right.

■

Example 10.5 Suppose an apothem of a regular pentagon has length 20. Find (a) the perimeter and (b) the area of the pentagon.

Solution The idea is to break the regular pentagon into five triangles, and use trigonometry to find the needed information.

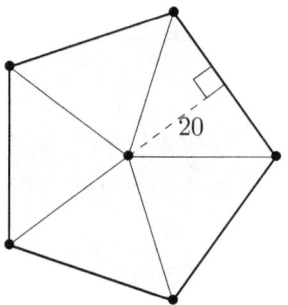

A complete circle has 360°, so the vertex angle of each isosceles triangle must have measure $350°/5 = 72°$.

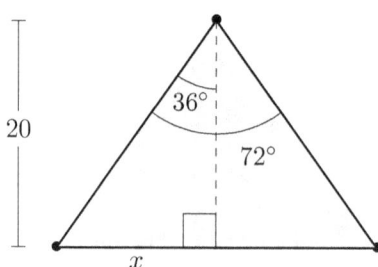

Let x be as shown in the diagram. Using trigonometry,
$$x = 20\tan 36°,$$
which implies the length of the base is
$$2(20\tan 36°) = 40\tan 36°.$$
Since the length of the triangle's base is also the side length of the pentagon, we are ready to answer our questions.

(a) The regular pentagon has five sides, so the perimeter is
$$5\left(40\tan 36°\right) = 200\tan 36° \approx 145.309.$$

(b) The area of each triangles is
$$\frac{1}{2}\left(40\tan 36°\right)(20) = 400\tan 36° \approx 290.617.$$

There are a total of five triangles in the regular pentagon. Hence, the area of the pentagon is
$$5\left(400\tan 36°\right) = 2000\tan 36° \approx 1453.085.$$

Example 10.6 A circumradius of a regular 15-gon has length 8. Find (a) the area of the 15-gon and (b) the perimeter.

Solution

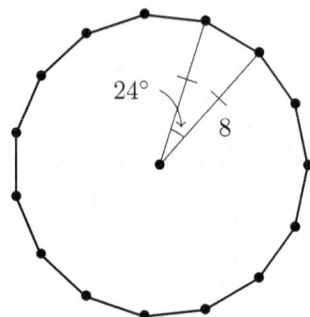

The figure above illustrates our situation. The vertex angle of each isosceles triangle has measure

$$\frac{360°}{15} = 24°.$$

(a) The two adjacent sides of the vertex angle both have length 8, because they are circumradii. Hence, Proposition 10.1 leads us to conclude each triangle has area

$$\frac{(8)(8)\sin 24°}{2} = 32\sin 24°.$$

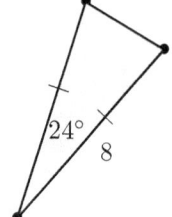

The 15-gon has fifteen triangles, so the total area is

$$15\left(32\sin 24°\right) = 480\sin 24° \approx 195.234.$$

(b) To find the perimeter, consider the isosceles triangle again. Use the apothem to form two congruent right triangles as shown on the next page.

294

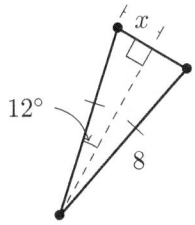

Let x be as shown in the diagram. Then

$$x = 8 \sin 12°.$$

Since the side length is double x, it must equal

$$2\left(8\sin 12°\right) = 16 \sin 12°.$$

The 15-gon has fifteen sides, so the perimeter is

$$15\left(16 \sin 12°\right) = 240 \sin 12° \approx 49.899.$$

Example 10.7 Imagine that the area of a regular hexagon is $150\sqrt{3}$. Find the lengths of its (a) circumradii, (b) apothems, and (c) sides.

Solution Divide the hexagon into six congruent triangles. Each triangle has area $150\sqrt{3}/6 = 25\sqrt{3}$, and each triangle's vertex angle has measure $360°/6 = 60°$.

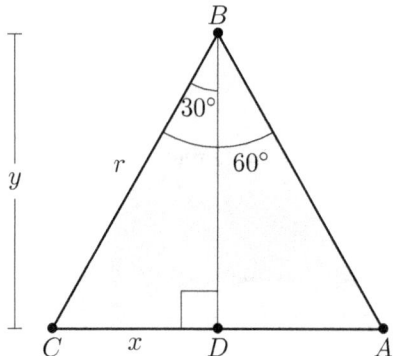

The hexagon's circumradii have length r, its apothems have length y, and its sides have length $2x$.

(a) Proposition 10.1 tells us the area is
$$\frac{r^2 \sin 60°}{2} = \frac{r^2\sqrt{3}}{4}.$$
It follows that
$$\frac{r^2\sqrt{3}}{4} = 25\sqrt{3} \quad \text{implies} \quad r = 10.$$
Hence, its circumradii have length 10.

(b) We see $\triangle BCD$ is a $30° - 60° - 90°$ special right triangle, so $y = 5\sqrt{3}$. This means the hexagon's apothems have length $5\sqrt{3}$ as well.

(c) Using the $30° - 60° - 90°$ special right triangle, we see $x = 5$. Therefore, the regular hexagon has sides of length 10. ∎

10.3 Segments of Circles

Definition 10.4 A **chord** is a line segment connecting two points on a circle.

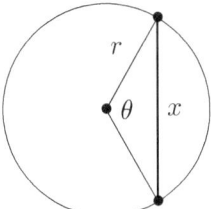

Proposition 10.3 *Let x be the length of a chord contained on a circle whose radius is length r. Suppose the chord subtends a central angle of measure θ. Then the length of the chord is*

$$x = 2r \sin \frac{\theta}{2}.$$

Proof The Law of Cosines tells us

$$x^2 = r^2 + r^2 - 2(r)(r) \cos \theta = 2r^2(1 - \cos \theta).$$

Proposition 7.3 (i) says

$$\sin^2 \frac{\theta}{2} = \frac{1 - \cos \theta}{2}.$$

It follows that

$$x^2 = 2r^2(1 - \cos \theta)$$
$$= 4r^2 \left(\frac{1 - \cos \theta}{2} \right)$$
$$= 4r^2 \sin^2 \frac{\theta}{2}.$$

Since $0 < \theta \leq 180°$ and r is a length, we know

$$\sin \frac{\theta}{2} > 0 \quad \text{and} \quad r > 0.$$

This implies
$$x = \sqrt{4r^2 \sin^2 \frac{\theta}{2}} = 2r \sin \frac{\theta}{2}.$$

∎

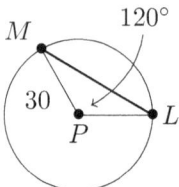

Example 10.8 Suppose $\overline{LM}$ is a chord of circle P. If $MP = 30$, and $m\angle LPM = 120°$, what is the length of $\overline{LM}$?

Solution Let $x = LM$, $r = 30$, and $\theta = 120°$. Then Proposition 10.3 tells us the length of chord $\overline{LM}$ is

$$x = 2r \sin \frac{\theta}{2}$$
$$= 2(30) \sin \frac{120°}{2}$$
$$= 30\sqrt{3}.$$

∎

Definition 10.5 A **segment of a circle** is the region bounded between a chord and the arc with the same endpoints.

To denote a segment of a circle, we write "segment ABC". The point written in the middle denotes the center of the circle.

The remainder of this section is dedicated to finding the perimeter and area of the segments of circles.

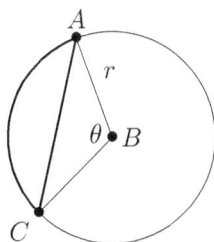

Consider the figure above. Suppose $m\angle ABC = \theta$ radians and the radius of circle B is r.

To find the area of segment ABC proceed as follows.

1. Find the area of sector ABC via Proposition 3.3, which says the area is
$$\frac{r^2\theta}{2}.$$

2. Find the area of $\triangle ABC$ via Proposition 10.1, which says the area is
$$\frac{r^2 \sin\theta}{2}.$$

3. The area of segment ABC is
$$\frac{r^2\theta}{2} - \frac{r^2 \sin\theta}{2}.$$

The idea behind this procedure is to calculate the area of the entire sector and the subtract the area of the isosceles triangle contained within the circle. The result is the area of the segment.

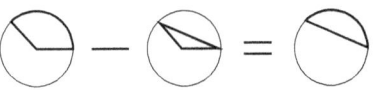

To find the perimeter of segment ABC proceed as follows.

1. Find the length of arc AC via Proposition 3.2, which says the arc length is
$$r\theta.$$

2. Find the length of chord AC via Proposition 10.3, which says the length is
$$2r\sin\frac{\theta}{2}.$$

3. The perimeter of sector ABC is
$$r\theta + 2r\sin\frac{\theta}{2}.$$

Example 10.9 Suppose circle Q has radius of length 15, and $m\angle PQR = 135°$. Find (a) the area of and (b) the perimeter of segment PQR.

Solution
We need the radian measure of $135°$ for (a) and (b). It is
$$135°\left(\frac{\pi}{180°}\right) = \frac{3\pi}{4}.$$

(a) The area of sector PQR is
$$\frac{15^2(3\pi/4)}{2} = \frac{675\pi}{8},$$
and the area of $\triangle PQR$ is
$$\frac{(15)^2 \sin 135°}{2} = \frac{225\sqrt{2}}{4}.$$
Hence, the area of segment PQR is
$$\frac{675\pi}{8} - \frac{225\sqrt{2}}{4} \approx 185.522.$$

(b) The length of arc PR and chord $\overline{PR}$ are
$$15\left(\frac{3\pi}{4}\right) = \frac{75\pi}{4} \quad \text{and} \quad 2(15)\sin\frac{135°}{2} = 30\sin\frac{135°}{2},$$

respectively. Therefore, the perimeter of segment PQR is

$$\frac{75\pi}{4} + 30\sin\frac{135°}{2} \approx 86.621.$$

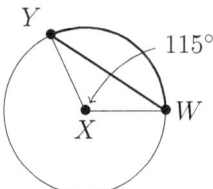

Example 10.10 Suppose the area of segment WXY is 79.350, and $m\angle X = 115°$. What is the length of the radius of circle X rounded to the nearest whole number?

Solution To find the radius's length r, we will find a formula for the area of segment WXY in terms of r, and then set it equal to 79.350.

Since $115° = 23\pi/36$ radians, the area of sector WXY is

$$\frac{r^2(23\pi/36)}{2} \approx 1.004r^2.$$

Proposition 10.1 tells us the area of $\triangle WXY$ is

$$\frac{r^2\sin 115°}{2} \approx 0.453r^2.$$

It follows that

$$\begin{aligned} & 1.004r^2 - 0.453r^2 &= 79.350 \\ \Rightarrow & 0.551r^2 &= 79.350 \\ \Rightarrow & r^2 &\approx 144.011 \\ \Rightarrow & r &\approx \sqrt{144.011} \\ & &\approx 12.000 \end{aligned}$$

We conclude that the radius of circle X is about 12.

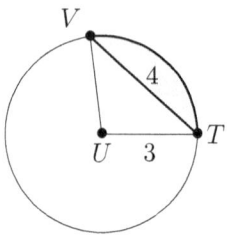

Example 10.11 Suppose the radius of circle U is 3, and the length of chord TV is 4. Find (a) the area of segment TUV and (b) the perimeter of segment TUV.

Solution Let us find $m\angle U$. Proposition 10.3 tells us

$$2(3)\sin\left(\frac{m\angle U}{2}\right) = 4.$$

It follows that

$$m\angle U = 2\arcsin\frac{2}{3}.$$

We will suppose $m\angle U$ is a radian measure and turn our calculator to radian mode. We are ready to find our solutions.

(a) The areas of sector TUV and $\triangle TUV$ are

$$\frac{3^2\left(2\arcsin(2/3)\right)}{2} \approx 6.568 \quad \text{and} \quad \frac{3^2\sin\left(2\arcsin(2/3)\right)}{2} \approx 4.472,$$

respectively. We conclude that the area of segment TUV is about

$$6.568 - 4.472 = 2.096.$$

(b) The length of arc TV is

$$3\left(2\arcsin\frac{2}{3}\right) \approx 4.378.$$

It follows that the perimeter of segment TUV is about

$$4.378 + 4 = 8.378.$$

∎

10.4 Exercises

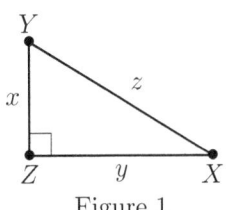

Figure 1

* Exercise 1

Consider Figure 1. Find the area and the perimeter of △XYZ.

(a) Suppose $x = 10$ and $z = 15$.

(b) Let $m\angle X = 20°$ and $x = 12$.

(c) Say $m\angle X = 35°$ and $y = 17$.

(d) Assume $m\angle X = 50°$ and $z = 11$.

(e) Given that $m\angle Y = 44°$ and $x = 32$

(f) Imagine that $m\angle Y = 65°$ and $x = 5$.

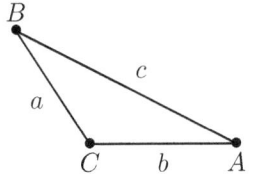

** Exercise 2

Use Figure 2 to obtain the area and the perimeter of △ABC. Suppose ...

(a) ... $m\angle A = 30°$, $b = 7$, and $c = 10$.

(b) ... $a = 5$, $b = 6$, and $c = 7$.

(c) ... $m\angle C = 105°$, $a = 10$, and $b = 15$.

(d) ... $a = 40$, $b = 52$, and $c = 90$.

** Exercise 3

Consider Figure 2. What is the area and the perimeter of △ABC?

(a) Suppose $a = 3$, $b = 5$, and $c = 7$.

(b) Let $m\angle A = 30°$, $m\angle B° = 55°$, and $a = 11$.

(c) Say $m\angle B = 45°$, $m\angle C = 100°$, and $a = 60$.

(d) Assume $m\angle C = 155°$, $b = 5$, and $c = 10$.

(e) Given that $m\angle A = 33°$, $b = 19$, and $c = 25$.

** Exercise 4

Use Proposition 10.1 to prove the Law of Sines.

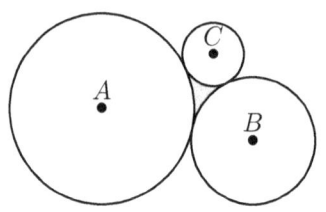

Figure 3

*** Exercise 5

In Figure 3, circles A, B, and C have radii 3, 2, and 1, respectively. Find the area of the gray region. Hint: Form an oblique triangle with the circles' centers.

** Exercise 6

Consider a regular n-gon. Suppose it has an apothem of length a, a circumradius of length r, and a side of length s. Calculate its area and perimeter using the information given.

(a) $n = 6$ and $s = 10$.

(b) $n = 30$ and $a = 15$.

(c) $n = 8$ and $r = 15$.

(d) $n = 10$ and $a = 4$.

(e) $n = 12$ and $r = 5$.

(f) $n = 15$ and $s = 40$.

** Exercise 7

Suppose the area of a regular hexagon is $600\sqrt{3}$. Find the length of its

(a) apothem,

(b) circumradius, and

(c) side.

** Exercise 8

Assume the area of a regular octagon is $450\sqrt{2}$. What is length of its

(a) apothem,

(b) circumradius, and

(c) side?

*** Exercise 9

A circle of radius 10 circumscribes a regular n-gon. Use the given values of n to compute the area and the perimeter of the n-gon.

(a) $n = 3$ (d) $n = 6$

(b) $n = 4$ (e) $n = 8$

(c) $n = 5$ (f) $n = 12$

** Exercise 10

In Exercise 9, what values are the areas and perimeters of the regular n-gons approaching as n grows?

*** Exercise 11

A circle of radius 10 inscribes a regular n-gon. Calculate the area and the perimeter of the triangle, for n equal to each of the following.

(a) $n = 3$
(b) $n = 4$
(c) $n = 5$
(d) $n = 6$
(e) $n = 8$
(f) $n = 12$

** Exercise 12

In Exercise 11, what values are the areas and perimeters of the regular n-gons approaching as n grows?

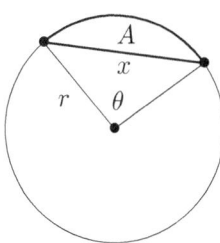

Figure 4

** Exercise 13

Consider Figure 4. In particular, consider the variables x, r, and θ. Find the third of the three variables using the givens information about the other two.

(a) $r = 5$ and $\theta = 120°$.
(b) $r = 12$ and $x = 12\sqrt{2}$.
(c) $x = 17/2$ and $\theta = 32°$.
(d) $r = 16$ and $\theta = 90°$.
(e) $r = 100$ and $x = 25$.
(f) $x = 10$ and $\theta = 100°$.

** Exercise 14

Suppose the variables A, r, and θ are as shown in Figure 4. Use the given values of the two variables to find the third variable.

(a) $A = 8\pi$ and $\theta = 180°$.
(b) $r = 9$ and $\theta = 150°$.
(c) $A = 0.352$ and $\theta = 45°$.
(d) $r = 20$ and $\theta = 60°$.
(e) $A = 100\pi - 200$ and $\theta = 90°$.
(f) $r = 19$ and $\theta = 45°$.

** Exercise 15

Consider Figure 4.

(a) Suppose $r = 9$ and $x = 10$. Find A.

(b) Let $A = 15$ and $\theta = 30°$. What is the value of x?

(c) Assume $x = 5$ and $\theta = 120°$. Calculate A.

*** Exercise 16

In Figure 5, suppose circles D and E have radii 5 and 3, respectively. Find the gray area.

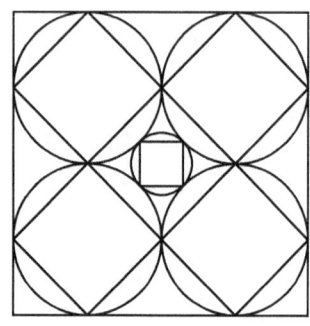

Figure 6

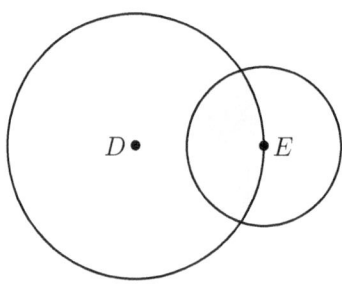

Figure 5

*** Exercise 17

Consider Figure 6. Assume the outer square has a side length of four. Calculate the ratio of the gray area to white area within the square.

Chapter 11

Vectors

In this chapter, we will study vectors. Vectors are of tremendous importance in mathematics, and they have many applications in physics and engineering. Our study will be limited to vectors in two dimensions, but it is not difficult to generalize many of the ideas.

The assumed knowledge is minimal. We will be evaluating trigonometric functions, so the reader needs to have familiarity with Chapter 5. Section 1.4.1 is needed for Section 11.3. One proof requires the Law of Cosines, so Chapter 9 is necessary for a complete theoretical understanding. We will use scientific calculators extensively.

11.1 The Basics

Definition 11.1 A nonzero **vector** is a mathematical expression that shows magnitude and direction.

Usually vectors are denoted by bold letters such as v or a letter with an arrow over it such as $\vec{v}$.

Vectors of the same magnitude and direction are equivalent. In other words, if u is obtained by shifting v, then $u = v$. All the vectors to the right equivalent.

To denote the vector with starting point A and ending point B, we write $\overrightarrow{AB}$. Usually, we refer to the starting point as the "tail" and the ending point the "tip".

Since we can shift vectors without losing equality, we can always assume that the tail of a vector is located at the origin. This practice occurs so frequently that we will introduce a definition to reference vectors whose tails are assumed to be at the origin.

Definition 11.2 A **position vector** is the representation of a vector which has its tail at the origin.

Position vectors allow us a means to reference vectors numerically.

Definition 11.3 Suppose v has a position vector whose tip is at (a, b). Then the **coordinate vector** of v is

$$\begin{pmatrix} a \\ b \end{pmatrix}.$$

Since shifts do not affect equivalence,

$$v = \begin{pmatrix} a \\ b \end{pmatrix}.$$

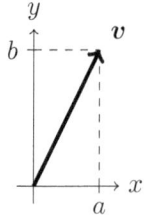

We say that a and b are the "first component" and "second component", respectively. For typesetting purposes, we will sometimes write $(a, b)^T$ to indicate the coordinate vector of v. The T denotes "transpose"; the transpose switches rows to columns and vice versa.

In general, if $A = (a_1, a_2)$ and $B = (b_1, b_2)$, then

$$\overrightarrow{AB} = \begin{pmatrix} b_1 - a_1 \\ b_2 - a_2 \end{pmatrix}.$$

Example 11.1 Find the coordinate vectors of (a) $\overrightarrow{PQ}$ and (b) $\overrightarrow{QP}$, if $P = (-2, 5)$ and $Q = (-7, 3)$.

308

Solution

(a) The coordinate vector of $\overrightarrow{PQ}$ is
$$\begin{pmatrix} -7-(-2) \\ 3-5 \end{pmatrix} = \begin{pmatrix} -5 \\ -2 \end{pmatrix}.$$

(b) The coordinate vector of $\overrightarrow{QP}$ is
$$\begin{pmatrix} -2-(-7) \\ 5-3 \end{pmatrix} = \begin{pmatrix} 5 \\ 2 \end{pmatrix}.$$

∎

Example 11.2 Find the angle that the position vector of
$$u = \begin{pmatrix} 2 \\ 3 \end{pmatrix}$$
makes with the positive x-axis.

Solution The first step is to draw the position vector of u.

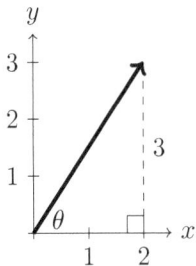

Then $\tan\theta = 3/2$ implies $\theta = \arctan(3/2) \approx 56.310°$. ∎

Definition 11.4 The **magnitude** of a vector $v = (a,b)^T$ is defined to be its length. To indicate the magnitude of a vector v we write $|v|$.

Proposition 11.1 *The magnitude of* $v = (a,b)^T$ *is*
$$|v| = \sqrt{a^2 + b^2}.$$

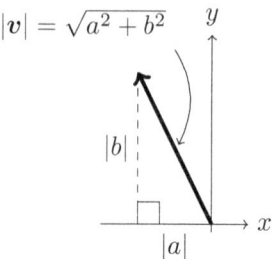

The proof of this follows via an application of the Pythagorean Theorem. For $v = (a, b)^T$, the diagram above indicates the idea.

Example 11.3 Find $|u|$, when

$$u = \begin{pmatrix} 3 \\ 4 \end{pmatrix}.$$

Solution Proposition 11.1 tells us

$$|u| = \sqrt{3^2 + 4^2} = 5.$$

■

Definition 11.5 The **zero vector**, denoted by **0**, is the vector with no magnitude or direction.

In component form,

$$\mathbf{0} = \begin{pmatrix} 0 \\ 0 \end{pmatrix}.$$

When graphed as a position vector, the zero vector is a point at the origin.

11.1.1 Arithmetic of Vectors

Suppose we have vectors u and v, and we want to find their sum. There are two methods to do this pictorially.

Parallelogram method: Draw vectors **u** and **v** with their tails at the origin. Then draw a parallelogram with **u** and **v** as sides. The diagonal of the parallelogram is **u** + **v**.

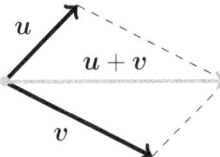

Triangle method: Draw vector **u** with its tail at the origin. Draw **v** with its tail at the tip of **u**. The vector that goes from the tail of **u** to the tip of **v** is **u** + **v**.

Another way to find the sum of two vectors is via coordinate vectors. Suppose $\boldsymbol{u} = (u_1, u_2)^T$ and $\boldsymbol{v} = (v_1, v_2)^T$. Then

$$\boldsymbol{u} + \boldsymbol{v} = \begin{pmatrix} u_1 + v_1 \\ u_2 + v_2 \end{pmatrix}.$$

Example 11.4 Assume

$$\boldsymbol{u} = \begin{pmatrix} -1 \\ 2 \end{pmatrix} \quad \text{and} \quad \boldsymbol{v} = \begin{pmatrix} 3 \\ 1 \end{pmatrix}.$$

(a) Use the graphs of **u** and **v** to find **u** + **v**.

(b) Use the coordinate vectors of **u** and **v** to find the coordinate vector of **u** + **v**.

Solution

(a) We will use the parallelogram method.

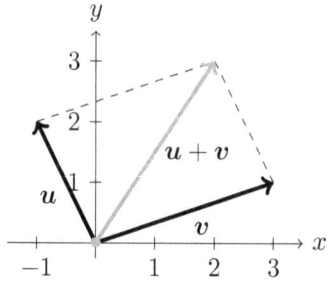

(b) Using coordinate vectors, we have

$$u + v = \begin{pmatrix} -1 \\ 2 \end{pmatrix} + \begin{pmatrix} 3 \\ 1 \end{pmatrix}$$
$$= \begin{pmatrix} -1 + 3 \\ 2 + 1 \end{pmatrix}$$
$$= \begin{pmatrix} 2 \\ 3 \end{pmatrix}.$$

■

Definition 11.6 Within the context of vectors, real numbers are called **scalars**.

Definition 11.7 Suppose c is a scalar, and $v = (a, b)^T$. Then define

$$cv = \begin{pmatrix} ca \\ cb \end{pmatrix}.$$

Example 11.5 Let

$$v = \begin{pmatrix} 3 \\ 3/2 \end{pmatrix}.$$

Graph the position vectors of v, $-2v$, and $\frac{1}{2}v$ using the same set of axes.

Solution

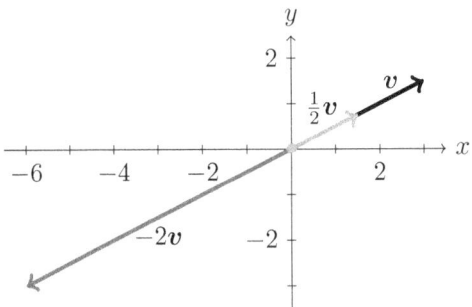

Definition 11.8 We say the vectors u and v are **parallel** if there exists a nonzero scalar c such that

$$cu = v.$$

Notice this definition implies that vectors going in opposite directions are parallel. For example, the vectors $(1,2)^T$ and $(-1,-2)^T$ are parallel, because

$$-1\begin{pmatrix} 1 \\ 2 \end{pmatrix} = \begin{pmatrix} -1 \\ -2 \end{pmatrix}.$$

Proposition 11.2 *The following hold for all vectors u, v, and w and all scalars c and d.*

(i)	$u + (v+w)$	$= (u+v)+w$	(v)	$c(du)$	$= (cd)u$
(ii)	$u + v$	$= v + u$	(vi)	$1u$	$= u$
(iii)	$u + 0$	$= 0 + u = u$	(vii)	$c(u+v)$	$= cu + cv$
(iv)	$u + (-u)$	$= -u + u = 0$	(iix)	$(c+d)u$	$= cu + du$

Proof We will provide the proofs of (i) and (ii) and leave the others as exercises. Let

$$u = \begin{pmatrix} u_1 \\ u_2 \end{pmatrix}, \quad v = \begin{pmatrix} v_1 \\ v_2 \end{pmatrix}, \quad \text{and} \quad w = \begin{pmatrix} w_1 \\ w_2 \end{pmatrix}.$$

(i)
$$\begin{aligned}
\boldsymbol{u} + (\boldsymbol{v} + \boldsymbol{w}) &= (u_1, u_2)^T + \left((v_1, v_2)^T + (w_1, w_2)^T\right) \\
&= (u_1, u_2)^T + (v_1 + w_1, v_2 + w_2)^T \\
&= \left(u_1 + (v_1 + w_1), u_2 + (v_2 + w_2)\right)^T \\
&= \left((u_1 + v_1) + w_1, (u_2 + v_2) + w_2\right)^T \\
&= (u_1 + v_1, u_2 + v_2)^T + (w_1, w_2)^T \\
&= \left((u_1, u_2)^T + (v_1, v_2)^T\right) + (w_1, w_2)^T \\
&= (\boldsymbol{u} + \boldsymbol{v}) + \boldsymbol{w}.
\end{aligned}$$

(ii)
$$\begin{aligned}
\boldsymbol{u} + \boldsymbol{v} &= (u_1, u_2)^T + (v_1, v_2)^T \\
&= (u_1 + v_1, u_2 + v_2)^T \\
&= (v_1 + u_1, v_2 + u_2)^T \\
&= (v_1, v_2)^T + (u_1, u_2)^T \\
&= \boldsymbol{v} + \boldsymbol{u}.
\end{aligned}$$

∎

To simplify our notation a bit. Let us introduce a convention.

For all vectors $\boldsymbol{u}$ and $\boldsymbol{v}$, assume

$$\boldsymbol{u} - \boldsymbol{v} = \boldsymbol{u} + (-\boldsymbol{v}).$$

This is analogous to the convention that $x - y$ means $x + (-y)$ for real numbers x and y.

Example 11.6 Suppose $\boldsymbol{u} = (-8, 2)^T$ and $\boldsymbol{v} = (-3, -4)^T$. Compute
$$\frac{3}{2}\boldsymbol{u} - 3(-\boldsymbol{v} + \boldsymbol{0}).$$

Solution

$$\begin{aligned}
\frac{3}{2}\boldsymbol{u} - 3(-\boldsymbol{v} + \boldsymbol{0}) &= \frac{3}{2}\boldsymbol{u} - 3(-\boldsymbol{v}) \\
&= \frac{3}{2}\boldsymbol{u} + 3\boldsymbol{v} \\
&= \frac{3}{2}(-8, 2)^T + 3(-3, -4)^T \\
&= \left(\frac{3}{2}(-8), \frac{3}{2}(2)\right)^T + (3(-3), 3(-4))^T \\
&= (-12, 3)^T + (-9, -12)^T \\
&= (-21, -9)^T.
\end{aligned}$$

■

Definition 11.9 A **unit vector** is a vector of magnitude is 1.

Example 11.7 Verify that $(-3/5, -4/5)^T$ is a unit vector.

Solution To verify that something is a unit vector, we will prove it has magnitude 1.

$$\begin{aligned}
\left|(-3/5, -4/5)^T\right| &= \sqrt{(-3/5)^2 + (-4/5)^2} \\
&= \sqrt{9/25 + 16/25} \\
&= \sqrt{25/25} \\
&= \sqrt{1} \\
&= 1.
\end{aligned}$$

■

Proposition 11.3 *Consider vector $\boldsymbol{v}$ and scalar c. Then*

$$|c\boldsymbol{v}| = |c|\,|\boldsymbol{v}|.$$

Proof Suppose $v = (v_1, v_2)^T$. Then
$$\begin{aligned}|cv| &= \left|(cv_1, cv_2)^T\right| \\ &= \sqrt{(cv_1)^2 + (cv_2)^2} \\ &= \sqrt{c^2 v_1^2 + c^2 v_2^2} \\ &= \sqrt{c^2(v_1^2 + v_2^2)} \\ &= \sqrt{c^2}\sqrt{v_1^2 + v_2^2} \\ &= |c|\,|v|.\end{aligned}$$

∎

Proposition 11.4 *The unit vector in the direction of $v \neq 0$ is*
$$\frac{1}{|v|}v.$$

Proof Proposition 11.3 tells us
$$\left|\frac{1}{|v|}v\right| = \left|\frac{1}{|v|}\right|\,|v|.$$
Since $1/|v| > 0$,
$$\left|\frac{1}{|v|}\right|\,|v| = \frac{1}{|v|}|v| = 1.$$

∎

Example 11.8 Find the unit vector in the direction of
$$w = \begin{pmatrix} -5 \\ 12 \end{pmatrix}.$$

Solution The magnitude
$$|w| = \sqrt{(-5)^2 + 12^2} = 13.$$
So, the unit vector in the direction of w is
$$\frac{1}{13}w = \begin{pmatrix} -5/13 \\ 12/13 \end{pmatrix}.$$

Let us introduce two popular unit vectors. We will use them as an alternative way to express vectors numerically.

Definition 11.10
$$i = \begin{pmatrix} 1 \\ 0 \end{pmatrix} \quad \text{and} \quad j = \begin{pmatrix} 0 \\ 1 \end{pmatrix}.$$

Definition 11.11 Suppose $v = (a,b)^T$. Then the **component form** of v is
$$ai + bj.$$

Example 11.9

(a) Write $(-4, 7)^T$ in component form.

(b) Find the coordinate vector of $-3i + 5j$.

Solution

(a) The component form of $(-4, 7)^T$ is
$$-4i + 7j.$$

(b) The coordinate vector of $-3i + 5j$ is
$$\begin{pmatrix} -3 \\ 5 \end{pmatrix}.$$

11.2 Bearing

Bearing refers to the degrees east or west of north or south. For example, $N30°W$ means $30°$ west of north.

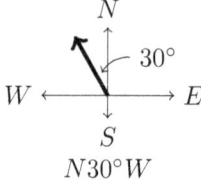

$N30°W$

The bearing $S25°E$ means $25°$ east of south.

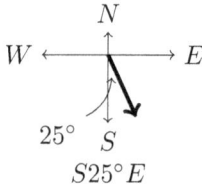

$S25°E$

Example 11.10 Suppose v has bearing $N60°E$ and $|v| = 20$. Find the coordinate vector of v, when the x- and y-axes are directed east and north, respectively.

Solution Using basic trigonometry, we conclude x- and y-coordinates of v are

$20 \sin 60° = 10\sqrt{3}$ and $20 \cos 60° = 10$,

respectively. Hence,

$$v = \begin{pmatrix} 10\sqrt{3} \\ 10 \end{pmatrix}.$$

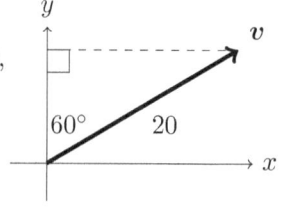

■

Example 11.11 A sailboat leaves Katsuura, Japan. It maintains a constant speed of 40 km/h. The sailboat's bearing is $S70°E$ for the first two hours. It then changes course and continues sailing at a bearing of $N30°E$ for three hours. Find the bearing of the straight path from the sail boat's starting to ending location.

Solution The distance the sailboat travels in the first two hours is $40(2) = 80$ kilometers, and it travels $40(3) = 120$ kilometers during the following three hours.

Suppose u is vector whose tail is the sailboat's original position and whose tip is its position after two hours. Let v be the vector whose tail is the position of the boat immediately after the first two hours and whose tip is the position of the boat three hours later.

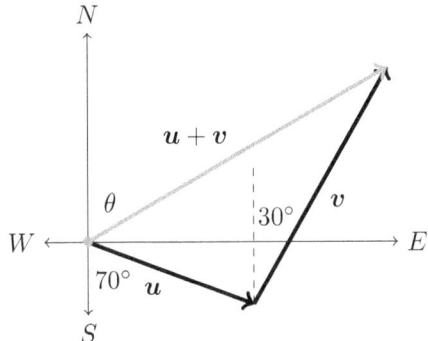

We have
$$u = \begin{pmatrix} 80\sin 70° \\ -80\cos 70° \end{pmatrix} \approx \begin{pmatrix} 75.175 \\ -27.362 \end{pmatrix},$$
and
$$v = \begin{pmatrix} 120\sin 30° \\ 120\cos 30° \end{pmatrix} \approx \begin{pmatrix} 60 \\ 103.923 \end{pmatrix}.$$

The vector which represents a straight path between the initial and final position is
$$u + v \approx \begin{pmatrix} 75.175 \\ -27.362 \end{pmatrix} + \begin{pmatrix} 60 \\ 103.923 \end{pmatrix}$$
$$= \begin{pmatrix} 135.175 \\ 76.561 \end{pmatrix}.$$

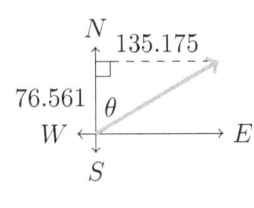

Let θ be the angle the resulting position vector makes with north. Then
$$\theta \approx \arctan\left(\frac{135.175}{76.561}\right) \approx 60.473°.$$
Therefore, the bearing of the straight path from the boat's starting to ending position is about $N60.473°E$. ■

Example 11.12 A pilot sets her plane's instruments so that it will head $S20°E$ at a speed of 550 mph in still air. However, there is a 15 mph East Wind. Compute the speed and bearing of the plane in the wind. (An East Wind blows from east to west.)

Solution Let w be the wind's vector, and a be the airplane's. Consider our diagram which is not to scale.

We see
$$w = \begin{pmatrix} -15 \\ 0 \end{pmatrix},$$
and
$$a = \begin{pmatrix} 550\sin 20° \\ -550\cos 20° \end{pmatrix} \approx \begin{pmatrix} 188.111 \\ -516.831 \end{pmatrix}.$$

Hence, the resulting vector is
$$a+w \approx \begin{pmatrix} 188.111 - 15 \\ -516.831 \end{pmatrix} = \begin{pmatrix} 173.111 \\ -516.831 \end{pmatrix}.$$

It follows that the speed of the plane in the wind is
$$|a+w| = \sqrt{(173.111)^2 + (-516.831)^2} \approx 545.052 \text{ mph}.$$

To find the bearing, we note that the vector $a+w$ makes a counterclockwise angle measure of
$$\arctan\left(\frac{173.111}{516.831}\right) \approx 18.518°$$
with south. Hence, the bearing is about $S18.518°E$ ■

11.3 Force

Newton's Laws of Mechanics give us the fundamental framework to understand force.

Newton's Laws of Mechanics

(i) Every object remains at rest or moves at a constant velocity unless an external force acts upon the object. In other words, if there is no external force acting upon an object, then the magnitude of acceleration is 0.

(ii) If forces $\boldsymbol{F}_1, \boldsymbol{F}_2, \ldots, \boldsymbol{F}_n$ act on an object of constant mass m, then their sum is

$$\boldsymbol{F}_1 + \boldsymbol{F}_2 + \ldots + \boldsymbol{F}_n = m\boldsymbol{a},$$

where $\boldsymbol{a}$ is the object's acceleration vector.

(iii) The forces that two interacting objects exert on each other are equal in magnitude and opposite in direction. In other words, if object A exerts a force of $\boldsymbol{F}$ on object B, then object B exerts a force of $-\boldsymbol{F}$ on object A.

There are an innumerable number of forces in nature. However, our studies will be restricted to the three below and forces from unnamed sources.

Definition 11.12

- **Gravitational force** is the force exerted by gravity. Denote it by $\boldsymbol{G}$. Occasionally, we use subscripts to indicate different gravitational forces.

- **Normal force** is the force perpendicular to a surface. We denote it by $\boldsymbol{N}$. We sometimes need a subscript when there are multiple normal forces.

- **Tension** is the force exerted by a cable. Tension travels along the cable and is obtained via the tightness of the cable. The vector $\boldsymbol{T}$ will represent tension in our calculations. When more than one cable exerts force or the same cable exerts force on different objects, we use subscripts to denote the various tensions.

The table below shows the most common units of measure in the metric and Old English system.

System	Force	Mass	Acceleration
Metric	newton (N)	kilogram (kg)	meters per square second (m/s^2)
Old English	pound (lb)	slug	feet per square second (ft/s^2)

As is customary in most science publications, we will use the metric system exclusively.

Within this text, we will assume that objects are close to the surface of the Earth. With this assumption in mind, let us introduce the following convention.

Unless otherwise stated, approximate the magnitude of gravitational g acceleration as 9.81 m/s^2.

For an object of mass m, the gravitational force has magnitude mg. Calculating the direction of gravitational force will rely on the context of the problem.

Example 11.13 A box of mass 20 kilograms rests on a surface. Calculate the component form of the normal force. Assume that one Newton of force right is i, and one Newton of force up is j.

Solution Gravitational force is

$$G = 20(-gj) \approx 20(-9.81j) = -196.2j.$$

Since the box is at rest, net force is 0. So,

$$N + G = 0 \quad \text{implies} \quad N = -G = 196.2j.$$

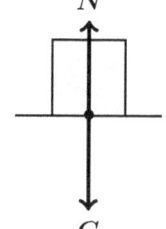

Example 13 illustrates Newton's Laws of Mechanics (iii). Gravitational force acts on the ground, and the ground exerts a force of equal magnitude in the opposite direction.

Example 11.14 A box of mass 50 kilograms hangs from a cable. Calculate the magnitude of tension.

Solution Say that gravitational force goes in the direct $(0, -1)^T$. Then

$$G = 50 \begin{pmatrix} 0 \\ -g \end{pmatrix} \approx \begin{pmatrix} 0 \\ -490.5 \end{pmatrix}.$$

Since the box is at rest,

$$T + G = 0 \quad \text{implies} \quad T = -G = \begin{pmatrix} 0 \\ 490.5 \end{pmatrix}.$$

It follows that the magnitude of T is about 491 newtons. ∎

In Examples 13 and 14, we drew all of the forces with their tails at a single point. Since vectors are equal when their magnitude and direction are the same, we can place them at any location. What matters is an understanding of what object the forces are acting on.

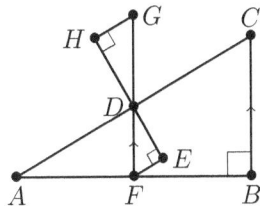

The next few examples will involve inclined planes. As a result, many problems require diagrams like the one above.

Suppose $\overline{BC}$ is parallel to $\overline{DF}$. Then it is helpful to note that

$$\angle A \cong \angle EDF \cong \angle GDH.$$

Subsection 1.4.1 discusses the reasoning behind this claim.

Let us introduce a convention before we begin our inclined plane examples:

> Assume that all inclined planes are frictionless surfaces, and there is no air resistance.

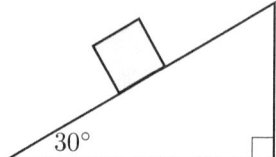

Example 11.15 A box slides down a plane inclined at an angle of 30° from horizontal. Suppose the mass of the box is 10 kilograms. Calculate the magnitude of the box's acceleration.

Solution Let us examine our forces. Gravitational force $\boldsymbol{G}$ acts on the box vertically. Because the magnitude of gravitational force is mass times acceleration,

$$|\boldsymbol{G}| = 10g.$$

Normal force $\boldsymbol{N}$ acts on the box and counteracts the portion of gravitational force perpendicular to the box.

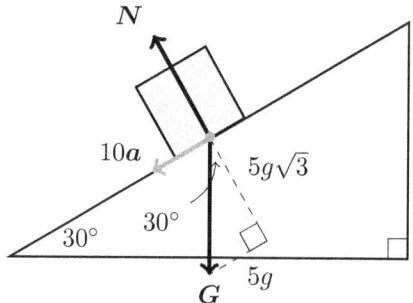

Newton's Laws of Mechanics (ii) tells us

$$\boldsymbol{G} + \boldsymbol{N} = m\boldsymbol{a},$$

where $\boldsymbol{a}$ is the acceleration of the box and $m = 10$ kilograms is its mass.

Suppose $m\boldsymbol{a}$ goes in the direction of $(-1, 0)^T$ and $\boldsymbol{N}$ goes in the direction of $(0, 1)^T$. Then

$$m\boldsymbol{a} = \begin{pmatrix} -10|\boldsymbol{a}| \\ 0 \end{pmatrix} \quad \text{and} \quad \boldsymbol{G} = \begin{pmatrix} -5g \\ -5g\sqrt{3} \end{pmatrix}.$$

324

Because normal force N counteracts the portion of gravitational force G perpendicular to the ramp,
$$N = \begin{pmatrix} 0 \\ 5g\sqrt{3} \end{pmatrix}.$$
It follows that
$$G + N = 10a \quad \text{implies} \quad \begin{pmatrix} -5g \\ 0 \end{pmatrix} = \begin{pmatrix} -10|a| \\ 0 \end{pmatrix}.$$
This leads us to conclude $-5g = -10|a|$. This gives
$$|a| = g/2 \approx 4.91 \text{ m/s}^2.$$
Hence, the box accelerates down the inclined plane at an acceleration of about 4.91 meters per square second. ∎

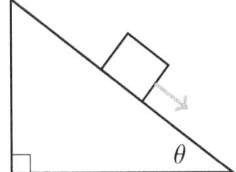

Example 11.16 A box of mass m slides down a plane inclined at an angle of θ from horizontal. If the box accelerates down the incline at 5.95 meters per square second, find θ.

Solution Consider our forces. Gravitation force G of magnitude mg acts on the box vertically. The portion of gravitational force perpendicular to the surface is equal in magnitude and opposite in direction to normal force N.

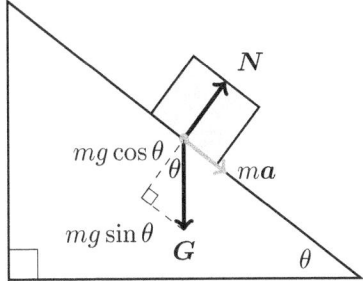

Suppose $\boldsymbol{a}$ goes in the direction of $(1,0)^T$, and $\boldsymbol{N}$ goes in the direction of $(0,1)^T$. Because $|\boldsymbol{a}| = 5.95$,

$$m\boldsymbol{a} = \begin{pmatrix} 5.95m \\ 0 \end{pmatrix}.$$

Using right triangle trigonometry, we have

$$\boldsymbol{G} = \begin{pmatrix} mg\sin\theta \\ -mg\cos\theta \end{pmatrix}.$$

Newton's Laws of Mechanics (ii) tells us

$$\boldsymbol{G} + \boldsymbol{N} = m\boldsymbol{a}.$$

Since normal force $\boldsymbol{N}$ cancels with the second component of $\boldsymbol{G}$, it follows that

$$\begin{pmatrix} mg\sin\theta \\ 0 \end{pmatrix} = \begin{pmatrix} 5.95m \\ 0 \end{pmatrix}.$$

This implies $5.95m = mg\sin\theta$. Using a bit of algebra, we see that $\sin\theta = 5.95/g$. The angle θ is within the range of arc sine. Ergo,

$$\theta = \arcsin\frac{5.95}{g} \approx 37.3°.$$

■

Our next example examines a problem involving a cable. Before we begin, we will introduce another convention:

Assume cables have no mass.

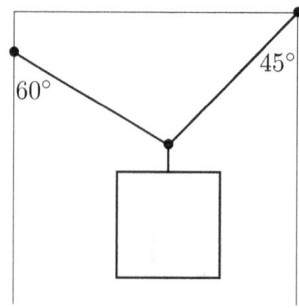

Example 11.17 Suppose the mass of the box is 10 kilograms. Calculate the magnitude of tension in each rope.

Solution Let us take account of our forces. Gravitational force G of magnitude $10g$ acts on the box vertically. Tension vectors are parallel to their corresponding ropes and pull away from the box; call the tension in the left rope T_1 and the tension in the right rope T_2.

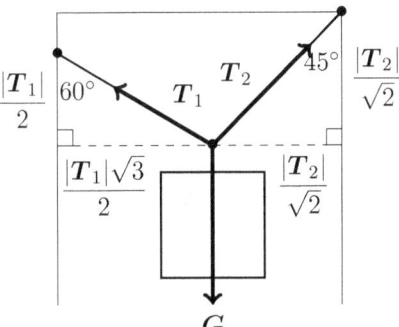

We need to define our coordinate system. Let G be parallel $(0, 1)^T$ and go in the opposite direction. Then

$$G = \begin{pmatrix} 0 \\ -10g \end{pmatrix} \approx \begin{pmatrix} 0 \\ -98.1 \end{pmatrix}.$$

Suppose $(1, 0)^T$ goes right and is perpendicular to G. Then special right triangles tell us

$$T_1 = \begin{pmatrix} -|T_1|\sqrt{3}/2 \\ |T_1|/2 \end{pmatrix} \approx \begin{pmatrix} -0.866|T_1| \\ 0.5|T_1| \end{pmatrix}$$

and

$$T_2 = \begin{pmatrix} |T_2|/\sqrt{2} \\ |T_2|/\sqrt{2} \end{pmatrix} \approx \begin{pmatrix} 0.707|T_2| \\ 0.707|T_2| \end{pmatrix}.$$

The box is at rest, which means the sum of forces is equal to **0**. So, we have

$$T_1 + T_2 + G = 0 \quad \text{implies} \quad \begin{pmatrix} -0.866|T_1| + 0.707|T_2| \\ 0.5|T_1| + 0.707|T_2| - 98.1 \end{pmatrix} \approx 0.$$

It follows that

$-0.866|T_1| + 0.707|T_2| \approx 0$ and $0.5|T_1| + 0.707|T_2| - 98.1 \approx 0$.

Solving the first equation for $|T_2|$ yields $|T_2| \approx 1.225|T_1|$. Substituting the right side of this equation for $|T_2|$ into our other equation gives
$$0.5|T_1| + 0.866|T_1| - 98.1 \approx 0$$
$$\Rightarrow \quad 1.366|T_1| \approx 98.1$$
$$\Rightarrow \quad |T_1| \approx 71.816.$$

It follows that

$$|T_2| \approx 1.225(71.816) \approx 87.974.$$

We conclude

$$|T_1| \approx 71.8 \text{ N} \quad \text{and} \quad |T_2| \approx 88.0 \text{ N}.$$

■

In our last example of this section, we will examine an inclined plane problem which contains a pulley and a cable. To simplify this type of problem, we will introduce yet another convention:

Assume pulleys are frictionless, which means that tension from cables are not reduced by the pulleys.

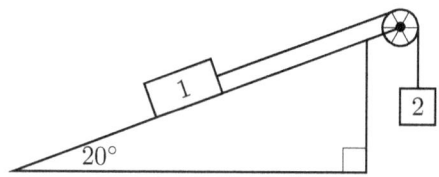

Example 11.18 Consider the diagram above. If the mass of box 1 is 7 kilograms, and the system is at rest, calculate the mass of box 2.

Solution Let us examine our forces. The forces acting on box 1 are gravitational force G_1, normal force N, and tension T_1. The forces acting on box 2 are gravitational force G_2 and tension T_2.

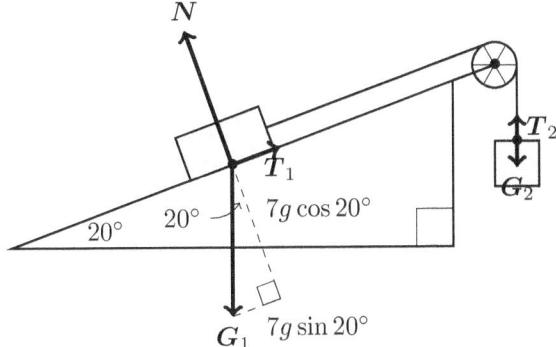

The system is at rest, which means the forces acting on each box add to $\mathbf{0}$. So,

$$G_1 + N + T_1 = \mathbf{0} \quad \text{and} \quad G_2 + T_2 = \mathbf{0}.$$

Let us put the forces acting on box 1 into coordinate form and make some calculations. Suppose T_1 goes in the direction of $(1,0)^T$ and N goes in the direction of $(0,1)^T$. Since $|G_1| = 7g$, right triangle trigonometry tells us

$$G_1 = \begin{pmatrix} -7g\sin 20° \\ -7g\cos 20° \end{pmatrix}.$$

Because T_1 is exclusively in the first component, N is exclusively in the second component, and $G_1 + N + T_1 = \mathbf{0}$,

$$N = \begin{pmatrix} 0 \\ 7g\cos 20° \end{pmatrix} \quad \text{and} \quad T_1 = \begin{pmatrix} 7g\sin 20° \\ 0 \end{pmatrix}.$$

It follows that $|T_1| = 7g\sin 20°$.

Now we will put the forces acting on box 2 into coordinate form and do a few more calculations to finish the example. For convenience, we will switch coordinate systems. Suppose T_2 goes in the direction $(0,1)^T$, and $(1,0)^T$ is a vector perpendicular which goes either left or right (we will not use this coordinate in our calculations so it does not matter). Then

$$G_2 = \begin{pmatrix} 0 \\ -mg \end{pmatrix},$$

where m is the mass of box 2.

Newton's Laws of Mechanics (iii) tells us that

$$|\boldsymbol{T}_1| = |\boldsymbol{T}_2|.$$

It follows that $|\boldsymbol{T}_2| = 7g\sin 20°$. Therefore,

$$\boldsymbol{T}_2 = \begin{pmatrix} 0 \\ 7g\sin 20° \end{pmatrix}.$$

We have

$$\boldsymbol{G}_2 + \boldsymbol{T}_2 = \boldsymbol{0} \quad \text{implies} \quad \begin{pmatrix} 0 \\ -mg + 7g\sin 20° \end{pmatrix} = \boldsymbol{0}.$$

So, $mg = 7g\sin 20°$. Dividing by g yields

$$m = 7\sin 20° \approx 2.39 \text{ kg}.$$

We conclude the mass of box 2 is about 2.39 kilograms. ∎

11.4 The Dot Product

Definition 11.13 Suppose $\boldsymbol{u} = (u_1, u_2)^T$ and $\boldsymbol{v} = (v_1, v_2)^T$. The **dot product** of $\boldsymbol{u}$ and $\boldsymbol{v}$ is

$$\boldsymbol{u} \bullet \boldsymbol{v} = u_1 v_1 + u_2 v_2.$$

Notice that the result of the dot product of two vectors is a scalar, *not* a vector.

Example 11.19 Suppose

$$\boldsymbol{u} = \begin{pmatrix} -1/2 \\ \pi \end{pmatrix}, \quad \boldsymbol{v} = \begin{pmatrix} 4 \\ 5 \end{pmatrix}, \quad \text{and} \quad \boldsymbol{w} = -3\boldsymbol{i} + 2\boldsymbol{j}.$$

Compute (a) $\boldsymbol{u} \bullet \boldsymbol{v}$, (b) $\boldsymbol{v} \bullet \boldsymbol{w}$, and (c) $\boldsymbol{u} \bullet (\boldsymbol{v} + \boldsymbol{w})$.

Solution

(a) Using the definition,
$$\boldsymbol{u}\cdot\boldsymbol{v} = -\frac{1}{2}(4) + \pi(5) = 5\pi - 2.$$

(b) We know
$$\boldsymbol{w} = -3\boldsymbol{i} + 2\boldsymbol{j} = \begin{pmatrix} -3 \\ 2 \end{pmatrix}.$$
This implies
$$\boldsymbol{v}\cdot\boldsymbol{w} = 4(-3) + 5(2) = -2.$$

(b) Parentheses tell us to do the operations contained within them first. So,
$$\boldsymbol{u}\cdot(\boldsymbol{v}+\boldsymbol{w}) = \left(-\frac{1}{2},\pi\right)^T \cdot \left((4,5)^T + (-3,2)^T\right)$$
$$= \left(-\frac{1}{2},\pi\right)^T \cdot (1,7)^T$$
$$= \left(-\frac{1}{2}\right)(1) + \pi(7)$$
$$= \frac{14\pi - 1}{2}.$$

■

Proposition 11.5 (Dot Product Properties) *Suppose $\boldsymbol{u}$, $\boldsymbol{v}$, and $\boldsymbol{w}$ are vectors, and suppose c is a scalar.*

(i) $\boldsymbol{u}\cdot\boldsymbol{v} = \boldsymbol{v}\cdot\boldsymbol{u}$ (iii) $\boldsymbol{u}\cdot(c\boldsymbol{v}) = c(\boldsymbol{u}\cdot\boldsymbol{v}) = (c\boldsymbol{u})\cdot\boldsymbol{v}$

(ii) $\boldsymbol{u}\cdot(\boldsymbol{v}+\boldsymbol{w}) = \boldsymbol{u}\cdot\boldsymbol{v} + \boldsymbol{u}\cdot\boldsymbol{w}$ (iv) $\boldsymbol{0}\cdot\boldsymbol{u} = \boldsymbol{u}\cdot\boldsymbol{0} = 0$

Proof We will prove (i) and (ii), and leave the rest as exercises. Let
$$\boldsymbol{u} = \begin{pmatrix} u_1 \\ u_2 \end{pmatrix}, \quad \boldsymbol{v} = \begin{pmatrix} v_1 \\ v_2 \end{pmatrix}, \quad \text{and} \quad \boldsymbol{w} = \begin{pmatrix} w_1 \\ w_2 \end{pmatrix}.$$

(i) Then
$$u \cdot v = u_1 v_1 + u_2 v_2$$
$$= v_1 u_1 + v_2 u_2$$
$$= v \cdot u.$$

(ii) Let us compute both sides of
$$u \cdot (v + w) \stackrel{?}{=} u \cdot v + u \cdot w$$
and prove they are equal.

$$u \cdot (v + w) = (u_1, u_2)^T \cdot \left((v_1, v_2)^T + (w_1, w_2)^T \right)$$
$$= (u_1, u_2)^T \cdot (v_1 + w_1, v_2 + w_2)^T$$
$$= u_1(v_1 + w_1) + u_2(v_2 + w_2)$$
$$= u_1 v_1 + u_1 w_1 + u_2 v_2 + u_2 w_2.$$

$$u \cdot v + u \cdot w = (u_1, u_2)^T \cdot (v_1, v_2)^T + (u_1, u_2)^T \cdot (w_1, w_2)^T$$
$$= u_1 v_1 + u_2 v_2 + u_1 w_1 + u_2 w_2$$
$$= u_1 v_1 + u_1 w_1 + u_2 v_2 + u_2 w_2.$$
Ergo,
$$u \cdot (v + w) = u \cdot v + u \cdot w.$$
■

Proposition 11.6 *For all vectors* u,
$$u \cdot u = |u|^2.$$
Proof Suppose $u = (u_1, u_2)^T$. Then
$$u \cdot u = u_1 u_1 + u_2 u_2$$
$$= u_1^2 + u_2^2$$
$$= \left(\sqrt{u_1^2 + u_2^2} \right)^2$$
$$= |u|^2.$$

Theorem 11.1 *Suppose u and v are vectors. Let θ be the angle between u and v of smaller measure. Then*

$$u \cdot v = |u||v| \cos \theta.$$

Proof Let us place u and v at the origin. The vector that starts at the tip of u and goes to the tip of v is $v - u$. Using these three vectors we can form a triangle.

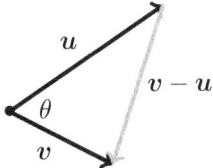

The Law of Cosines (Theorem 9.1) tells us

$$|v - u|^2 = |u|^2 + |v|^2 - 2|u||v| \cos \theta.$$

Using Proposition 11.6,

$$\begin{aligned}
|v - u|^2 &= (v - u) \cdot (v - u) \\
&= (v - u) \cdot v - (v - u) \cdot u \\
&= |v|^2 - u \cdot v - v \cdot u + |u|^2 \\
&= |v|^2 - 2u \cdot v + |u|^2
\end{aligned}$$

So,

$$\begin{aligned}
|v|^2 - 2u \cdot v + |u|^2 &= |u|^2 + |v|^2 - 2|u||v| \cos \theta \\
\Rightarrow \quad -2u \cdot v &= -2|u||v| \cos \theta \\
\Rightarrow \quad u \cdot v &= |u||v| \cos \theta.
\end{aligned}$$

■

Example 11.20 Suppose the angle between u and v is 30°. If $|u| = 5$ and $|v| = 7$, find $u \cdot v$.

Solution Because of Theorem 11.1,

$$u \cdot v = |u||v| \cos 30°$$
$$= 5(7)\left(\frac{\sqrt{3}}{2}\right)$$
$$= \frac{35\sqrt{3}}{2}.$$

∎

In Theorem 11.1, the assumption that θ is the angle between u and v of smaller measure, leads us to the statement $0 \leq \theta \leq 180°$. This is helpful, because it allows us to use arc cosine without careful consideration of the measure of θ.

Example 11.21 Find the angle between $r = -3i + 2j$ and $s = 5i - j$.

Solution Theorem 11.1 tells us that

$$r \cdot s = |r||s| \cos \theta.$$

Let us find the pieces and then solve for θ:

$r \cdot s = -3(5) + 2(-1),\quad |r| = \sqrt{(-3)^2 + 2^2},\quad$ and $\quad |s| = \sqrt{5^2 + (-1)^2}$
$= -15 - 2 = \sqrt{9 + 4} = \sqrt{25 + 1}$
$= -17 = \sqrt{13} = \sqrt{26}.$

It follows that

$$-17 = \sqrt{13} \cdot \sqrt{26} \cos \theta \quad \text{implies} \quad \cos \theta = -\frac{17}{13\sqrt{2}}.$$

Hence,

$$\theta = \arccos\left(-\frac{17}{13\sqrt{2}}\right) \approx 157.620°.$$

∎

Definition 11.14 The vectors u and v are **orthogonal** if
$$u \cdot v = 0.$$

Since
$$u \cdot v = |u||v|\cos\theta,$$
there are two ways for vectors to be orthogonal. The angle between the two vectors, θ, could equal 90° or at least one of the vectors could be $\mathbf{0}$.

Example 11.22 Find a unit vector orthogonal to
$$a = \begin{pmatrix} -1 \\ 5 \end{pmatrix}.$$

Solution Suppose the vector b is a unit vector. Then
$$b = \begin{pmatrix} \cos\alpha \\ \sin\alpha \end{pmatrix}.$$
for some angle α. If b is orthogonal to a, then
$$a \cdot b = -\cos\alpha + 5\sin\alpha = 0.$$
This implies
$$\begin{aligned} 5\sin\alpha &= \cos\alpha \\ \Rightarrow \quad 5\tan\alpha &= 1 \\ \Rightarrow \quad \tan\alpha &= \frac{1}{5} \\ \Rightarrow \quad \alpha &= \arctan\left(\frac{1}{5}\right) \\ &\approx 11.310°. \end{aligned}$$
Thus,
$$b \approx \begin{pmatrix} \cos 11.310° \\ \sin 11.310° \end{pmatrix} \approx \begin{pmatrix} 0.981 \\ 0.196 \end{pmatrix}.$$

■

In Exercise 22, we simply found one vector. There is another solution. In particular,
$$\begin{pmatrix} -0.981 \\ -0.196 \end{pmatrix},$$

which corresponds to $\alpha = \arctan(1/5) + 180° \approx 191.310°$. Within the xy-plane, there are always two vectors orthogonal to a nonzero vector.

11.5 Projection

Perhaps the easiest way to understand the concept of projection is via examination of a few diagrams.

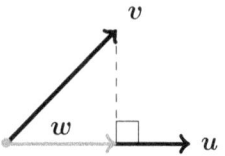

The vector $\boldsymbol{w}$ is the "projection" of $\boldsymbol{v}$ onto $\boldsymbol{u}$, and $|\boldsymbol{w}|$ is the "scalar projection" of $\boldsymbol{v}$ onto $\boldsymbol{u}$. We write

$$\operatorname{proj}_{\boldsymbol{u}}\boldsymbol{v} \quad \text{and} \quad \operatorname{comp}_{\boldsymbol{u}}\boldsymbol{v},$$

to denote the vector projection and scalar projection, respectively.

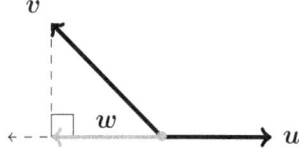

Sometimes the vector projection of $\boldsymbol{v}$ onto $\boldsymbol{u}$ does not lie on $\boldsymbol{u}$. This happens when the angle between $\boldsymbol{v}$ and $\boldsymbol{u}$ is obtuse. However, the projection of $\boldsymbol{v}$ onto $\boldsymbol{u}$ always lies on the line which contains the vector $\boldsymbol{u}$.

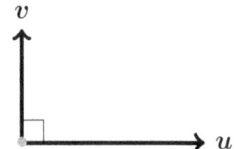

It is possible for the vector projection of v onto u to be $\mathbf{0}$; this corresponds to the scalar projection being 0. The vector project is $\mathbf{0}$ if and only if u is orthogonal to v. Using our notation, $\text{proj}_u v = \mathbf{0}$ and $\text{comp}_u v = 0$, if and only if $u \cdot v = 0$.

Proposition 11.7 *Suppose $u \neq 0$. Then*

$$\text{proj}_u v = \frac{u \cdot v}{|u|^2} u \quad \text{and} \quad \text{comp}_u v = \frac{u \cdot v}{|u|}.$$

Proof We will prove

$$\text{comp}_u v = \frac{u \cdot v}{|u|}.$$

Let θ be the angle between u and v. Suppose that u and v are position vectors on the xy-plane such that u lies on the positive x-axis. Notice that $\text{comp}_u v$ is the x-coordinate of the tip of v. Since v has length $|v|$, Theorem 5.1 tells us $\text{comp}_u v = |v| \cos \theta$.

To get rid of the $\cos \theta$ in the equation $\text{comp}_u v = |v| \cos \theta$, notice

$$u \cdot v = |u||v| \cos \theta \quad \text{implies} \quad \cos \theta = \frac{u \cdot v}{|u||v|}.$$

Hence,

$$\text{comp}_u v = |v| \cos \theta$$
$$= |v| \left(\frac{u \cdot v}{|u||v|} \right)$$
$$= \frac{u \cdot v}{|u|}.$$

Now, we want to find $\text{proj}_u v$. The magnitude of the vector projection is

$$\text{comp}_u v = \frac{u \cdot v}{|u|},$$

and it goes in the direction of u. The unit vector $u/|u|$ goes in the direction of u, so scaling it by $u \cdot v / |u|$ gives the vector projection. Therefore,

$$\text{proj}_u v = \left(\frac{u \cdot v}{|u|} \right) \frac{u}{|u|} = \frac{u \cdot v}{|u|^2} u.$$

Example 11.23 Suppose $c = -3i + 2j$ and $d = i - 5j$. Compute (a) $\text{comp}_d c$ and (b) $\text{proj}_d c$.

Solution

(a) Using Proposition 11.7,

$$\text{comp}_d c = \frac{d \cdot c}{|d|}$$

$$= \frac{1(-3) - 5(2)}{\sqrt{1^2 + (-5)^2}}$$

$$= -\frac{13}{\sqrt{26}}$$

$$= -\frac{13}{\sqrt{26}} \cdot \frac{\sqrt{26}}{\sqrt{26}}$$

$$= -\frac{\sqrt{26}}{2}.$$

(b) We could scale the unit vector $d/|d|$ by $-\sqrt{26}/2$ to find $\text{proj}_d c$. However, let us use the formula in Proposition 11.7:

$$\text{proj}_d c = \frac{d \cdot c}{|d|^2} d$$

$$= \frac{1(-3) - 5(2)}{1^2 + (-5)^2}(1, -5)^T$$

$$= -\frac{13}{26}(1, -5)^T$$

$$= -\frac{1}{2}(1, -5)^T$$

$$= \left(-\frac{1}{2}, \frac{5}{2}\right)^T.$$

11.6 Work

Definition 11.15 The **work** done by a force $\boldsymbol{F}$ which moves an object from point P to point Q is

$$W = \boldsymbol{F} \cdot \overrightarrow{PQ}.$$

The vector $\overrightarrow{PQ}$ is referred to as the "displacement vector".

The standard measure of work in the metric system is Joules, which is abbreviated J.

$$1 \text{ J} = \text{N·m} = \text{kg·m}^2/\text{s}^2.$$

In the Old English system, the unit pound-feet is used. We will use the metric system whenever units are utilized. However, some problems formulate work as a purely abstract phenomenon. In such cases, units are omitted.

Example 11.24 The force vector $\boldsymbol{F} = 4\boldsymbol{i} - 2\boldsymbol{j}$ moves an object from the origin to the point $(5, -3)$. Find the work done.

Solution Since the displacement vector is

$$(5-0)\boldsymbol{i} + (-3-0)\boldsymbol{j} = 5\boldsymbol{i} - 3\boldsymbol{j},$$

the work done must be

$$W = \boldsymbol{F} \cdot (5\boldsymbol{i} - 3\boldsymbol{j}) = 4(5) - 2(-3) = 26.$$

∎

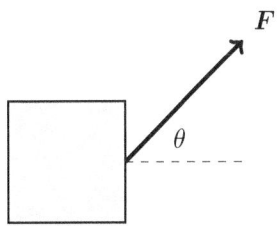

Example 11.25 A force F of magnitude 10 Newtons is exerted on a box at an angle θ from horizontal. The box moves 7 meters right. Calculate the work done when (a) $\theta = 0$, (b) $\theta = 60°$, (c) $\theta = 90°$, and (d) $\theta = 120°$.

Solution Since $|\vec{PQ}| = 7$ and $|F| = 10$,

$$W = F \cdot \vec{PQ} = |F| \left| \vec{PQ} \right| \cos\theta = 70\cos\theta.$$

(a) If $\theta = 0$, then $W = 70\cos 0 = 70$ J.
(b) If $\theta = 60°$, then $W = 70\cos 60° = 35$ J.
(c) If $\theta = 90°$, then $W = 70\cos 90° = 0$.
(d) If $\theta = 120°$, then $W = 70\cos 120° = -35$ J. ∎

Parts (c) and (d) illustrate how the definition of work differs from the colloquial conception of "work". In part (c), force was exerted on the box, but no work was done because it was not in the direction of motion. In part (d), we see that the concept of negative work, in fact, makes sense given our definition; because the non-orthogonal component of force was applied counter to the direction of motion, a negative amount of work was done.

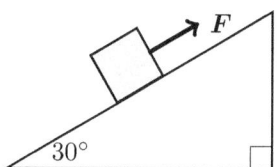

Example 11.26 Consider Example 15. How much work is required to pull the box 3 meters up the incline?

Solution From our calculations in Example 15, we found that the the gravitational force vector acting on the box is

$$G = \begin{pmatrix} -5g \\ -5g\sqrt{3} \end{pmatrix},$$

where $(1,0)^T$ is parallel to the incline and points up, and $(0,1)^T$ is perpendicular to the incline and points up. It follows that the force requires to pull the box up at a constant velocity is

$$\boldsymbol{F} = \begin{pmatrix} 5g \\ 0 \end{pmatrix}.$$

The displacement vector is $(3,0)^T$. Hence,

$$\begin{aligned} W &= \begin{pmatrix} 5g \\ 0 \end{pmatrix} \cdot \begin{pmatrix} 3 \\ 0 \end{pmatrix} \\ &= 5g(3) + 0 \\ &= 15g \\ &\approx 147 \text{ J}. \end{aligned}$$

Therefore, about 147 Joules of work are required to move the box 3 meters up the inclined plane. ■

11.7 Exercises

* Exercise 1

Find the coordinate vector of $\overrightarrow{AB}$.

(a) $A = (1, 2)$ and $B = (7, 2)$.

(b) $A = (3, 5)$ and $B = (-1, 7)$.

(c) $A = (-2, 1)$ and $B = (-2, -7)$.

(d) $A = (3, -5)$ and $B = (-7, -15)$.

* Exercise 2

Suppose v lies on the terminal side of a standard position angle θ. Find the degree measure of θ. Assume $0 \leq \theta < 360°$.

(a) $v = \left(\frac{\sqrt{3}}{2}, \frac{1}{2}\right)^T$

(b) $v = 3i - 1j$

(c) $v = \left(-\frac{1}{2}, \frac{\sqrt{3}}{2}\right)^T$

(d) $v = -3j$

(e) $v = (-7, -7)^T$

(f) $v = i$

(g) $v = (-5, -2)^T$

(h) $v = i - j$

** Exercise 3

Compute the magnitude of each vector.

(a) $-3i + 4j$

(b) $(3, 6)^T$

(c) $-12i + 5j$

(d) $\frac{2}{\sqrt{13}}i - \frac{3}{\sqrt{13}}j$

(e) $(-17, -51)^T$

(f) $\frac{11}{61}i + \frac{60}{61}j$

** Exercise 4

Suppose u lies on the terminal side of a standard position angle θ. Find the coordinate vector of u for the given values of $|u|$ and θ.

(a) $|u| = 22$ and $\theta = 0°$

(b) $|u| = 1$ and $\theta = -\pi/4$

(c) $|u| = 2.1$ and $\theta = 270°$

(d) $|u| = 11/5$ and $\theta = 2\pi/3$

(e) $|u| = 2\sqrt{5}$ and $\theta = 30°$

(f) $|u| = 10\sqrt{\pi}$ and $\theta = 7\pi/6$

* Exercise 5

For each pair of vectors u and v, use the parallelogram or triangle method to draw $u + v$ on the xy-plane.

(a) $u = (-3, 2)^T$ and $v = (1, -1)^T$

(b) $u = i - 2j$ and $v = -i - 4j$

(c) $u = (-2, -5)^T$ and $v = (2, 5)^T$.

* Exercise 6

Use the given pairs of coordinate vectors of u and v in Exercise 5 to find the coordinate vectors of $u + v$.

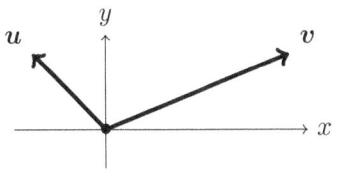

Figure 1

* Exercise 7

Consider Figure 1. Sketch each of the following.

(a) $2u$ (c) $u + v$

(b) $-\frac{1}{2}v$ (d) $2u + v$

* Exercise 8

For each vector v, draw v, $-2v$, $3v$ on the xy-plane.

(a) $v = (-1, 3)^T$

(b) $v = i + j$

(c) $v = (-3, -2)^T$

* Exercise 9

Use the given coordinate vectors of v in Exercise 8 to find the coordinate vectors of $-2v$ and $3v$.

* Exercise 10

Use the given values of u, v, and w to compute

(i) $u + v$,
(ii) $-5v$, and
(iii) $3u - 3(v - 2w)$.

(a) $u = -15i - 2j$, $v = -\frac{3}{5}i + 3j$, and $w = -4i + 13j$.

(b) $u = 2i + 4j$, $v = 3i + 3j$, and $w = \frac{1}{2}i$.

(c) $u = 15i - 5j$, $v = -2j$, and $w = \frac{1}{10}i - 14j$.

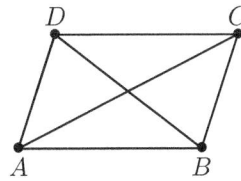

Figure 2

** Exercise 11

Consider Figure 2. Let $u = \overrightarrow{AB}$ and $v = \overrightarrow{AD}$. Write the vectors listed using the coordinates in the diagram. Remember that vectors are invariant under shifts.

(a) $-u$
(b) $u + v$
(c) $u - v$
(d) $v - u$

** Exercise 12

Prove (iii)-(iix) of Proposition 11.2.

** Exercise 13

Find the unit vector in the direction of each vector.

(a) $(-5, -12)^T$
(b) $7i + j$
(c) $(40, -9)^T$
(d) $\frac{3}{7}i - \frac{4}{7}j$
(e) $(3\sqrt{23}, 6)^T$
(f) $-\frac{5}{3}i + \frac{5}{3}j$

** Exercise 14

Compute the vector u which has the given magnitude and the same direction as v.

(a) $|u| = 5$ and $v = (-\sqrt{3}, 1)^T$
(b) $|u| = \frac{1}{2}$ and $v = \frac{6}{5}i - \frac{8}{5}j$
(c) $|u| = 13$ and $v = \left(-6, -\frac{5}{2}\right)^T$
(d) $|u| = \sqrt{2}$ and $v = -\sqrt{3}i - \sqrt{3}j$
(e) $|u| = 15$ and $v = (3, 4)^T$

** Exercise 15

Calculate the coordinate vector of v with the given bearing and magnitude. Suppose the x- and y-axes are directed east and north, respectively.

(a) $N30°E$ and $|v| = 20$
(b) $N45°W$ and $|v| = 6\sqrt{2}$
(c) $S30°W$ and $|v| = 100$
(d) $S60°E$ and $|v| = 5\sqrt{3}$

** Exercise 16

Find the bearing of each vector. Assume the x- and y-axes are directed east and north, respectively.

(a) $15i - 15j$
(b) $(-\sqrt{3}, 1)^T$
(c) $-8i - 8j$
(d) $(25, 25\sqrt{3})^T$
(e) $7i - 11j$
(f) $(\sqrt{2}, \sqrt{3})^T$

** Exercise 17

Point B is 100 meters due east of point A. Point C is 150 meters due north of B. Point D is 70 meters due west of point C.

(a) Calculate the bearing from A to C.

(b) What is the bearing from A to D?

(c) Find the bearing from B to D.

(d) Compute the bearing from C to A.

(e) Calculate the bearing from D to A.

(f) What is the bearing from D to B?

** Exercise 18

Ship A travels at a bearing of $S70°W$ and a rate of 25 kilometers per hour. Ship B travels at a bearing of $N30°E$ and a rate of 20 kilometers per hour. Suppose that the ships were initially at the same location. Calculate the distance and bearing from ship A to ship B after one hour.

** Exercise 19

An airplane travels along a path with a bearing of $N25°E$ for two hours. It changes course and heads $S80°E$ and lands an hour later. Suppose the plane flies at a constant speed. Find the bearing of a plane traveling at the same speed which takes-off and lands at the same locations, but travels along a straight path.

** Exercise 20

A boat sets its instruments so that it would travel $S10°W$ at a speed of 50 kilometers per hour in still water. However, there is an eastbound current which travels at a speed of 5 kilometers per hour. Compute (a) the speed of the boat in the water, and (b) the bearing of the boat in the water.

** Exercise 21

An East wind of speed 10 kilometers per hour, causes a plane to travel 850 kilometers per hour at a bearing of $N55°W$. What would be (a) the speed and (b) the bearing of the plane if it were in still air?

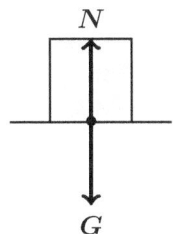

Figure 3

* Exercise 22

In Figure 3, suppose the mass of the box is m.

(a) Assume $|G| = 100$ N. Then $|N|$ must equal what value?

(b) Let $m = 20$ kg. Calculate $|G|$.

(c) If $|N| = 10$ N, what is m?

(d) Suppose $m = 5$ kg. Find $|N|$.

(e) What is m, if $|G| = 78.48$ N?

(e) What is m, if $|T| = 25$ N?

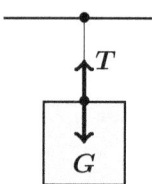

Figure 4

* Exercise 23

Consider Figure 4. Let the mass of the box be m.

(a) Assume $|G| = 100$ N. Then $|T|$ must equal what value?

(b) Let $m = 10$ kg. Calculate $|G|$.

(c) If $|T| = 25$ N, what is m?

(d) Suppose $m = 2$ kg. Find $|G|$.

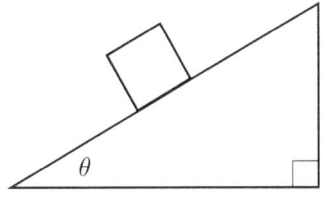

Figure 5

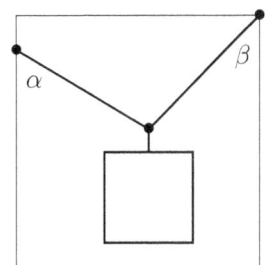

Figure 6

** Exercise 24

In Figure 5, let θ be as shown. Assume m is the mass of the box and a is the magnitude of acceleration the box. Calculate the missing variable using the other two.

(a) $m = 100$ kg and $\theta = 60°$

(b) $m = 50$ kg and $a = 6.94$ m/s^2

(c) $m = 75$ m/s^2 and $\theta = 20°$

(d) $m = 90$ kg and $a = 3.99$ m/s^2

** Exercise 25

Consider Figure 5. Let θ be as shown and a be the magnitude of acceleration. Prove

$$a = g \sin \theta.$$

** Exercise 26

In Figure 6, suppose the mass of the box is 100 kilograms.

(a) Let $\alpha = 45°$ and $\beta = 30°$. Calculate the tension in each cable.

(b) Suppose $\alpha = 50°$ and $\beta = 30°$. What is the magnitude of tension in each cable?

(c) Say $\beta = 20°$ and the left cable contains 25 percent less tension than the right. Find α.

(d) Assume $\alpha = 60°$ and the right cable contains 30 percent more tension than the left. Compute the tension in the right cable.

** Exercise 27

Consider Figure 6. Let the magnitude of tension in the left cable be 300 Newtons, the magnitude of tension in the right cable be 900 Newtons, and $\beta = 15°$. Determine the mass of the box.

*** Exercise 28

In Figure 6, assume the box has mass 200 kilograms. Use the given information to calculate α and β.

(a) Say the magnitude tension in the left cable is 981 Newtons and the magnitude tension in the right cable is $981\sqrt{3}$ Newtons.

(b) Suppose the magnitude of tension in left cable is 700 Newtons and the magnitude of tension in the right cable is 1400 Newtons.

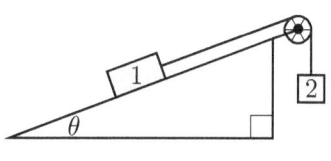

Figure 7

** Exercise 29

Consider Figure 7. Suppose the mass of box 1 is 100 kilograms.

(a) Let $\theta = 30°$. Assume the system is at rest. Calculate the mass of box 2.

(b) Say the mass of box 2 is 75 kilograms. If the system is at rest, what is the measure of θ?

(c) Suppose $\theta = 60°$ and box 1 accelerates down the ramp at 5 m/s². Find the mass of box 2.

(d) Assume the mass of box 2 is 80 kilograms and box 1 accelerates up the ramp at 3 m/s². Compute θ.

* Exercise 30

Let

$$u = -i + 2j, \quad v = \begin{pmatrix} 5 \\ -2 \end{pmatrix},$$

and $w = i - 4j$. Find each of the following.

(a) $u \cdot v$

(b) $v \cdot w$

(c) $u \cdot (v + w)$

(d) $2u \cdot (2v - 3w)$

* Exercise 31

Repeat Exercise 30, except suppose

$$u = \begin{pmatrix} -1/2 \\ 3 \end{pmatrix}, \quad v = 6i + j,$$

and $w = 15i$.

** Exercise 32

Prove properties (iii) and (iv) of Proposition 11.5.

* Exercise 33

Compute $u \cdot v$ using the given information. Suppose θ is the angle between u and v.

(a) $|u| = 3$, $|v| = 2/3$, and $\theta = 120°$

(b) $|u| = 5$, $|v| = 7$, and $\theta = \pi/6$

(c) $|u| = 5/7$, $|v| = 21$, and $\theta = 135°$

(d) $|u| = 13$, $|v| = 10$, and $\theta = \pi/2$

** Exercise 34

What is the measure of the angle between u and v?

(a) $u = i$ and $v = -5j$

(b) $u = (-3, 0)^T$ and $v = (2, 0)^T$

(c) $u = \frac{1}{2}i - 3j$ and $v = 2i + 2j$

(d) $u = \left(-3\sqrt{2}, \sqrt{6}\right)^T$ and $v = (0, 5)^T$

(e) $u = i - 1j$ and $v = 3i + j$

(f) $u = \frac{3}{5}i + \frac{4}{5}j$ and $v = (-12, 5)^T$

** Exercise 35

Determine whether the given vectors are parallel, orthogonal, or neither.

(a) $3i + 2j$ and $9i + 6j$

(b) $(2, -3)^T$ and $(-3, 2)^T$

(c) $\frac{5}{2}i - \frac{\sqrt{3}}{2}j$ and $-3\sqrt{3}i + \frac{9}{5}j$

(d) $(1, -2)^T$ and $2i + j$

(e) $(3, 4)^T$ and $(3, -4)^T$

(f) $-3\sqrt{2}i + 6j$ and $\left(10, 5\sqrt{2}\right)^T$

** Exercise 36

Find two vectors orthogonal to v.

(a) $v = 3i - 5j$

(b) $v = (2, 7)^T$

(c) $v = i + j$

(d) $v = \left(\cos 22.5°, \sin 22.5°\right)^T$

** Exercise 37

Compute $\text{comp}_u v$ for the given u and v.

(a) $u = (1, -2)^T$ and $v = (3, 1)^T$

(b) $u = 3i + 2j$ and $v = \frac{1}{3}i + j$

(c) $u = (5, -12)^T$ and $v = (12, 5)^T$

** Exercise 38

Use the vectors u and v in Exercise 37 to compute $\text{proj}_u v$.

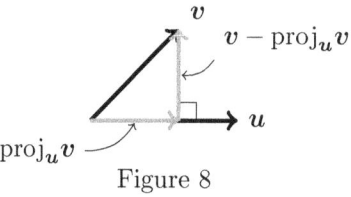

Figure 8

** Exercise 39

The component of v perpendicular to u is

$$v - \text{proj}_u v.$$

Figure 8 illustrates the idea. Calculate the component of v perpendicular to u.

(a) $u = (6,8)^T$ and $v = (-15, 5)^T$

(b) $u = -i - 2i$ and $v = 7i + 7j$

(c) $u = (12, -9)^T$ and $v = (-15, -45)^T$

(d) $u = \sqrt{3}i - j$ and $v = i + \sqrt{3}j$

** Exercise 40

A force F moves an object from point A to point B. Compute the work done.

(a) $A = (0,5)$, $B = (7,4)$, and $F = (1,2)^T$

(b) $A = (1,2)$, $B = (5,7)$, and $F = 2i - 5j$

(c) $A = (1,2)$, $B = (-5,0)$, and $F = (-1,-4)^T$

(d) $A = (-3,2)$, $B = (-1,5)$, and $F = (-3,2)^T$

(e) $A = (4,-2)$, $B = (3,-3)$, and $F = 7i - 11j$

(f) $A = (1, \sqrt{\pi})$, $B = (1/2, \sqrt{\pi}/3)$, and $F = (\sqrt{\pi}, -4)^T$

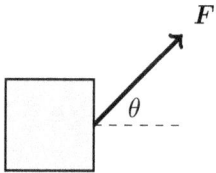

Figure 9

* Exercise 41

In Figure 9, a force F of magnitude F Newtons is exerted on a box at an angle θ from horizontal. The box moves d meters right. Calculate the work done using the given values.

(a) $d = 3$, $F = 12$, and $\theta = 0$

(b) $d = 20$, $F = 5$, and $\theta = \pi/4$

(c) $d = 10$, $F = 20$, and $\theta = 90°$

(d) $d = 1$, $F = 13$, and $\theta = 5\pi/4$

** **Exercise 42**

Consider Figure 5. Suppose the ramp is d meters long and the box has mass m kilograms. Use the given values to calculate the work required to pull the box up the ramp.

(a) $d = 16$, $m = 100$, and $\theta = 10°$

(b) $d = 15$, $m = 20$, and $\theta = \pi/6$

(c) $d = 8$, $m = 50$, and $\theta = 45°$

Chapter 12

Complex Numbers

In this chapter, we will study complex numbers. A solid understanding of Chapters 1, 3, 5, and 7 is required. Complex numbers use some of the same ideas as vectors. As a result, knowledge of Chapter 11 is helpful though not strictly necessary. We will not use calculators in this chapter.

12.1 The Basics

As you know from algebra class, there are certain polynomials with no real solution. For example,
$$z^2 + 1 = 0$$
has no real solution. Our goal is to "fill the gap" in the set of solvable polynomial equations.

Definition 12.1 Define i to be a solution of $z^2 + 1 = 0$. In other words, let
$$i^2 = -1 \quad \text{or} \quad i = \sqrt{-1}.$$

The existence of i allows us to construct a new number system.

Definition 12.2 The **complex numbers** are the set
$$= \{a + bi : a, b \in \mathbb{R}\}.$$

We would like to preform arithmetic operations on complex numbers, e.g. adding or multiplying two complex numbers. To do this, we need to introduce a standard representation of a complex number.

Definition 12.3 A complex number is in **standard form** when it is written $a + bi$ for a and b are real numbers.

So, for example, $7 + 3i$ is in standard form, but

$$\frac{1 + 4i}{2 - 3i}$$

is not. Expressions with a minus between the terms, like $7 - 3i$, are considered to be in standard form as well; note that $7 - 3i = 7 + (-3)i$. We consider expressions like 1 and $5i$ to be in standard form; note that $1 = 1 + 0i$ and $5i = 0 + 5i$.

We are ready to define addition and multiplication of complex numbers in standard form.

Definition 12.4 For complex numbers

$$z_1 = a + bi \quad \text{and} \quad z_2 = c + di,$$

define
$$z_1 + z_2 = (a + b) + (b + d)i$$
and
$$z_1 z_2 = (ac - bd) + (ad + bc)i.$$

The following is the natural extension of the corresponding properties of real numbers and will come as no surprise to most readers. We include them as a formality.

Proposition 12.1 *Suppose z_1, z_2, and z_3 are complex numbers.*

(i) $(z_1 + z_2) + z_3 = z_1 + (z_2 + z_3)$

(ii) $z_1 + z_2 = z_2 + z_1$

(iii) $(z_1 z_2) z_3 = z_1(z_2 z_3)$

(iv) $z_1 z_2 = z_2 z_1$

(v) $z_1(z_2 + z_3) = z_1z_2 + z_1z_2$ and $(z_1 + z_2)z_3 = z_1z_3 + z_2z_3$

Proposition 12.1 allows us to treat i like a variable in a linear polynomial with real coefficients. This approach alleviates the need for substantial amounts of memorization; simply remember that $i^2 = -1$.

Example 12.1 Let
$$z_1 = 2 + i \quad \text{and} \quad z_2 = 3 - 5i.$$
Compute (a) $z_1 + z_2$ and (b) $z_1 z_2$.

Solution

(a) Addition of complex numbers uses the same ideas that were used to combine like terms in algebra.
$$\begin{aligned} z_1 + z_2 &= (2 + i) + (3 - 5i) \\ &= (2 + 3) + (1 - 5)i \\ &= 5 - 4i. \end{aligned}$$

(b) Instead of using the multiplication formula, we will use expansion techniques from algebra.
$$\begin{aligned} z_1 z_2 &= (2 + i)(3 - 5i) \\ &= 6 - 10i + 3i - 5i^2 \\ &= 6 - 7i + 5 \\ &= 11 - 7i. \end{aligned}$$

■

Definition 12.5 The **complex conjugare** of $z = a + bi$ is
$$\overline{z} = a - bi.$$

Example 12.2 Assume
$$z_1 = 2 - 5i \quad \text{and} \quad z_2 = \frac{1 + 3i}{2}.$$
Find (a) $\overline{z_1}$ and (b) $\overline{z_2}$.

Solution

(a) We have, $\overline{z_1} = \overline{2 - 5i} = 2 + 5i$.

(b) Technically, we need to convert the complex number to standard form to use the formula. We will show these steps, but some readers may prefer to omit them because they are obvious.

$$\overline{z_2} = \overline{\frac{1 + 3i}{2}}$$
$$= \overline{\frac{1}{2} + \frac{3}{2}i}$$
$$= \frac{1}{2} - \frac{3}{2}i$$
$$= \frac{1 - 3i}{2}.$$

■

Proposition 12.2 *Suppose $z = a + bi$. Then*

$$z\bar{z} = a^2 + b^2.$$

Proof We will consider the definition of the conjugate and use expansion techniques from algebra:

$$z\bar{z} = (a + bi)(a - bi)$$
$$= a^2 - abi + abi - (bi)^2$$
$$= a^2 - b^2 i^2$$
$$= a^2 - (-1)b^2$$
$$= a^2 + b^2.$$

■

Example 12.3 Let $z = 2 - 3i$. Compute $z\bar{z}$.

Solution Proposition 12.2 tells us

$$z\bar{z} = 2^2 + (-3)^2 = 13.$$

■

Proposition 12.3 (Inverses and Identities) *Assume z is a complex number.*

(i) *The number $0 = 0 + 0i$ is such that*
$$0 + z = z + 0 = z.$$

(ii) *The number $1 = 1 + 0i$ is such that*
$$1z = z1 = z.$$

(iii) *There exists $-z$ such that*
$$z + (-z) = -z + z = 0.$$

(iv) *If $z \neq 0$, then there exists z^{-1} such that*
$$zz^{-1} = z^{-1}z = 1.$$

For complex numbers z_1 and z_2, we often write
$$z_1 - z_2 \quad \text{instead of} \quad z_1 + (-z_2).$$
Similarly, we will often write
$$\frac{z_1}{z_2} \quad \text{instead of} \quad z_1 z_2^{-1}.$$

In practice, finding the additive inverse of a complex number is no problem. The additive inverse of $a + bi$ is simply $-a - bi$.

Finding the multiplicative inverse is more challenging. The standard form of the inverse of $z = a + bi \neq 0$ is
$$z^{-1} = \frac{a}{a^2 + b^2} - \frac{b}{a^2 + b^2}i.$$

However, this formula is difficult to remember, and rarely used in practice.

Instead of using our formula, we will introduce a list of procedures to find the standard form of the inverse of $z \neq 0$:

1. Consider $\dfrac{1}{z}$.
2. Multiply the top and bottom by $\bar{z}$.
3. Divide term by term to write the result in standard form.

Example 12.4 Write the multiplicative inverse of $z = 3 - 4i$ in standard form.

Solution

$$\begin{aligned} z^{-1} &= \frac{1}{3-4i} \\ &= \frac{1}{3-4i} \cdot \frac{3+4i}{3+4i} \\ &= \frac{3+4i}{9+16} \\ &= \frac{3+4i}{25} \\ &= \frac{3}{25} + \frac{4}{25}i. \end{aligned}$$

∎

A similar process can be used to find the standard form of the quotient of two complex numbers.

Example 12.5 Write
$$\frac{3-2i}{1+3i}$$
in standard form.

Solution We will multiply the denominator by the conjugate and simplify:

$$\frac{3-2i}{1+3i} = \frac{3-2i}{1+3i}\frac{1-3i}{1-3i}$$
$$= \frac{3-11i+6i^2}{1+9}$$
$$= \frac{3-11i-6}{10}$$
$$= \frac{-3-11i}{10}$$
$$= -\frac{3}{10} - \frac{11}{10}i.$$

∎

Proposition 12.4 (Properties of the Conjugate) *Suppose z_1 and z_2 are complex numbers.*

(i) $\overline{z_1 + z_2} = \overline{z_1} + \overline{z_2}$

(ii) $\overline{z_1 z_2} = \bar{z}_1 \bar{z}_2$

(iii) $\overline{\dfrac{z_1}{z_2}} = \dfrac{\bar{z}_1}{\bar{z}_2}$

Proof We will prove (i) and (ii). Property (iii) will be left as an exercise. Let
$$z_1 = a+bi \quad \text{and} \quad z_2 = c+di.$$
(i)
$$\overline{z_1+z_2} = \overline{a+c+(b+d)i}$$
$$= a+c-(b+d)i$$
$$= a+c-bi-di$$
$$= a-bi+c-di$$
$$= \overline{a+bi} + \overline{c+di}$$
$$= \overline{z_1} + \overline{z_2}.$$

(ii) We will compute the left and right side of the equation to prove that they are equal. We have

$$\overline{z_1 z_2} = \overline{(a+bi)(c+di)}$$
$$= \overline{ac - bd + (ad + bc)i}$$
$$= ac - bd - (ad + bc)i,$$

and

$$\overline{z_1}\,\overline{z_2} = \overline{(a+bi)}\,\overline{(c+di)}$$
$$= (a-bi)(c-di)$$
$$= ac - bd + (-ad - bc)i$$
$$= ac - bd - (ad + bc)i.$$

∎

Definition 12.6 The **modulus** of a complex number $z = a + bi$ is

$$|z| = \sqrt{a^2 + b^2}.$$

The plural form of modulus is moduli.

Example 12.6 Compute $|5 - 12i|$.

Solution

$$|5 - 12i| = \sqrt{5^2 + (-12)^2}$$
$$= \sqrt{25 + 144}$$
$$= \sqrt{169}$$
$$= 13.$$

∎

Proposition 12.5 *For complex numbers z_1 and z_2,*

$$|z_1 z_2| = |z_1|\,|z_2| \quad \text{and} \quad \left|\frac{z_1}{z_2}\right| = \frac{|z_1|}{|z_2|}.$$

Before we begin the proof, we note a useful property which follows from Proposition 12.2:

$$|z| = \sqrt{z\bar{z}} \quad \text{or} \quad |z|^2 = z\bar{z}.$$

Proof We will prove

$$|z_1 z_2| = |z_1| |z_2|.$$

We will leave the proof of

$$\left|\frac{z_1}{z_2}\right| = \frac{|z_1|}{|z_2|}$$

as an exercise.

We have

$$\begin{aligned}
|z_1 z_2|^2 &= (z_1 z_2)(\overline{z_1 z_2}) \\
&= z_1 z_2 \bar{z}_1 \bar{z}_2 \\
&= (z_1 \bar{z}_1)(z_2 \bar{z}_2) \\
&= |z_1|^2 |z_2|^2.
\end{aligned}$$

Because moduli are always nonnegative real numbers,

$$|z_1 z_2|^2 = |z_2|^2 |z_1|^2 \quad \text{implies} \quad |z_1 z_2| = |z_2||z_1|.$$

∎

Example 12.7 Suppose $z_1 = 3 - 4i$ and $z_2 = 1 + i$. Compute (a) $|z_1 z_2|$ and (b) $|z_1/z_2|$.

Solution We will compute $|z_1|$ and $|z_2|$, and then use Proposition 12.5:

$$\begin{aligned}
|z_1| &= |3 - 4i| & \text{and} \quad |z_2| &= |1 + i| \\
&= \sqrt{3^2 + (-4)^2} & &= \sqrt{1^2 + 1^2} \\
&= \sqrt{25} & &= \sqrt{2} \\
&= 5
\end{aligned}$$

(a) $|z_1 z_2| = |z_1| |z_2| = 5\sqrt{2}$

(b) $\left|\dfrac{z_1}{z_2}\right| = \dfrac{|z_1|}{|z_2|} = \dfrac{5}{\sqrt{2}} = \dfrac{5\sqrt{2}}{2}$

∎

12.1.1 Exponents

Definition 12.7 For a complex number z, let
$$z^n = \underbrace{z \cdot z \cdot \ldots \cdot z}_{n \text{ times}}$$
for $n = 1, 2, 3, \ldots$. If $z \neq 0$, define
$$z^0 = 1.$$

Example 12.8 Compute
$$(3 - 2i)^n$$
when (a) $n = 0$, (b) $n = 2$, and (c) $n = 3$.

Solution

(a) Since $3 - 2i \neq 0$, we have $(3 - 2i)^0 = 1$.

(b) This is like our work in Example 1:
$$\begin{aligned}(3 - 2i)^2 &= (3 - 2i)(3 - 2i) \\ &= 9 - 12i + 4i^2 \\ &= 9 - 12i - 4 \\ &= 5 - 12i.\end{aligned}$$

(c) Since $(3 - 2i)^2 = 5 - 12i$,
$$\begin{aligned}(3 - 2i)^3 &= (3 - 2i)^2(3 - 2i) \\ &= (5 - 12i)(3 - 2i) \\ &= 15 - 46i + 24i^2 \\ &= 14 - 46i - 24 \\ &= -10 - 46i.\end{aligned}$$

An important observation to make is that
$$i^1 = i, \quad i^2 = -1, \quad i^3 = -i, \quad \text{and} \quad i^4 = 1.$$

Example 12.9 Compute i^5.

Solution We have
$$\begin{aligned} i^5 &= i^4 \cdot i \\ &= 1 \cdot i \\ &= i. \end{aligned}$$
∎

Example 12.10 Simplify i^{953}.

Solution Since i repeats itself every multiple of four, we observe
$$953 = 4(238) + 1.$$
So,
$$\begin{aligned} i^{953} &= i^{4(238)+1} \\ &= \left(i^4\right)^{238} \cdot i^1 \\ &= 1 \cdot i \\ &= i. \end{aligned}$$
∎

Example 12.11 Wirte
$$-4i^7 + \frac{3}{i^3} + 3i^{100}$$
in standard form.

Solution
$$\begin{aligned} -4i^7 + \frac{3}{i^3} + 3i^{100} &= -4i^{4+3} + \frac{3i}{i^4} + 3i^{4(25)} \\ &= -4i^4 i^3 + \frac{3i}{i^4} + 3\left(i^4\right)^{25} \\ &= -4(1)(-i) + \frac{3i}{1} + 3(1) \\ &= 3 + 7i. \end{aligned}$$

Example 12.12 Write
$$i + i^2 + i^2 + \ldots + i^{33}$$
in standard form

Solution Notice that the first four terms of the sum add to 0, i.e.
$$i + i^2 + i^3 + i^4 = i - 1 - i + 1 = 0.$$
Furthermore, all consecutive groups of four terms add to 0, e.g.
$$i^5 + i^6 + i^7 + i^8 = i^4(i + i^2 + i^3 + i^4)$$
$$= i(0)$$
$$= 0.$$

Since $33 = 4(8) + 1$, we arrange our sum into eight consecutive groups of four terms plus the last term. The eight groups of four terms are all zero, so the sum is simply equal to the last term. In other words,

$$i + i^2 + i^2 + \ldots + i^{33} = \left(i + i^2 + i^3 + i^4\right) + \left(i^5 + i^6 + i^7 + i^8\right)$$
$$+ \ldots + \left(i^{29} + i^{30} + i^{31} + i^{32}\right) + i^{33}$$
$$= 0 + 0 + \ldots + 0 + i^{33}$$
$$= i^{33}.$$

All that is left is to simplify i^{33}:
$$i^{33} = i^{4(8)+1}$$
$$= \left(i^4\right)^8 i^1$$
$$= 1 \cdot i$$
$$= i.$$

In summery,
$$i + i^2 + i^2 + \ldots + i^{33} = i.$$

12.1.2 The Complex Plane

Definition 12.8 Consider the complex number $z = a + bi$.

- The **real part** of z is the real number a.
- The **imaginary part** of z is the real number b.

Example 12.13 What are the real and imaginary parts of
$$z = \frac{1 - 5i}{3}?$$

Solution Since
$$\frac{1 - 5i}{3} = \frac{1}{3} - \frac{5}{3}i,$$
the real part of z is $1/3$ and the imaginary part of z is $-5/3$. ∎

The complex plane is a means of representing complex numbers graphically. The horizontal axis represents the real part of the complex number and the vertical axis represents the imaginary part. There is a simple correspondence between points on the xy-plane and numbers on the complex plane; the point (x, y) occupies the same location on the xy-plane as $z = x + iy$ does on the complex plane. So, the complex number $a + bi$ is plotted as shown.

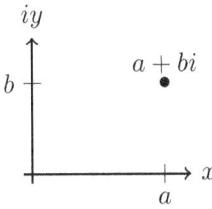

Example 12.14 Graph the complex numbers (a) $-3 + 2i$, (b) $2 - i$, (c) 0, and (d) $-2i$.

Solution

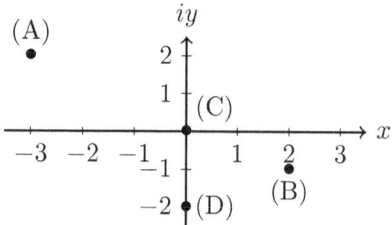

The point corresponding to $z = 0$ on the complex plane is called the "origin", just like on the xy-plane.

Using the Pythagorean Theorem, it is easy to see that the modulus of z is the distance between the origin and the point z on the complex plane.

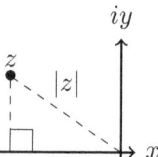

Example 12.15 Find the distance between z and the origin on the complex plane for (a) $z = 1 - i$ and (b) $z = 3 + 7i$.

Solution Since the distance between z and the origin on the complex plane is simply the modulus, we do not need to graph the complex numbers. We will simply compute their moduli.

(a) The distance between $1 - i$ and the origin on the complex plane is
$$\sqrt{1^2 + (-1)^2} = \sqrt{2}.$$

(b) The distance between $3 + 7i$ and the origin on the complex plane is
$$\sqrt{3^2 + 7^2} = \sqrt{58}.$$

12.2 Polar Form

The complex plane gives us a means to represent complex numbers geometrically. We can utilize this thinking to represent nonzero complex numbers using their moduli and standard position angles. Consider $z = a + bi$.

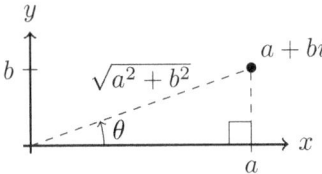

Theorem 5.1 tells us that there is an angle θ in standard position such that
$$\cos\theta = \frac{a}{r} \quad \text{and} \quad \sin\theta = \frac{b}{r}$$
where $r = \sqrt{a^2 + b^2}$. Hence,
$$\begin{aligned} z &= a + bi \\ &= r\left(\frac{a}{r} + \frac{b}{r}i\right) \\ &= r\left(\cos\theta + i\sin\theta\right). \end{aligned}$$

This can be further simplified using our next theorem. We omit its proof because it requires Calculus techniques.

Theorem 12.1 (Euler's Formula) *For all θ in $\mathbb{R}$,*
$$e^{i\theta} = \cos\theta + i\sin\theta.$$

Using Euler's Formula, we conclude
$$z = r\left(\cos\theta + i\sin\theta\right) = re^{i\theta}.$$

Definition 12.9

- The **polar form** of the complex number $z = a + bi \neq 0$ is

$$z = re^{i\theta},$$

where $r = |z|$ and θ is a standard position angle on the complex plane whose terminal side contains z.

- The angle θ is called an **argument** of z.

Two important remarks:

(i) If $z = 0$, it is impossible to write the expression in polar form, because θ cannot be determined.

(ii) Polar form is not unique. Indeed, if θ is an argument, so is $\theta + 2\pi n$, for $n = 1, -1, 2, -2, \ldots$.

Example 12.16 Find two polar forms of $z = -2 + 2i$.

Solution We know

$$|z| = \sqrt{(-2)^2 + 2^2} = 2\sqrt{2}.$$

So, an argument θ must satisfy

$$\cos\theta = -\frac{2}{2\sqrt{2}} = -\frac{\sqrt{2}}{2} \quad \text{and} \quad \sin\theta = \frac{2}{2\sqrt{2}} = \frac{\sqrt{2}}{2}.$$

The angle measure $\theta = 3\pi/4$ is an argument of z, because it satisfies our equation. Hence, $z = -2 + 2i$ can be written in polar form as

$$z = 2\sqrt{2}e^{3\pi i/4}.$$

Since $\theta = 3\pi/4$ is an argument of z, so is $3\pi/4 - 2\pi = -5\pi/4$. Thus, another polar form of z is

$$z = 2\sqrt{2}e^{-5\pi i/4}.$$

∎

Example 12.17 Write each expression in standard form.

(a) $z_1 = 2e^{\pi i/2}$

(b) $z_2 = e^{\pi i}$

(c) $z_3 = 6e^{\pi i/6}$

(d) $z_4 = 5e^{-2\pi i/3}$

Solution

(a) $\begin{aligned}z_1 &= 2e^{\pi i/2}\\ &= 2\left(\cos\frac{\pi}{2} + i\sin\frac{\pi}{2}\right)\\ &= 2(0+i)\\ &= 2i.\end{aligned}$

(b) $\begin{aligned}z_2 &= e^{\pi i}\\ &= \cos\pi + i\sin\pi\\ &= -1 + 0i\\ &= -1.\end{aligned}$

(c) $\begin{aligned}z_3 &= 6e^{\pi i/6}\\ &= 6\left(\cos\frac{\pi}{6} + i\sin\frac{\pi}{6}\right)\\ &= 6\left(\frac{\sqrt{3}}{2} + \frac{1}{2}i\right)\\ &= 3\sqrt{3} + 3i.\end{aligned}$

(d) $\begin{aligned}z_4 &= 5e^{-2\pi i/3}\\ &= 5\left(\cos\left(-\frac{2\pi}{3}\right) + i\sin\left(-\frac{2\pi}{3}\right)\right)\\ &= 5\left(-\frac{1}{2} - i\frac{\sqrt{3}}{2}\right)\\ &= -\frac{5}{2} - \frac{5i\sqrt{3}}{2}.\end{aligned}$

∎

12.3 More on Polar Form

Proposition 12.6 *Consider the complex numbers*

$$z_1 = r_1 e^{i\alpha} \quad \text{and} \quad z_2 = r_2 e^{i\beta}.$$

Then

$$z_1 z_2 = r_1 r_2 e^{i(\alpha+\beta)} \quad \text{and} \quad \frac{z_1}{z_2} = \frac{r_1}{r_2} e^{i(\alpha-\beta)}.$$

The proof of this follows due to properties of exponents.

Example 12.18 Assume

$$z_1 = \sqrt{2}e^{i\pi/2} \quad \text{and} \quad z_2 = 2e^{-i\pi/6}.$$

Compute (a) $z_1 z_2$ and (b) z_1/z_2. Write the result in standard form.

Solution

(a) Using Proposition 12.6,

$$\begin{aligned} z_1 z_2 &= 2\sqrt{2} e^{i(\pi/2 + (-\pi/6))} \\ &= 2\sqrt{2} e^{i\pi/3} \\ &= 2\sqrt{2} \left(\cos \frac{\pi}{3} + i \sin \frac{\pi}{3} \right) \\ &= 2\sqrt{2} \left(\frac{1}{2} + i \frac{\sqrt{3}}{2} \right) \\ &= \sqrt{2} + i\sqrt{6}. \end{aligned}$$

(b) Proposition 12.6 tells us

$$\begin{aligned} \frac{z_1}{z_2} &= \frac{\sqrt{2}}{2} e^{i(\pi/2 - (-\pi/6))} \\ &= \frac{\sqrt{2}}{2} e^{2i\pi/3} \\ &= \frac{\sqrt{2}}{2} \left(\cos \frac{2\pi}{3} + i \sin \frac{2\pi}{3} \right) \\ &= \frac{\sqrt{2}}{2} \left(-\frac{1}{2} + i \frac{\sqrt{3}}{2} \right) \\ &= -\frac{\sqrt{2}}{4} + i \frac{\sqrt{6}}{4}. \end{aligned}$$

■

Proposition 12.7 (De Moivre's Formula) *Suppose*

$$z = r \left(\cos \theta + i \sin \theta \right),$$

where $r > 0$. Then

$$z^n = r^n \left(\cos n\theta + i \sin n\theta \right)$$

for all n.

Proof The proof is a simple application of exponent rules and Euler's Formula:

$$z = r \left(\cos \theta + i \sin \theta \right) = r e^{i\theta}.$$

Hence,
$$\begin{aligned} z^n &= \left(re^{i\theta}\right)^n \\ &= r^n \left(e^{i\theta}\right)^n \\ &= r^n e^{ni\theta} \\ &= r^n \left(\cos n\theta + i \sin n\theta\right). \end{aligned}$$

∎

Example 12.19 Compute the standard form of
$$\left(1 - i\sqrt{3}\right)^5.$$

Solution Let us find the modulus and an argument of $1 - i\sqrt{3}$, and then use Proposition 12.7. We have
$$\begin{aligned} \left|1 - i\sqrt{3}\right| &= \sqrt{1^2 + \left(-\sqrt{3}\right)^2} \\ &= \sqrt{4} \\ &= 2. \end{aligned}$$

It follows that
$$\cos\theta = \frac{1}{2} \quad \text{and} \quad \sin\theta = -\frac{\sqrt{3}}{2}.$$

This implies that $\theta = -\pi/3$ is an argument of $1 - i\sqrt{3}$.

So,
$$1 - i\sqrt{3} = 2\left(\cos\left(-\frac{\pi}{3}\right) + i\sin\left(-\frac{\pi}{3}\right)\right).$$

Thus, De Moivre's Formula tells us
$$\begin{aligned} \left(1 - i\sqrt{3}\right)^5 &= 2^5 \left(\cos\left(-\frac{\pi}{3}(5)\right) + i\sin\left(-\frac{\pi}{3}(5)\right)\right) \\ &= 2^5 \left(\cos\left(-\frac{5\pi}{3}\right) + i\sin\left(-\frac{5\pi}{3}\right)\right) \\ &= 32\left(\frac{1}{2} + i\frac{\sqrt{3}}{2}\right) \\ &= 16 + 16i\sqrt{3}. \end{aligned}$$

Example 12.20 Use Euler's Formula to prove

$$\cos(\alpha + \beta) = \cos\alpha\cos\beta - \sin\alpha\sin\beta$$

and

$$\sin(\alpha + \beta) = \sin\alpha\cos\beta + \cos\alpha\sin\beta.$$

Solution Using Euler's Formula, we have

$$e^{i\alpha} = \cos\alpha + i\sin\alpha \quad \text{and} \quad e^{i\beta} = \cos\beta + i\sin\beta.$$

Furthermore,

$$e^{i(\alpha+\beta)} = \cos(\alpha + \beta) + i\sin(\alpha + \beta).$$

It follows that

$$\begin{aligned}
\cos(\alpha + \beta) &+ i\sin(\alpha + \beta) \\
&= e^{i(\alpha+\beta)} \\
&= e^{i\alpha}e^{i\beta} \\
&= \Big(\cos\alpha + i\sin\alpha\Big)\Big(\cos\beta + i\sin\beta\Big) \\
&= \cos\alpha\cos\beta + i\cos\alpha\sin\beta + i\sin\alpha\cos\beta + i^2\sin\alpha\sin\beta \\
&= \Big(\cos\alpha\cos\beta - \sin\alpha\sin\beta\Big) + i\Big(\sin\alpha\cos\beta + \cos\alpha\sin\beta\Big).
\end{aligned}$$

Since $a + bi = c + di$ implies $a = c$ and $b = d$, we conclude

$$\cos(\alpha + \beta) = \cos\alpha\cos\beta - \sin\alpha\sin\beta$$

and

$$\sin(\alpha + \beta) = \sin\alpha\cos\beta + \cos\alpha\sin\beta.$$

12.3.1 Finding Roots

Let us turn our attention to solving equations like

$$z^n + a = 0.$$

It is useful to know the number of solutions that such an equation has. So, we introduce the following theorem.

Theorem 12.2 (Fundamental Theorem of Algebra) *Consider*

$$f(z) = a_n z^n + a_{n-1} z^{n-1} + \ldots + a_1 z + a_0,$$

where a_n, a_{n-1}, $\ldots$, and a_0 are complex numbers. Counting multiplicity, the function f has n roots.

The Fundamental Theorem of Algebra is more robust than necessary for our purposes. The following corollary emphasizes the needed properties.

Corollary 12.1 *An equation of the form $z^n + a = 0$ has n unique solutions if $a \neq 0$.*

To find the n roots of $z^n + a$ proceed as follows:

1. Solve for z^n, which yields $z^n = -a$.

2. Write $-a$ in polar form with the general form of the argument in the exponent, i.e. write $-a$ as

$$|-a|e^{(\theta + 2\pi k)i},$$

where θ is an argument of $-a$.

3. Take the n-th root, which gives

$$z = \sqrt[n]{|-a|} e^{(\theta + 2\pi k)i/n}.$$

4. Evaluate the expression for $k = 0, 1, 2, \ldots, n-1$. This gives the n unique solutions.

Example 12.21 Compute the standard form solutions of $z^3 + 8 = 0$.

Solution We know $z^3 = -8$. The polar form of -8 is $8e^{(\pi+2\pi k)i}$. It follows that

$$z = \sqrt[3]{8}e^{(\pi+2\pi k)i/3} = 2e^{(\pi+2\pi k)i/3}.$$

All that is left is to evaluate for $k=0$, 1, and 2.

$k = 0$:

$$\begin{aligned} z &= 2e^{(\pi+2\pi(0))i/3} \\ &= 2\left(\cos\frac{\pi}{3} + i\sin\frac{\pi}{3}\right) \\ &= 2\left(\frac{1}{2} + i\frac{\sqrt{3}}{2}\right) \\ &= 1 + i\sqrt{3}. \end{aligned}$$

$k = 1$:

$$\begin{aligned} z &= 2e^{(\pi+2\pi(1))i/3} \\ &= 2\left(\cos\pi + i\sin\pi\right) \\ &= 2(-1 + 0i) \\ &= -2. \end{aligned}$$

$k = 2$:

$$\begin{aligned} z &= 2e^{(\pi+2\pi(2))i/3} \\ &= 2\left(\cos\frac{5\pi}{3} + i\sin\frac{5\pi}{3}\right) \\ &= 2\left(\frac{1}{2} - i\frac{\sqrt{3}}{2}\right) \\ &= 1 - i\sqrt{3}. \end{aligned}$$

■

Example 12.22 Find all possible values of z, given that

$$z^3 = -125 + 125i.$$

Solution The modulus of $-125+125i$ is

$$\begin{aligned}|-125+125i| &= \sqrt{(-125)^2+125^2}\\ &= \sqrt{2(125)^2}\\ &= 125\sqrt{2}.\end{aligned}$$

So,

$$\begin{aligned}-125+125i &= 125\sqrt{2}\left(-\frac{125}{125\sqrt{2}}+i\frac{125}{125\sqrt{2}}\right)\\ &= 125\sqrt{2}\left(-\frac{1}{\sqrt{2}}+i\frac{1}{\sqrt{2}}\right)\\ &= 125\sqrt{2}\left(-\frac{\sqrt{2}}{2}+i\frac{\sqrt{2}}{2}\right)\end{aligned}$$

Because

$$\cos\theta = -\frac{\sqrt{2}}{2} \quad\text{and}\quad \sin\theta = \frac{\sqrt{2}}{2},$$

the general argument of $-125+125i$ is

$$\frac{3\pi}{4}+2\pi k = \frac{(3+8k)\pi}{4}.$$

It follows that

$$-125+125i = 125\sqrt{2}e^{(3+8k)\pi i/4}.$$

Hence,

$$\begin{aligned}z &= \sqrt[3]{125\sqrt{2}}e^{(3+8k)\pi i/12}\\ &= \sqrt[3]{125}\sqrt[3]{\sqrt{2}}e^{(3+8k)\pi i/12}\\ &= 5\sqrt[6]{2}e^{(3+8k)\pi i/12}.\end{aligned}$$

We need to evaluate for $k=1,2,$ and 3. To do this, we will use the Half-Angle Identities

$$\cos\frac{\theta}{2} = \pm\sqrt{\frac{1+\cos\theta}{2}} \quad\text{and}\quad \sin\frac{\theta}{2} = \pm\sqrt{\frac{1-\cos\theta}{2}},$$

where the sign in front is determined by what quadrant the terminal side of $\theta/2$ lies. Let us evaluate.

$k = 0$:

$$z = 5\sqrt[6]{2}e^{(3+8(0))\pi i/12}$$
$$= 5\sqrt[6]{2}\left(\cos\frac{\pi}{4} + i\sin\frac{\pi}{4}\right)$$
$$= 5\sqrt[6]{2}\left(\frac{\sqrt{2}}{2} + i\frac{\sqrt{2}}{2}\right)$$
$$= \frac{5\sqrt[6]{2}\sqrt{2}}{2} + i\frac{5\sqrt[6]{2}\sqrt{2}}{2}$$
$$= \frac{5\sqrt[6]{2}\sqrt[6]{8}}{2} + i\frac{5\sqrt[6]{2}\sqrt[6]{8}}{2}$$
$$= \frac{5\sqrt[6]{16}}{2} + i\frac{5\sqrt[6]{16}}{2}$$

$k = 1$:

$$z = 5\sqrt[6]{2}e^{(3+8(1))\pi i/12}$$
$$= 5\sqrt[6]{2}\left(\cos\frac{11\pi}{12} + i\sin\frac{11\pi}{12}\right)$$
$$= 5\sqrt[6]{2}\left(\cos\left(\frac{11\pi/6}{2}\right) + i\sin\left(\frac{11\pi/6}{2}\right)\right)$$
$$= 5\sqrt[6]{2}\left(-\sqrt{\frac{1+\cos(11\pi/6)}{2}} + i\sqrt{\frac{1-\cos(11\pi/6)}{2}}\right)$$
$$= 5\sqrt[6]{2}\left(-\sqrt{\frac{1+\sqrt{3}/2}{2}} + i\sqrt{\frac{1-\sqrt{3}/2}{2}}\right)$$
$$= 5\sqrt[6]{2}\left(-\sqrt{\frac{2+\sqrt{3}}{4}} + i\sqrt{\frac{2-\sqrt{3}}{4}}\right)$$
$$= \frac{5\sqrt[6]{2}}{2}\left(-\sqrt{2+\sqrt{3}} + i\sqrt{2-\sqrt{3}}\right)$$

$k = 2$:

$$z = 5\sqrt[6]{2}e^{(3+8(2))\pi i/12}$$

$$= 5\sqrt[6]{2}\left(\cos\frac{19\pi}{12} + i\sin\frac{19\pi}{12}\right)$$

$$= 5\sqrt[6]{2}\left(\cos\left(\frac{19\pi/6}{2}\right) + i\sin\left(\frac{19\pi/6}{2}\right)\right)$$

$$= 5\sqrt[6]{2}\left(\sqrt{\frac{1+\cos(19\pi/6)}{2}} - i\sqrt{\frac{1-\cos(19\pi/6)}{2}}\right)$$

$$= 5\sqrt[6]{2}\left(\sqrt{\frac{1-\sqrt{3}/2}{2}} - i\sqrt{\frac{1+\sqrt{3}/2}{2}}\right)$$

$$= 5\sqrt[6]{2}\left(\sqrt{\frac{2-\sqrt{3}}{4}} - i\sqrt{\frac{2+\sqrt{3}}{4}}\right)$$

$$= \frac{5\sqrt[6]{2}}{2}\left(\sqrt{2-\sqrt{3}} - i\sqrt{2+\sqrt{3}}\right).$$

∎

Definition 12.10 An n-th **root of unity** is a solution of

$$z^n = 1.$$

Example 12.23 Find the sixth roots of unity.

Solution Since
$$1 = e^{2\pi k i},$$
we have
$$z^6 = 1 \quad \text{implies} \quad z = e^{2\pi k i/6} = e^{\pi k i/3}.$$

Hence, the sixth roots of unity are

$z = e^{\pi(0)i/3}$ $z = e^{\pi(1)i/3}$ $z = e^{\pi(2)i/3}$

$= 1,$ $= \cos\dfrac{\pi}{3} + i\sin\dfrac{\pi}{3}$ $= \cos\dfrac{2\pi}{3} + i\sin\dfrac{2\pi}{3}$

 $= \dfrac{1}{2} + i\dfrac{\sqrt{3}}{2},$ $= -\dfrac{1}{2} + i\dfrac{\sqrt{3}}{2},$

$z = e^{\pi(3)i/3}$ $z = e^{\pi(4)i/3}$

$= \cos\pi + i\sin\pi$ $= \cos\dfrac{4\pi}{3} + i\sin\dfrac{4\pi}{3}$

$= -1,$ $= -\dfrac{1}{2} - i\dfrac{\sqrt{3}}{2},$

and

$$z = e^{\pi(5)i/3}$$
$$= \cos\frac{5\pi}{3} + i\sin\frac{5\pi}{3}$$
$$= \frac{1}{2} - i\frac{\sqrt{3}}{2}.$$

■

Example 12.24 Let $z = e^{2\pi i/7}$. What is the standard form of

$$1 + z^2 + z^3 + 2z^4 + 2z^5 + 2z^6 + 2z^7 + z^8 + z^9 + z^{10}?$$

Solution The complex number $z = e^{2\pi i/7}$ is a seventh root of unity and $z \neq 1$. This means

$$z^7 - 1 = (z-1)(z^6 + z^5 + z^4 + z^3 + z^2 + z + 1) = 0.$$

Since $z \neq 1$, it follows that

$$z^6 + z^5 + z^4 + z^3 + z^2 + z + 1 = 0.$$

Hence,

$$
\begin{aligned}
&1 + z^2 + z^3 + 2z^4 + 2z^5 + 2z^6 + 2z^7 + z^8 + z^9 + z^{10} \\
&= \left(1 + z^2 + z^3 + z^4 + z^5 + z^6\right) + \left(z^4 + z^5 + z^6 + z^7 + z^8 + z^9 + z^{10}\right) + z^7 \\
&= \left(1 + z^2 + z^3 + z^4 + z^5 + z^6\right) + z^4\left(1 + z^2 + z^3 + z^4 + z^5 + z^6\right) + z^7 \\
&= 0 + z^4(0) + z^7 \\
&= z^7 \\
&= \left(e^{2\pi i/7}\right)^7 \\
&= \cos 2\pi + i \sin 2\pi \\
&= 1.
\end{aligned}
$$

∎

12.4 Exercises

* Exercise 1

Suppose

$z_1 = 3 - 5i$ and $z_2 = 5 + 12i$.

Find each of the following. Write the results in standard form.

(a) $z_1 + z_2$
(b) $2z_1 - 3z_2$
(c) $z_1 z_2$
(d) $i z_1$

* Exercise 2

Assume

$z_1 = 7 - 2i$ and $z_2 = 1 + i$.

Compute. Write the answer in standard form.

(a) $\dfrac{z_1}{2} + z_2$
(b) $5z_1 - 9z_2$
(c) $z_1 z_2$
(d) $i z_2$

* Exercise 3

Let

$z_1 = 2 - 3i$, $z_2 = -11 + 7i$,

and $z_3 = 5 - i$. What is the standard form of each of the following?

(a) $z_1(2z_2 - 3z_3)$
(b) $z_1(z_2 z_3)$
(c) $(z_1 z_2) z_3$
(d) $(5z_2 + z_1) i z_3$
(e) $5i z_2 z_3 + i z_1 z_3$

* Exercise 4

Suppose

$z_1 = \dfrac{1}{2} - 3i$ and $z_2 = -2 + 4i$.

Write standard form of each expression.

(a) $\overline{z_1 + z_2}$
(b) $\overline{z_1} + \overline{z_2}$
(c) $\dfrac{z_1 + \overline{z_1}}{2}$
(d) $\dfrac{z_2 - \overline{z_2}}{2}$
(e) $\overline{z_1 z_2}$
(f) $\overline{z_1}\, \overline{z_2}$

* Exercise 5

Say

$z_1 = 2 - \dfrac{3}{2}i$ and $z_2 = 5 + 3i$.

Determine the standard form of each expression.

(a) z_1^{-1}
(b) $\dfrac{1}{z_2}$
(c) $\dfrac{z_2}{z_1}$
(d) $\dfrac{\overline{3z_1}}{z_2}$

380

* Exercise 6

Calculate $z\bar{z}$ given $z=$

(a) 2

(b) $1-i$

(c) $\dfrac{3+i}{2}$

(d) $-i\sqrt{3}$

(e) $5+2i$

(f) $\sqrt{3}-i$

* Exercise 7

What is the modulus of each complex number?

(a) -3

(b) $11i$

(c) $1+i$

(d) $20-15i$

(e) $-\sqrt{3}+i$

(f) $\dfrac{5}{13}-\dfrac{12}{13}i$

(g) $6+3i$

(h) $\dfrac{1}{2}-i\sqrt{11}$

* Exercise 8

Assume

$z_1 = 2.4-i$ and $z_2 = -1-2i$.

Write the expressions in standard form.

(a) $|z_1|$

(b) $|z_2|$

(c) $\left|-\dfrac{5z_1}{2}\right|$

(d) $|z_1 z_2|$

(e) $\left|\dfrac{z_1}{z_2}\right|$

(f) $\left|\dfrac{z_1}{2z_1 - z_2}\right|$

** Exercise 9

Find
$$(1-3i)^n$$
for each n.

(a) $n=-1$

(b) $n=0$

(c) $n=2$

(d) $n=3$

** Exercise 10

Compute
$$(2+i)^n$$
for each n.

(a) $n=-2$

(b) $n=-1$

(c) $n=0$

(d) $n=2$

** Exercise 11

Simplify.

(a) i^{423}

(b) $\dfrac{2}{i^{107}}$

(c) $i^{52} i^{74}$

(d) $\dfrac{17 i^{893}}{i^{977}}$

** Exercise 12

Write in standard form.

(a) $-5 + i^{33} + \dfrac{7i^{34}}{i^{15}}$

(b) $i^{22}\left(2 + i^{88} i^{13} + \dfrac{55}{i^{567}}\right)$

(c) $\dfrac{i^{135} - 7i^{134}}{i^{23} - 12}$

** Exercise 13

Write the sum as a complex number in standard form.

(a) $1 + i + i^2 + i^3 + i^4 + i^5$

(b) $2i + 2i^2 + \ldots + 2i^{10} + 2i^{11}$

(c) $-5i^{15} - 5i^{16} - \ldots - 5i^{52}$

* Exercise 14

Find the real and imaginary parts of each complex number.

(a) $2 - 3i$ (c) -7

(b) $4i$ (d) $\dfrac{1 + 2i}{3}$

* Exercise 15

Graph on the complex plane.

(a) $-3 + i$ (d) $-2i$

(b) $1 + i$ (e) 0

(c) 5 (f) $-\dfrac{3 + 5i}{2}$

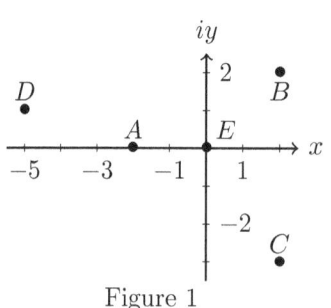

Figure 1

* Exercise 16

Consider Figure 1. Find the standard form of the complex numbers represented by the points in the complex plane.

* Exercise 17

Compute the distance between each complex number and the origin on the complex plane.

(a) $3 + 4i$ (d) $1 - i$

(b) $-12 + 5i$ (e) $7i$

(c) $-\dfrac{4}{5} + \dfrac{3}{5}i$ (f) $\dfrac{1 - 2i}{2}$

** Exercise 18

Use the definition of complex number addition, complex number multiplication, and the corresponding properties for real numbers to prove Proposition 12.1.

** Exercise 19

Prove Proposition 12.4 (iii).

** Exercise 20

For complex numbers z_1 and $z_2 \neq 0$, prove

$$\left| \frac{z_1}{z_2} \right| = \frac{|z_1|}{|z_2|}.$$

** Exercise 21

Find the polar form of each complex number given that the argument θ is such that $0 \leq \theta < 2\pi$.

(a) $3 - 3i\sqrt{3}$

(b) $-\dfrac{\sqrt{15}}{7} + \dfrac{i\sqrt{5}}{7}$

(c) $-i$

(d) $5 + 5i$

(e) $-\sqrt{3} - i\sqrt{3}$

(f) -2

(g) $42i$

** Exercise 22

Calculate the polar form of each complex number given that the argument θ is such that $-\pi < \theta \leq \pi$.

(a) $-i\sqrt{15}$

(b) $-\sqrt{2} + i\sqrt{2}$

(c) $-\dfrac{\sqrt{3}}{2} - \dfrac{i}{2}$

(d) $\dfrac{3}{2} - \dfrac{i\sqrt{3}}{2}$

(e) $\dfrac{\pi}{4} + \dfrac{\pi i\sqrt{3}}{4}$

(f) -7

(g) $1 + i$

** Exercise 23

Write each complex number in standard form.

(a) $\pi e^{-5\pi i/3}$

(b) $4e^{11\pi i/6}$

(c) $\sqrt{6} e^{-\pi i/3}$

(d) $3e^{11\pi i/4}$

(e) $\sqrt{2} e^{-2\pi i/3}$

(f) $e^{2\pi i}$

(g) $\dfrac{\pi\sqrt{2}}{3} e^{\pi i/4}$

(h) $12 e^{5\pi i/6}$

(i) $3 e^{5\pi i/2}$

(j) $\sqrt{5} e^{-7\pi i/6}$

** Exercise 24

Suppose

$$z_1 = 3e^{7\pi i/6} \text{ and } z_2 = \dfrac{1}{2} e^{-5\pi i/3}.$$

What is the standard form of each expression?

(a) $z_1 z_2$

(b) $\dfrac{z_1}{z_2}$

(c) z_1^3

(d) z_2^{-5}

** Exercise 25

Repeat Exercise 24, but assume

$$z_1 = \dfrac{3}{2} e^{\pi i/3} \text{ and } z_2 = \sqrt{2} e^{-\pi i/6}.$$

** Exercise 26

Let $z = 1 + i$. Use polar form to compute the following.

Write the final answer in standard form.

(a) z^5
(b) z^{-6}
(c) z^{-12}
(d) z^7

** Exercise 27

Suppose $z = 1 - i\sqrt{3}$. Use polar form to find the standard form of each expression.

(a) z^4
(b) z^{-5}
(c) z^{-7}
(d) z^6

*** Exercise 28

Use Euler's Formula to prove

$$\cos(\alpha-\beta)=\cos\alpha\cos\beta+\sin\alpha\sin\beta$$

and

$$\sin(\alpha-\beta)=\sin\alpha\cos\beta-\cos\alpha\sin\beta.$$

*** Exercise 29

Utilize Euler's Formula to prove

$$\cos(2\theta) = \cos^2\theta - \sin^2\theta$$

and

$$\sin(2\theta) = 2\sin\theta\cos\theta.$$

*** Exercise 30

Use Euler's Formula to prove

$$\cos(-\theta) = \cos\theta$$

and

$$\sin(-\theta) = -\sin\theta.$$

** Exercise 31

Find the standard form solutions.

(a) $z^2 = i$
(b) $z^3 = 125$
(c) $z^4 + 2 = 1$
(d) $z^6 = 64$

*** Exercise 32

Find all solutions.

(a) $z^6 = -1$
(b) $z^3 = -\dfrac{\sqrt{2}}{2} + \dfrac{i\sqrt{2}}{2}$
(c) $z^4 = 1 - i\sqrt{3}$
(d) $z^6 = -64i$

** Exercise 33

What is the standard form of (a) the second, (b) the third, and (c) the fourth roots of unity?

** Exercise 34

Suppose $z = e^{2\pi i/9}$. Simplify.

$$z^9 + z^8 + \ldots + z + 1$$

*** Exercise 35

Assume $z = e^{\pi i/3}$. Write in standard form.

$$z^5 + 2z^4 + z^3 + 2z^2 + z + 2$$

Chapter 13

Polar Coordinates and Equations

In this chapter, we will study an alternative means to describe points on a plane. Previously, we studied the xy-coordinate system (also called the rectangular coordinate system) and the complex plane. Now we will study the polar coordinate system. Much likes how the rectangular coordinate system is a good way to describe geometric objects like lines, the polar coordinate system is an excellent means to describe other types of geometric objects, like circles.

A solid understanding of Chapter 5 is necessary. Polar coordinates use many of the same ideas as vectors and complex numbers. As a result, knowledge of Chapters 11 and 12 is helpful but not essential. Theorem 7.1 will be needed for some exercises. This chapter requires a scientific calculator.

13.1 Polar Coordinates

Definition 13.1

- The **pole** is the central point of the polar coordinate system.

- The **polar axis** is the horizontal ray with endpoint the pole. By convention, the polar axis points to the right.

Polar coordinates (r, θ) reference a point via a signed distance r and an angle measure θ. The value r is the signed distance between the point and the pole O. The value θ is the directed angle whose initial side is the polar axis and whose terminal side contains the point.

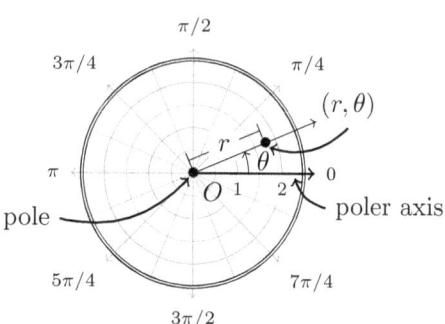

Example 13.1 Graph the polar coordinates.

(a) $(2, 180°)$ (c) $(0.5, 60°)$ (e) $(9/4, 315°)$
(b) $(2.5, 90°)$ (d) $(3/2, 225°)$ (f) $(\pi/2, 45°)$

Solution

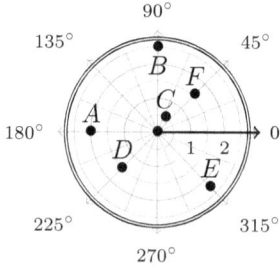

Let us examine how the signs of θ and r affect points' locations. To illustrate this, consider point P which has polar coordinates $(1, 0)$, i.e. it lies on the polar axis and is one unit away from the pole O.

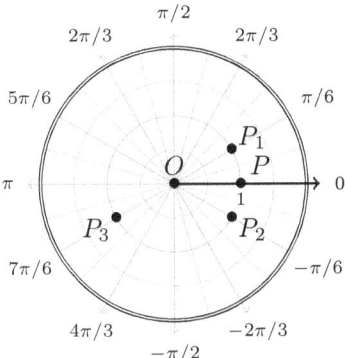

Consider P_1. It has polar coordinates $(1, \pi/6)$. We obtain P_1 by rotating initial side $\overline{OP}$ to terminal side $\overline{OP_1}$. This rotation corresponds to a directed angle of measure $\pi/6$ radians.

Consider P_2 which has polar coordinates $(1, -\pi/6)$. We obtain P_2 via rotating initial side $\overline{OP}$ to terminal side $\overline{OP_2}$. This rotation corresponds to a directed angle of measure $-\pi/6$ radians.

Consider P_3, which has polar coordinates $(-1, \pi/6)$. To obtain P_3 rotate initial side $\overline{OP}$ to terminal side $\overline{OP_1}$. We then reflect P_1 about the pole. The result is P_3. Notice that another representation of P_3 is the polar coordinates $(1, 7\pi/6)$.

Proposition 13.1 *For all polar coordinates*

$$(-r, \theta) = (r, \theta + \pi),$$

or, more generally,

$$(-r, \theta) = (r, \theta + \pi k)$$

for $k = 1, -1, 3, -3, \ldots$.

Example 13.2 Consider the polar coordinate $(2, \pi/4)$. Find another representation such that ...

(a) ... the radius and angle measure are positive.

(b) ... the radius is positive and the angle measure is negative.

(d) ... the radius is negative and the angle measure is positive.

(c) ... the radius and angle measure are negative.

Solution

(a) Since an additional counterclockwise revolution places the point at the same location, the point can also be written as
$$\left(2, \frac{\pi}{4} + 2\pi\right) = \left(2, \frac{9\pi}{4}\right).$$

(b) A complete clockwise rotation puts the point at the identical position as well. Hence, the point also has polar coordinates
$$\left(2, \frac{\pi}{4} - 2\pi\right) = \left(2, -\frac{7\pi}{4}\right).$$

(c) Proposition 13.1 tells us that adding π to the angle measure negates the radius. Ergo, a representation with a negative radius and positive angle measure is
$$\left(-2, \frac{\pi}{4} + \pi\right) = \left(-2, \frac{5\pi}{4}\right).$$

(d) We will use our work from (c). Since a complete clockwise rotation places the point at the same spot, the point is described by polar coordinates
$$\left(-2, \frac{5\pi}{4} - 2\pi\right) = \left(-2, -\frac{3\pi}{4}\right).$$
∎

We will now develop a means to convert between the polar coordinate system and the xy-coordinate systems. Assume:

The pole has the same location as the origin, the polar axis contains the positive x-axis, and the ray defined by $\theta = \pi/2$ and $r \geq 0$ contains the positive y-axis.

Proposition 13.2 *Suppose the rectangular coordinates (x, y) and polar coordinates (r, θ) describe the same point. Then*
$$x = r\cos\theta, \quad y = r\sin\theta, \quad \tan\theta = \frac{y}{x}, \quad \text{and} \quad x^2 + y^2 = r^2.$$

These conversions are fairly clear in quadrant I. Proposition 13.2 follows generally due to Theorem 5.1 and a bit of tinkering with the signs. We will skip the proof because it is tedious and adds very little insight.

Example 13.3 Convert the polar coordinates to xy-coordinates.

(a) $\left(\dfrac{3}{2}, \dfrac{\pi}{3}\right)$, (b) $\left(-2, \dfrac{5\pi}{6}\right)$, and (c) $\left(1, \dfrac{3\pi}{2}\right)$

Solution

(a) Since $r = 3/2$ and $\theta = \pi/3$,

$$x = \frac{3}{2}\cos\frac{\pi}{3} = \frac{3}{4} \quad \text{and} \quad y = \frac{3}{2}\sin\frac{\pi}{3} = \frac{3\sqrt{3}}{4}.$$

Therefore, the xy-coordinate form is $(3/4, 3\sqrt{3}/4)$.

(b) Because $r = -2$ and $\theta = 5\pi/6$,

$$x = -2\cos\frac{5\pi}{6} = \sqrt{3} \quad \text{and} \quad y = -2\sin\frac{5\pi}{6} = -1.$$

Thus, the xy-coordinates are $(\sqrt{3}, -1)$.

(c) Due to the fact that $r = 1$ and $\theta = 3\pi/2$,

$$x = 1\cos\frac{3\pi}{2} = 0 \quad \text{and} \quad y = 1\sin\frac{3\pi}{2} = -1.$$

Ergo, the xy-coordinate form of our point is $(0, -1)$. ∎

Example 13.4 Convert the xy-coordinates to polar coordinates.

(a) $(3, 0)$
(b) $(2, 2)$
(c) $(0, -1)$
(d) $\left(-2, -2\sqrt{3}\right)$
(e) $\left(\sqrt{3}, -1\right)$
(f) $(-3, 4)$

Solution Let us graph these points, so that we know where they lie. This will help us find θ later.

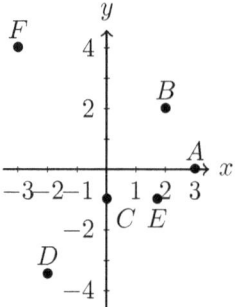

We will assume r is positive.

(a) Using Proposition 13.2,
$$r^2 = 3^2 + 0^2 \quad \text{implies} \quad r = 3.$$
By inspection of the graph, we conclude $\theta = 0$. Hence, a polar representation of our point is $(3, 0)$.

(b) Due to Proposition 13.2,
$$r^2 = 2^2 + 2^2 \quad \text{implies} \quad r = 2\sqrt{2}.$$
Since $(2, 2)$ is in quadrant I,
$$\tan \theta = \frac{2}{2} = 1 \quad \text{implies} \quad \theta = \arctan 1 = \frac{\pi}{4}.$$
Thus, a polar representation is $(2\sqrt{2}, \pi/4)$.

(c) Proposition 13.2 tells us
$$r^2 = 0^2 + (-1)^2 \quad \text{implies} \quad r = 1.$$
Since $(0, -1)$ is on the negative y-axis, $\theta = 3\pi/2$. Hence, $(1, 3\pi/2)$ is a polar coordinate representation of our point.

(d) Because of Proposition 13.2,
$$r^2 = (-2)^2 + (-2\sqrt{3})^2 \quad \text{implies} \quad r = 4.$$

We have
$$\tan\theta = \frac{-2\sqrt{3}}{-2} = \sqrt{3}.$$
This implies that the reference angle is $\pi/3$. Since θ is in quadrant III, it follows that
$$\theta = \frac{\pi}{3} + \pi = \frac{4\pi}{3}.$$
We conclude that a solution is $(4, 4\pi/3)$.

(e) We utilize Proposition 13.2 to conclude
$$r^2 = (\sqrt{3})^2 + (-1)^2 \quad \text{implies} \quad r = 2.$$
Since
$$\tan\theta = -\frac{1}{\sqrt{3}} = -\frac{\sqrt{3}}{3},$$
the reference angle is $\pi/6$. Because θ is in quadrant IV,
$$\theta = 2\pi - \frac{\pi}{6} = \frac{11\pi}{6}.$$
Ergo, a solution is $(2, 11\pi/6)$.

(f) Due to Proposition 13.2,
$$r^2 = (-3)^2 + 4^2 \quad \text{implies} \quad r = 5.$$
We know $\tan\theta = -4/3$. This implies the reference angle is $\arctan(4/3) \approx 0.927$ radians. Since we are in quadrant II, we conclude
$$\theta = \pi - \arctan\frac{4}{3} \approx 2.214 \text{ rad}.$$
A polar representation of our point is
$$\left(5, \pi - \arctan\frac{4}{3}\right) \approx (5, 2.214).$$

■

Since the polar representation of a point is not unique, there are other correct answers. We chose to have $r > 0$ and $0 \leq \theta < 2\pi$. However, this was simply our preference. There are many other options. For example, we could let $\theta = \arctan(y/x)$ and place the point in the correct quadrant via changing the sign of r.

13.2 Basic Polar Graphs

Proposition 13.3

- *The graph of $\theta = a$ is a line through the pole with slope $\tan a$ for a any degree or radian measure.*

- *The graph of $r = b$, where b is a real number, is a circle with center at the pole and radius $|b|$.*

Example 13.5 Graph $\theta = 60°$. What is its slope?

Solution The graph includes all of the points of the form $(r, 60°)$, where r is a real number. This describes a line. Its slope is $\tan 60° = \sqrt{3}$.

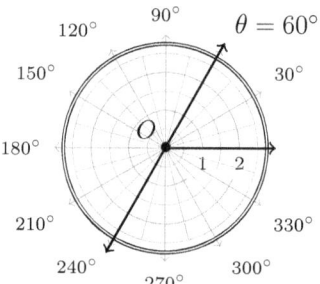

■

Example 13.6 Graph $r = 3/2$. What is its radius?

Solution The graph contains all the points of the form $(3/2, \theta)$, where θ is any radian measure. So, it is a circle. Its radius is $3/2$.

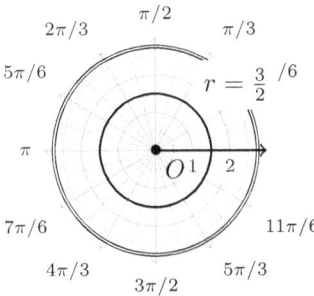

Some graphs require a fair amount of point plotting.

Example 13.7 Graph $r = \theta/4$, where $0 \leq \theta < 2\pi$.

Solution Our first task is to construct a table of values. We will start at $\theta = 0$ and go in increments of $\pi/2$. The point corresponding to $\theta = 2\pi$ will be in our table, though this value is not included in the graph; we will draw an open circle there.

θ	0	$\dfrac{\pi}{2}$	π	$\dfrac{3\pi}{2}$	2π
r	0	$\dfrac{\pi}{8}$	$\dfrac{\pi}{4}$	$\dfrac{3\pi}{8}$	$\dfrac{\pi}{2}$

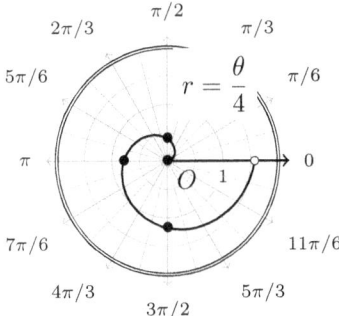

395

Example 13.8 Graph $r = \dfrac{3}{2} - \sin\theta$.

Solution We will start our table at 0 and go in increments of a fourth the period. That is, our increments will be $360°/4 = 90°$.

θ	0	90°	180°	270°	360°
r	$\dfrac{3}{2}$	$\dfrac{1}{2}$	$\dfrac{3}{2}$	$\dfrac{5}{2}$	$\dfrac{3}{2}$

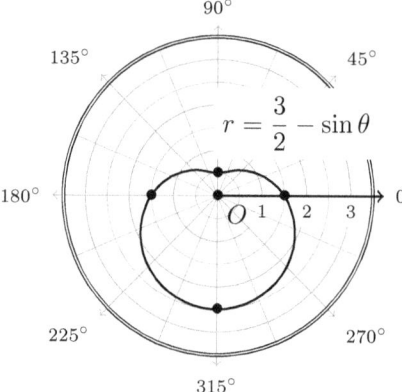

Proposition 13.4

- The graph of $r = a\cos\theta$ is a circle of radius $|a|/2$ and center $(a/2, 0)$.

- The graph of $r = a\sin\theta$ is a circle of radius $|a|/2$ and center $(0, a/2)$.

Proof We will prove the graph of $r = a\cos\theta$ is a circle, and leave $r = a\sin\theta$ as an exercise. Recall that the equation of a circle with radius $|a|/2$ and center $(a/2, 0)$ is

$$\left(x - \frac{a}{2}\right)^2 + y^2 = \frac{a^2}{4}.$$

Our task is to prove $r = a\cos\theta$ converts to the equation above. Using Proposition 13.2, we have

$$r = a\cos\theta$$

$$\Rightarrow \qquad r^2 = ar\cos\theta$$

$$\Rightarrow \qquad x^2 + y^2 = ax$$

$$\Rightarrow \qquad x^2 - ax + y^2 = 0$$

$$\Rightarrow \qquad x^2 - ax + \frac{a^2}{4} + y^2 = \frac{a^2}{4}$$

$$\Rightarrow \qquad \left(x - \frac{a}{2}\right)^2 + y^2 = \frac{a^2}{4}.$$

∎

Example 13.9 Graph $r = 2\cos\theta$.

Solution This is a circle of radius 1 and center $(1, 0)$.

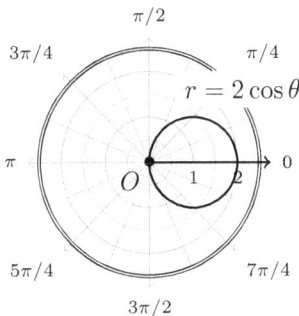

∎

397

Proposition 13.5

- The polar equation $r = a\sec\theta$ has a corresponding rectangular equation $x = a$.
- The polar equation $r = a\csc\theta$ has a corresponding rectangular equation $y = a$.

Proof We will prove $r = a\sec\theta$ is equivalent to $x = a$, and leave the rest of the proposition as an exercise.

$$\begin{aligned} r &= a\sec\theta \\ \Rightarrow \quad r &= \frac{a}{\cos\theta} \\ \Rightarrow \quad r\cos\theta &= a \\ \Rightarrow \quad x &= a \end{aligned}$$

■

Example 13.10 Graph $r = \sec\theta$ and $r = -2\csc\theta$.

Solution Due to Proposition 13.5, the graph of $r = \sec\theta$ is equivalent to $x = 1$ and the graph of $r = -2\csc\theta$ is equivalent to $y = -2$.

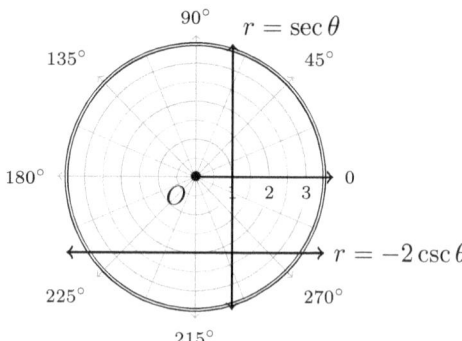

■

In our last two examples, we will convert between equations in the xy-coordinate and polar coordinate systems. Many of the ideas have already been used in proofs, but an ability to convert between the two systems is important, so we will provide an ample amount of material to help students learn.

Example 13.11 Convert xy-coordinate system equation
$$y = x^2$$
into a polar equation.

Solution Since
$$x = r\cos\theta \quad \text{and} \quad y = r\sin\theta$$
we have
$$y = x^2$$
$$\Rightarrow \quad r\sin\theta = (r\cos\theta)^2$$
$$\Rightarrow \quad r\sin\theta = r^2\cos^2\theta$$
$$\Rightarrow \quad r = \frac{\sin\theta}{\cos^2\theta}$$
$$= \frac{1}{\cos\theta} \cdot \frac{\sin\theta}{\cos\theta}$$
$$= \sec\theta\tan\theta.$$
So, the corresponding polar equation is
$$r = \sec\theta\tan\theta.$$
∎

Example 13.12 Convert
$$r^2 = 4\sin\theta$$
into an equation in the xy-coordinate system.

Solution Since
$$r^2 = x^2 + y^2 \quad \text{and} \quad \sin\theta = \frac{y}{\sqrt{x^2 + y^2}},$$

399

it follows that

$$r^2 = 4\sin\theta$$

$$\Rightarrow \quad x^2 + y^2 = \frac{4y}{\sqrt{x^2+y^2}}$$

$$\Rightarrow \quad (x^2+y^2)\sqrt{x^2+y^2} = 4y$$

$$\Rightarrow \quad (x^2+y^2)^{3/2} = 4y.$$

∎

13.3 Intermediate Polar Graphs

Proposition 13.6 (Symmetry Tests) *Consider an equation in polar form.*

- *Its graph is symmetric about the line $\theta = \pi/2$ if replacing (r,θ) by $(-r,-\theta)$ or $(r,\pi-\theta)$ produces an equivalent equation.*

- *Its graph is symmetric about the polar axis if replacing (r,θ) by $(r,-\theta)$ produces an equivalent equation.*

- *Its graph is symmetric about the pole if replacing (r,θ) by $(-r,\theta)$ or $(r,\pi+\theta)$ produces an equivalent equation.*

Example 13.13 Determine the symmetry of each equation's graph.

(a) $r = \cos 2\theta$

(b) $r = \dfrac{1}{2} + \sin\theta$

(c) $r^2 = \sin\theta$

(d) $r = \sin 3\theta$

Solution

(a) If we replace (r,θ) with $(r,-\theta)$, then

$$r = \cos(-2\theta) = \cos 2\theta,$$

because cosine is even. That means (a) is symmetric about the polar axis.

(b) We will replace (r, θ) with $(r, \pi - \theta)$, and use Theorem 7.1 to prove the resulting expression is equivalent to the original. This will prove (b) is symmetric about the line $\theta = \pi/2$:

$$r = \frac{1}{2} + \sin(\pi - \theta)$$
$$= \frac{1}{2} + \sin\pi\cos\theta - \sin\theta\cos\pi$$
$$= \frac{1}{2} + \sin\theta.$$

(c) Replacing (r, θ) with $(-r, \theta)$ yields

$$(-r)^2 = \sin\theta \quad \text{implies} \quad r^2 = \sin\theta.$$

Hence, (c) is symmetric about the pole.

(d) Let us replace (r, θ) with $(-r, -\theta)$. Since sine is odd,

$$-r = \sin(-3\theta)$$
$$\Rightarrow \quad -r = -\sin 3\theta$$
$$\Rightarrow \quad r = \sin 3\theta.$$

We conclude that (d) is symmetric about the line $\theta = \pi/2$. ■

Example 13.14 Graph $r = \cos 2\theta$.

Solution Example 13 (a) proved the graph of $r = \cos 2\theta$ is symmetric about the polar axis, so we only to plot points between 0 and π. The period of $\cos 2\theta$ is π, so let us go in increments of $\pi/4$.

θ	0	$\frac{\pi}{4}$	$\frac{\pi}{2}$	$\frac{3\pi}{4}$	π
r	1	0	-1	0	1

As of now, we have the following graph.

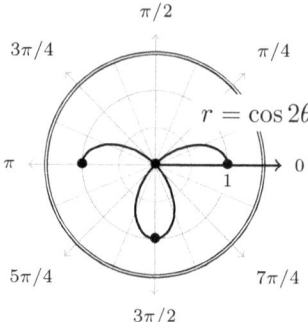

Symmetry about the polar axis allows us to complete the graph without plotting any more points.

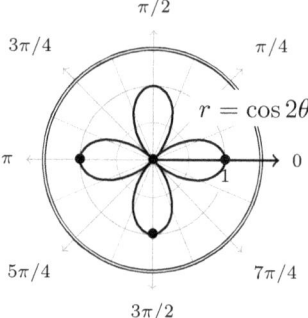

Example 13.15 Graph $r = 1/2 + \sin\theta$.

Solution Because replacing (r, θ) with $(r, \pi-\theta)$ gives an equivalent equation, the graph is symmetric about $\theta = \pi/2$. So, we only need to graph the equation from $-\pi/2$ to $\pi/2$.

Negative values of r make the graph more challenging. As a result, checking the zeros of the equation is helpful. We have

$$\frac{1}{2} + \sin\theta = 0 \quad \text{implies} \quad \sin\theta = -\frac{1}{2}.$$

Since we are considering values of θ in the interval $[-\pi/2, \pi/2]$, the only zero of interest is $\theta = -\pi/6$.

Let us plot some points. We will start at $-\pi/2$. Normally, we go in increments of a fourth the period, which would be $\pi/2$ for $r = 1/2 + \sin\theta$. However, we will go in increments of $\pi/4$ to be safe.

θ	$-\dfrac{\pi}{2}$	$-\dfrac{\pi}{4}$	0	$\dfrac{\pi}{4}$	$\dfrac{\pi}{2}$
r	$-\dfrac{1}{2}$	$\dfrac{1-\sqrt{2}}{2} \approx -0.207$	$\dfrac{1}{2}$	$\dfrac{1+\sqrt{2}}{2} \approx 1.207$	$\dfrac{3}{2}$

Thus far we have the following graph.

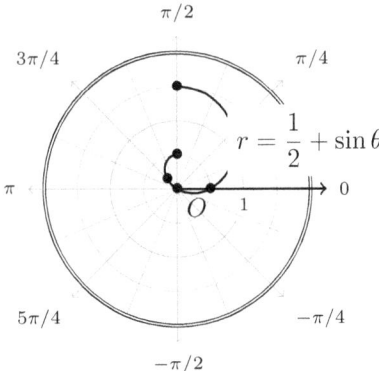

Symmetry about the line $\theta = \pi/2$ allows us to quickly sketch the rest of the graph.

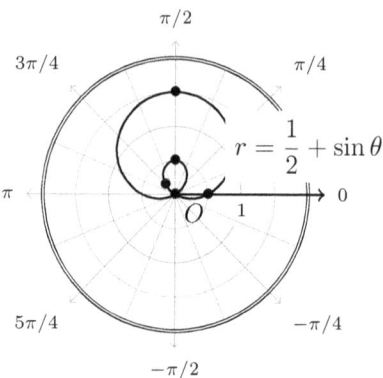

Example 13.16 Graph $r^2 = \sin\theta$.

Solution Example 13 (c) proved the graph of $r^2 = \sin\theta$ is symmetric about the pole. Knowing this, we will only consider values of θ such that $0 \leq \theta \leq \pi$.

It is clear that the only zeros of $\sin\theta$ within our interval occur at $\theta = 0$ and π.

Let us plot some points. We will go in increments of $\pi/2$ and start at 0.

θ	0	$\dfrac{\pi}{2}$	π
r	0	± 1	0

Plotting points and connecting the dots gives us the following.

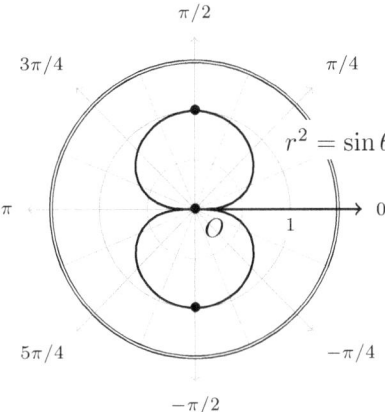

This graph is already symmetric about the pole, so we are done. ∎

Example 13.17 Graph $r = \sin 3\theta$.

Solution Because replacing (r, θ) with $(-r, -\theta)$ produces an equivalent equation, the graph is symmetric about the line $\theta = \pi/2$. As a result, we will suppose $-\pi/2 \leq \theta \leq \pi/2$. Symmetry will allow us to complete the graph.

Let us check the zeros. If
$$\sin 3\theta = 0 \quad \text{and} \quad -\frac{\pi}{2} \leq \theta \leq \frac{\pi}{2},$$
then
$$\theta = -\frac{\pi}{3}, \ 0, \ \text{or} \ \frac{\pi}{3}.$$

We will start at $-\pi/2$ and go in increments of a fourth of the period. Since the period is $2\pi/3$ our increment will be
$$\frac{2\pi/3}{4} = \frac{\pi}{6}.$$

θ	$-\dfrac{\pi}{2}$	$-\dfrac{\pi}{3}$	$-\dfrac{\pi}{6}$	0	$\dfrac{\pi}{6}$	$\dfrac{\pi}{3}$	$\dfrac{\pi}{2}$
r	1	0	-1	0	1	0	-1

This gives us the graph below, which is already symmetric about $\theta = \pi/2$ so we are done.

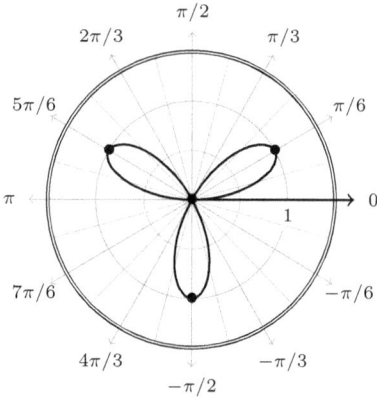

■

The graphs in Examples 14 and 17 are called "rose curves" of four and three "petals", respectively. There are rules that can be used to graph rose curves easily.

Proposition 13.7 *The graph of*

$$r = a\cos n\theta \quad \text{and} \quad r = a\sin n\theta$$

are rose curves, when n is an integer greater than or equal to 2. If n is even, then there are $2n$ petals. If n is odd, then there are n petals. Each petal has length $|a|$.

Example 13.18 Graph $r = \cos 4\theta$.

Solution This is a rose curve with eight petals. The first petal occurs between the positive and negative zeros of r closest to $\theta = 0$.

These zeros are $\theta = -12.5°$ and $\theta = 12.5°$. We can find subsequent petals using similar reasoning or via the use of symmetry. For example, the next petal occurs in the interval $[12.5°, 57.5°]$.

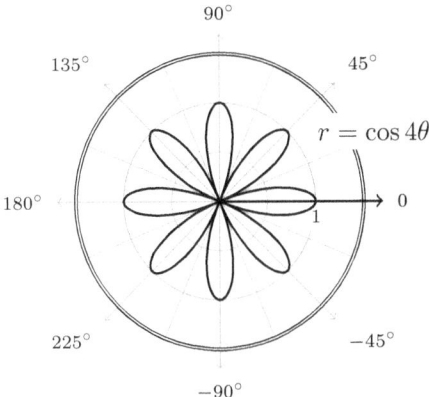

13.4 Classification of Polar Graphs

Our last section addresses the classification of polar graphs. We have already covered circles, lines, and rose curves, but they are included here for completeness.

Lines The graph of the polar equations $\theta = a$ is a line for a any degree or radian measure. The slope of the line is $\tan a$.

$$\theta = a$$

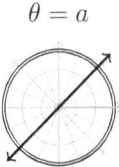

The equations of vertical and horizontal lines in rectangular coordinates are $x = a$ and $y = a$, respectively. The corresponding polar equations are $r = a\sec\theta$ and $r = a\csc\theta$, respectively.

Circles The graph of the polar equation $r = a$ is a circle with center at the pole and radius $|a|$.

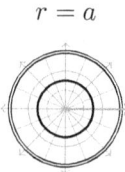

A circle of radius $|a|/2$ centered at $(a/2, 0)$ has polar equation

$$r = a\cos\theta.$$

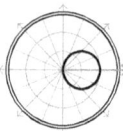

A circle of radius $|a|/2$ centered at $(0, a/2)$ has polar equation

$$r = a\sin\theta.$$

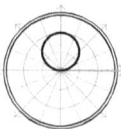

Limaçons A limaçon graph is obtain via the polar equation

$$r = a + b\cos\theta \quad \text{or} \quad r = a + b\sin\theta.$$

Our graphs suppose $r = a + b\cos\theta$, $a > 0$, and $b > 0$.
An inner loop limaçon occurs when $|a/b| < 1$.

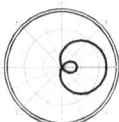

A "cardioid" (heart shape) occurs when $|a/b| = 1$.

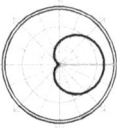

A dimpled limaçon occurs when $1 < |a/b| < 2$.

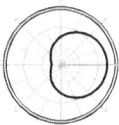

A convex limaçon occurs when $|a/b| \geq 2$.

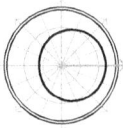

Rose curve The polar equations which produce rose curves are

$$r = a\cos n\theta \quad \text{and} \quad r = a\sin n\theta,$$

where n is an integer greater than or equal to 2.

If n is even, the rose curve has $2n$ petals of length $|a|$. If n is odd, the rose curve has n petals of length $|a|$. We suppose $a > 0$ in our graphs.

$r = \cos 2\theta$ $\qquad$ $r = \cos 3\theta$

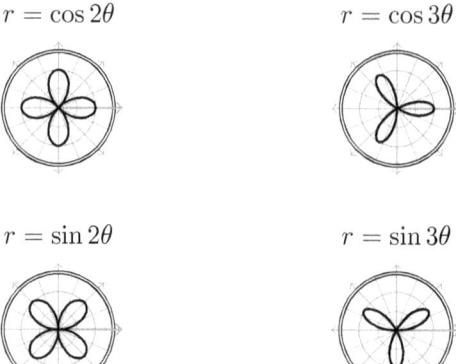

$r = \sin 2\theta$ $\qquad$ $r = \sin 3\theta$

Lemniscates A polar equation which produces a graph of a lemniscate is of the form

$$r^2 = a^2 \cos 2\theta \quad \text{or} \quad r^2 = a^2 \sin 2\theta,$$

where $a \neq 0$. The length of the entire lemniscate in either case is $2|a|$.

$r^2 = a^2 \cos 2\theta$ $\qquad$ $r^2 = a^2 \sin 2\theta$

13.5 Exercises

* Exercise 1

Graph the polar coordinates.

(a) $\left(1, \dfrac{\pi}{6}\right)$ (e) $\left(\pi, \dfrac{3\pi}{2}\right)$

(b) $\left(\dfrac{1}{3}, 300°\right)$ (f) $(4, 180°)$

(c) $\left(\dfrac{3}{4}, \dfrac{\pi}{4}\right)$ (g) $\left(0.2, \dfrac{\pi}{2}\right)$

(d) $(2, 135°)$ (h) $(3, 210°)$

** Exercise 2

Graph the polar coordinates.

(a) $(-2, -315°)$ (e) $(\sqrt{2}, -210°)$

(b) $\left(-1.5, \dfrac{3\pi}{4}\right)$ (f) $\left(-\dfrac{3}{4}, \dfrac{2\pi}{3}\right)$

(c) $(5, -360°)$

(d) $\left(-\dfrac{1}{2}, \dfrac{11\pi}{3}\right)$ (g) $(6, -495°)$

(h) $(-\pi, 2.5\pi)$

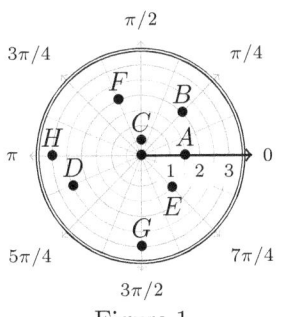

Figure 1

* Exercise 3

Consider Figure 1. Write polar coordinates for each point.

** Exercise 4

Consider each polar coordinate. Find another representation such that ...

(i) ... the radius and angle measure are positive.

(ii) ... the radius is positive and the angle measure is negative.

(iii) ... the radius is negative and the angle measure is positive.

(iv) ... the radius and angle measure are negative.

(a) $(1, 330°)$ (d) $\left(2.5, \dfrac{4\pi}{3}\right)$

(b) $\left(\dfrac{1}{2}, 0\right)$ (e) $(-2, -345°)$

(c) $(5, 135°)$ (f) $\left(-1, -\dfrac{\pi}{2}\right)$

** Exercise 5

Convert the polar coordinates to xy-coordinates.

(a) $(3, 270°)$

(b) $\left(\sqrt{6}, -\dfrac{5\pi}{6}\right)$

(c) $(-2, 90°)$

(d) $\left(7\sqrt{2}, \dfrac{3\pi}{4}\right)$

(e) $(-4, 420°)$

(f) $\left(\dfrac{3}{2}, -\dfrac{13\pi}{4}\right)$

** Exercise 6

Convert the xy-coordinates to polar.

(a) $(-2, 2)$

(b) $\left(-\dfrac{\sqrt{3}}{2}, -\dfrac{3}{2}\right)$

(c) $(0, 4)$

(d) $(3, -\sqrt{3})$

(e) $(-1, 0)$

(f) $(3\sqrt{2}, 3\sqrt{2})$

** Exercise 7

Convert the xy-coordinate system equations to polar equations.

(a) $x^2 + y^2 = 1$

(b) $x^2 + (y+2)^2 = 4$

(c) $y = 5$

(d) $y = x$

(e) $y = -2x^2$

(f) $y = x + 1$

(g) $(x-3)^2 + (y+1)^2 = 10$

(h) $y = \dfrac{1}{x}$

(i) $y^2 - x^2 = 1$

** Exercise 8

Convert the polar equations to equations in the xy-coordinate system.

(a) $r = 2$

(b) $r = 2\cos\theta$

(c) $\theta = \dfrac{5\pi}{6}$

(d) $r^2 = \sin\theta$

(e) $r = \theta$

(f) $r = -3\sec\theta$

(g) $r^2 = \sin 2\theta$

(h) $r = \dfrac{1}{1 - \sin\theta}$

(i) $r^2 = \sec 2\theta$

** Exercise 9

Prove the graph of the polar equation $r = a\csc\theta$ is equivalent to the rectangular equation $y = a$.

** Exercise 10

Prove the graph of polar equation $r = a\sin\theta$ is a circle of radius $|a|/2$ and center $(0, a/2)$.

*** Exercise 11

Prove the distance between the polar coordinates (r_1, θ_1) and (r_2, θ_2) is

$$d = \sqrt{r_1^2 + r_2^2 - 2r_1 r_2 \cos(\theta_1 - \theta_2)}.$$

Hint: Convert the points to their rectangular form and then use the xy-coordinate system distance formula

$$d = \sqrt{(x_2 - x_1)^2 + (y_2 - y_1)^2}.$$

** Exercise 12

Use the result of Exercise 11 to compute the distance between the polar coordinates.

(a) $(8, 30°)$ and $(6, 120°)$

(b) $\left(-15\sqrt{3}, \dfrac{11\pi}{10}\right)$ and $\left(30, \dfrac{4\pi}{15}\right)$

(c) $(4, -110°)$ and $(2, 190°)$

** Exercise 13

Graph.

(a) $\theta = \pi/2$
(b) $r = 2\csc\theta$
(c) $r = -3\sec\theta$
(d) $\theta = -45°$
(e) $r - \sec\theta = 0$
(f) $\theta = \dfrac{5\pi}{3}$

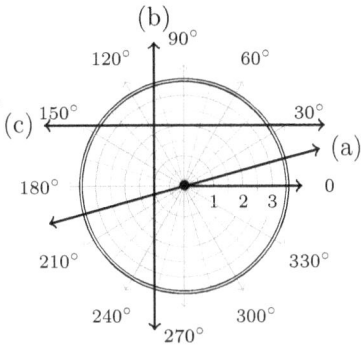

Figure 2

** Exercise 14

Consider Figure 2. Find a polar equations for each line.

** Exercise 15

Graph.

(a) $r = 5$
(b) $r = 2\sin\theta$
(c) $r = -1$
(d) $r = 4\cos\theta$
(e) $r = \sin\theta$
(f) $r = -\dfrac{3}{2}$

** Exercise 16

Graph.

(a) $r = \theta/2$

(b) $r = e^\theta$

(c) $r = 1 - \cos\theta$

(d) $r = 3 - \sin 2\theta$

** Exercise 17

Use Proposition 13.6 to determine the symmetry of the following polar equations.

(a) $r = \cos\theta$

(b) $r^2 = \sin 2\theta$

(c) $r = \cos 2\theta$

(d) $r = 2 + \sin\theta$

(e) $r = 2 + 3\cos\theta$

(f) $r = 10\csc\theta$

(g) $r^2 = \sin 3\theta$

(h) $r = 3$

* Exercise 18

Determine the number of petals contained in each polar equation's rose curve.

(a) $r = 3\cos 10\theta$

(b) $r = -5\sin 7\theta$

(c) $r = -2\cos(-3\theta)$

(d) $r = \pi \sin 22\theta$

** Exercise 19

Graph.

(a) $r = -2\sin 3\theta$

(b) $r = 3\cos 2\theta$

(c) $r = 2\sin 4\theta$

(d) $r = -\cos 3\theta$

** Exercise 20

(i) Classify the graphs on pages 415 and 417 based in the categories in Section 13.4.

(ii) Write corresponding polar equations for each graph.

** Exercise 21

Graph.

(a) $r^2 = \sin 2\theta$
(b) $r = 1 - 3\cos\theta$
(c) $r = 3 - \sin\theta$
(d) $r^2 = \cos 2\theta$
(e) $r = 3 + 2\cos\theta$
(f) $r^2 = 4\cos 2\theta$
(g) $r^2 = 9\sin 2\theta$
(h) $r = 2 + 2\sin\theta$

** Exercise 22

Graph.

(a) $r^2 = 9\cos 2\theta$
(b) $r = -1.5\cos 4\theta$
(c) $r = \dfrac{5}{2} + \cos\theta$
(d) $r^2 = \sin 2\theta$
(e) $r = 2 - 4\sin\theta$
(f) $r = \dfrac{5\sin 3\theta}{2}$
(g) $r = \cos\theta$
(h) $r = 10\csc\theta$

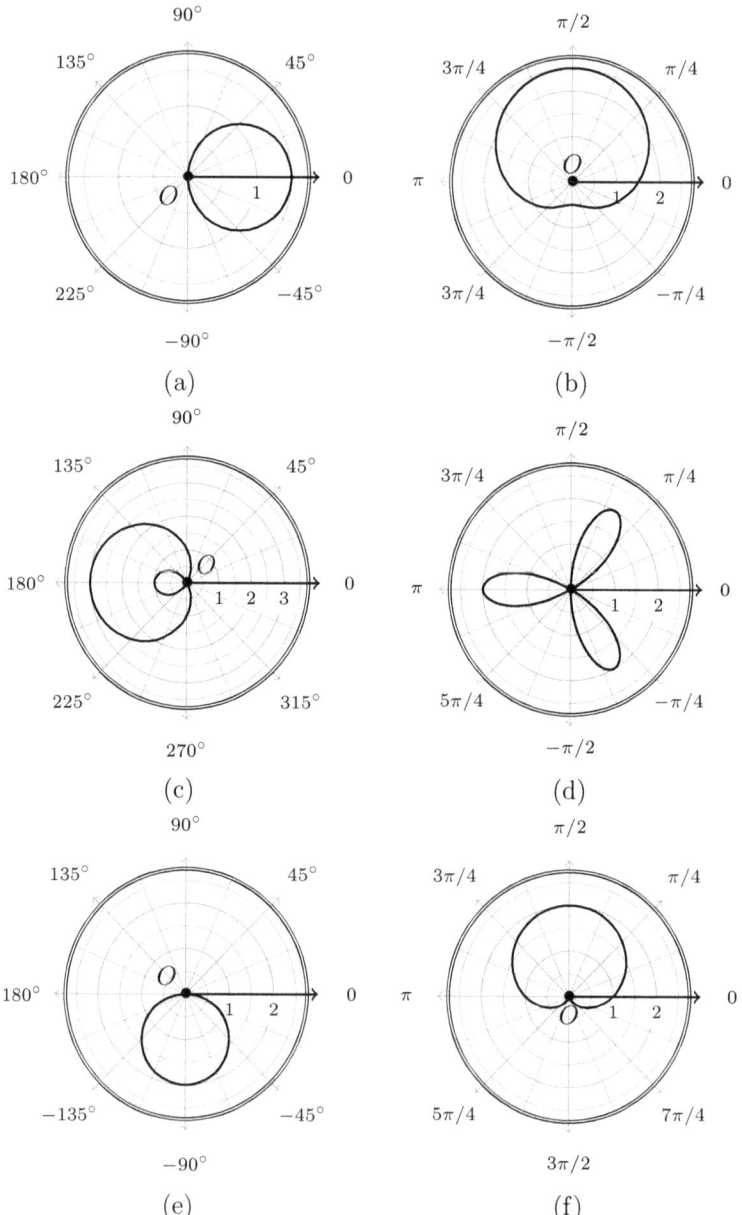

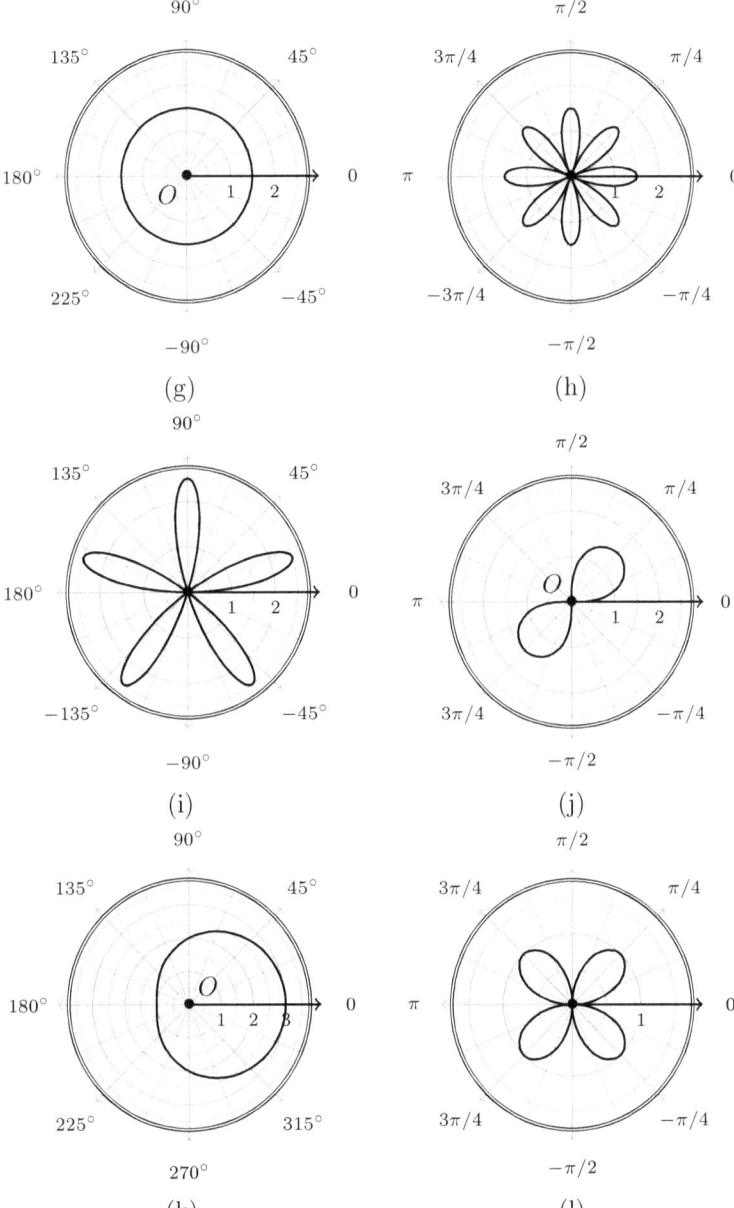

(g)

(h)

(i)

(j)

(k)

(l)

Appendices

Appendix A

Rational Expressions and Equations

Definition A.1 A **rational expression** is any expression of the form
$$\frac{a_m x^m + a_{m-1} x^{m-1} + \ldots + a_1 x + a_0}{b_n x^n + b_{n-1} x^{n-1} + \ldots + b_1 x + b_0},$$
where $a_m, a_{m-1}, \ldots, a_0, b_n, b_{n-1}, \ldots,$ and b_0 are constants.

Definition A.2 Consider an expression of the form
$$\frac{p(x)}{q(x)}.$$
The top function $p(x)$ is the **numerator** and the bottom function $q(x)$ is the **denominator**.

Example A.1 Simplify when possible.

(a) $\dfrac{x^2+x}{2x-x^2}$, (b) $\dfrac{x^2+x}{x+1}$, (c) $\dfrac{x-2}{2-x}$, and (d) $\dfrac{x+1}{2-x}$.

Solution To simplify a rational expression, factor the numerator and denominator. Factors held in common between the top and bottom cancel.

(a) We have
$$\dfrac{x^2+x}{2x-x^2} = \dfrac{x(x+1)}{x(2-x)}.$$
In this case, a factor of x is held in common between the numerator and denominator, so it cancels:
$$\dfrac{\cancel{x}(x+1)}{\cancel{x}(2-x)} = \dfrac{x+1}{2-x}.$$

(b) First, we factor:
$$\dfrac{x^2+x}{x+1} = \dfrac{x(x+1)}{x+1}.$$
The $x+1$ in the numerator and denominator cancel:
$$\dfrac{x\cancel{(x+1)}}{\cancel{x+1}} = \dfrac{x}{1} = x.$$

(c) Sometimes factoring out a negative allows us to cancel:
$$\dfrac{x-2}{2-x} = \dfrac{x-2}{-(-2+x)} = \dfrac{x-2}{-(x-2)}.$$
The $x-2$ in the numerator and denominator cancel:
$$\dfrac{\cancel{x-2}}{-\cancel{(x-2)}} = \dfrac{1}{-1} = -1.$$

(d) The numerator and denominator of
$$\dfrac{x+1}{2-x}$$
have no common factors. Hence, there is no cancelation.

Complex fractions sometimes arise in problems. As a result, it is important to know how to simplify them. The following identity is useful
$$\frac{a/b}{c/d} = \frac{a}{b} \cdot \frac{d}{c} = \frac{ad}{bc}.$$

Example A.2 Simplify
$$\frac{\dfrac{2w^2 - 7w - 15}{w^2 - 1}}{\dfrac{w^2 - 25}{w^2 + 2w + 1}}.$$

Solution The first step is to factor:
$$\frac{\dfrac{2w^2 - 7w - 15}{w^2 - 1}}{\dfrac{w^2 - 25}{w^2 + 2w + 1}} = \frac{\dfrac{(2w+3)(w-5)}{(w-1)(w+1)}}{\dfrac{(w-5)(w+5)}{(w+1)^2}}.$$

We are ready to flip, cancel, and multiply:
$$\frac{\dfrac{(2w+3)(w-5)}{(w-1)(w+1)}}{\dfrac{(w-5)(w+5)}{(w+1)^2}} = \frac{(2w+3)\cancel{(w-5)}}{(w-1)\cancel{(w+1)}} \cdot \frac{(w+1)^{\cancel{2}}}{\cancel{(w-5)}(w+5)}$$
$$= \frac{2w+3}{w-1} \cdot \frac{w+1}{w+5}$$
$$= \frac{(2w+3)(w+1)}{(w-1)(w+5)}.$$

Since any algebraic expression can be written as the ratio of itself and 1, the identity
$$\frac{a/b}{c/d} = \frac{ad}{bc}$$
can be used to simplify ratios of the form
$$\frac{a/b}{c} \quad \text{and} \quad \frac{a}{b/c}.$$

In particular, we have

$$\frac{a/b}{c} = \frac{a/b}{c/1} \quad \text{and} \quad \frac{a}{c/d} = \frac{a/1}{c/d}$$

$$= \frac{a}{b} \cdot \frac{1}{c} \qquad\qquad\qquad = \frac{a}{1} \cdot \frac{d}{c}$$

$$= \frac{a}{bc}, \qquad\qquad\qquad = \frac{ad}{c}.$$

Example A.3 Simplify

$$\frac{1 - \dfrac{2x+3}{x^2+x+1}}{x+1}.$$

Solution We will convert the numerator into a single fraction, and then factor:

$$1 - \frac{2x+3}{x^2+x+1} = \frac{x^2+x+1}{x^2+x+1} - \frac{2x+3}{x^2+x+1}$$
$$= \frac{x^2+x+1-(2x+3)}{x^2+x+1}$$
$$= \frac{x^2-x-2}{x^2+x+1}$$
$$= \frac{(x-2)(x+1)}{x^2+x+1}.$$

We put the denominator over one, flip, cancel, and multiply:

$$\frac{\dfrac{(x-2)(x+1)}{x^2+x+1}}{\dfrac{x+1}{1}} = \frac{(x-2)\cancel{(x+1)}}{x^2+x+1} \cdot \frac{1}{\cancel{x+1}} = \frac{x-2}{x^2+x+1}.$$

∎

Definition A.3 A **rational equation** is an equation which contains one or more rational expressions.

Cross-multiplication is a useful technique for solving rational equations. It says

$$\frac{a}{b} = \frac{c}{d} \quad \text{is equivalent to} \quad ad = bc.$$

Example A.4 Solve

$$\frac{t+6}{t-2} = \frac{t+10}{t-6}.$$

Solution Using cross-multiplication,

$$\frac{t+6}{t-2} = \frac{t+10}{t-6} \quad \text{implies} \quad (t+6)(t-6) = (t-2)(t+10).$$

All that is left is to expand and solve:

$$\begin{aligned} (t+6)(t-6) &= (t-2)(t+10) \\ \Rightarrow \quad t^2 - 36 &= t^2 + 8t - 20 \\ \Rightarrow \quad -16 &= 8t \\ \Rightarrow \quad t &= -2. \end{aligned}$$

■

Appendix B

Radical Expressions

Our goal is to provide a sketch of the prerequisite knowledge of radicals required for the main text. Of particular emphasis are square roots, though radicals of an arbitrary index are studied as well.

B.1 Square Roots

Definition B.1 The **square root** of x, denoted $\sqrt{x}$, is the function defined by the relationship

$$y = \sqrt{x} \quad \text{if} \quad y^2 = x$$

for $x \geq 0$ and $y \geq 0$.

In Chapter 12, we will expand this definition to include values of x less than zero.

Example B.1 Evaluate (a) $\sqrt{64}$ and (b) $\sqrt{0}$ without a calculator.

Solution

(a) We know $8^2 = 64$ and $8 \geq 0$, so we conclude
$$\sqrt{64} = 8.$$

(b) Because $0^2 = 0$ and $0 \geq 0$, we have
$$\sqrt{0} = 0.$$

■

Proposition B.1 *Suppose $a \geq 0$ and $b \geq 0$. Then*
$$\sqrt{ab} = \sqrt{a} \cdot \sqrt{b}.$$
Furthermore, if $b \neq 0$,
$$\sqrt{\frac{a}{b}} = \frac{\sqrt{a}}{\sqrt{b}}.$$

Example B.2 Simplify. (a) $\sqrt{48}$ and (b) $\sqrt{\dfrac{5}{16}}$.

Solution

(a) Notice that $48 = 16 \cdot 3$. Hence,
$$\sqrt{48} = \sqrt{16 \cdot 3}$$
$$= \sqrt{16} \cdot \sqrt{3}$$
$$= 4\sqrt{3}.$$

(b) We have
$$\sqrt{\frac{5}{16}} = \frac{\sqrt{5}}{\sqrt{16}} = \frac{\sqrt{5}}{4}.$$

■

Definition B.2 A number is **rational** if it can be written in the form
$$\frac{a}{b},$$
where $a = 0, 1, -1, 2, -2, \ldots$ and $b = 1, -1, 2, -2, \ldots$. When a real number is not rational, we say that it is **irrational**.

Some square root expressions are not rational numbers. For example,
$$\sqrt{2} = 1.4142135\ldots$$
is an irrational number because it cannot be written as a ratio of two integers.

Expressions with rational denominators are considered to be more simple. As a result, it is important to know how to rationalize denominators.

Example B.3 Simplify (a) $\dfrac{6}{\sqrt{3}}$ and (b) $\dfrac{\sqrt{14}}{\sqrt{21}}$.

Solution

(a) Because $\sqrt{3} \cdot \sqrt{3} = 3$, we can rationalize the denominator by multiplying the top and bottom of that ratio by $\sqrt{3}$:
$$\frac{6}{\sqrt{3}} = \frac{6}{\sqrt{3}} \cdot \frac{\sqrt{3}}{\sqrt{3}} = \frac{6\sqrt{3}}{3} = 2\sqrt{3}.$$

(b) Since $\sqrt{21} \cdot \sqrt{21} = 21$, we will multiply the top and bottom of the ratio by $\sqrt{21}$. So,
$$\frac{\sqrt{14}}{\sqrt{21}} = \frac{\sqrt{14}}{\sqrt{21}} \cdot \frac{\sqrt{21}}{\sqrt{21}}$$
$$= \frac{\sqrt{49 \cdot 6}}{21}$$
$$= \frac{7\sqrt{6}}{21}$$
$$= \frac{\sqrt{6}}{3}.$$

Sometime we need to multiple the top and bottom of a ratio by the "conjugate radical" to simplify. The conjugate radical of

$$\sqrt{a} + \sqrt{b} \quad \text{is} \quad \sqrt{a} - \sqrt{b}$$

and the conjugate radical of

$$\sqrt{a} - \sqrt{b} \quad \text{is} \quad \sqrt{a} + \sqrt{b}.$$

Multiplying the sum or difference of two radicals by the conjugate radical is effective because it utilizes the difference of two squares identity

$$(x - y)(x + y) = x^2 - y^2.$$

In particular, with radicals this identity tells us

$$\left(\sqrt{a} - \sqrt{b}\right)\left(\sqrt{a} + \sqrt{b}\right) = \left(\sqrt{a}\right)^2 - \left(\sqrt{b}\right)^2 = a - b.$$

Example B.4 Simplify.

(a) $\dfrac{7}{\sqrt{5} - \sqrt{2}}$ and (b) $\dfrac{\sqrt{15}}{\sqrt{3} - 2}$.

Solution

(a) The conjugate of $\sqrt{5} - \sqrt{2}$ is $\sqrt{5} + \sqrt{2}$. So,

$$\frac{7}{\sqrt{5} - \sqrt{2}} = \frac{7}{\sqrt{5} - \sqrt{2}} \cdot \frac{\sqrt{5} + \sqrt{2}}{\sqrt{5} + \sqrt{2}}$$
$$= \frac{7\left(\sqrt{5} + \sqrt{2}\right)}{5 - 2}$$
$$= \frac{7\left(\sqrt{5} + \sqrt{2}\right)}{3}.$$

(b) The conjugate of $\sqrt{3}-2$ is $\sqrt{3}+2$. Hence,

$$\frac{\sqrt{15}}{\sqrt{3}-2} = \frac{\sqrt{15}}{\sqrt{3}-2} \cdot \frac{\sqrt{3}+2}{\sqrt{3}+2}$$
$$= \frac{\sqrt{15}\left(\sqrt{3}+2\right)}{3-4}$$
$$= \frac{\sqrt{45}+2\sqrt{15}}{-1}$$
$$= -3\sqrt{5}-2\sqrt{15}.$$

■

Proposition B.2 *Suppose x is a real number. Then*

$$\left(\sqrt{x}\right)^2 = x \quad \text{and} \quad \sqrt{x^2} = |x|$$

whenever the expressions are defined.

Example B.5 Solve $x^2 - 23 = 2$.

Solution The first step is to isolate the term with x on one side:

$$x^2 - 23 = 2 \quad \text{implies} \quad x^2 = 25.$$

To solve for x, take the square root of each side of the equation:

$$x^2 = 25 \quad \text{implies} \quad \sqrt{x^2} = \sqrt{25} = 5.$$

Proposition B.2 says that $\sqrt{x^2} = |x|$, so

$$|x| = 5 \quad \text{implies} \quad x = \pm 5.$$

■

Example B.6 Expand and simplify

$$\left(\sqrt{2}-\sqrt{3}\right)^2.$$

Solution We will use the expansion formula

$$(x-y)^2 = x^2 - 2xy + y^2.$$

Readers unfamiliar with this formula can also use expansion techniques from algebra.

We have

$$\left(\sqrt{2} - \sqrt{3}\right)^2 = \left(\sqrt{2}\right)^2 - 2\left(\sqrt{2}\right)\left(\sqrt{3}\right) + \left(\sqrt{3}\right)^2.$$

Proposition B.2 tells us $\left(\sqrt{2}\right)^2 = 2$ and $\left(\sqrt{3}\right)^2 = 3$. Furthermore, Proposition B.1 implies $\left(\sqrt{2}\right)\left(\sqrt{3}\right) = \sqrt{6}$. It follows that

$$\left(\sqrt{2} - \sqrt{3}\right)^2 = 2 - 2\sqrt{6} + 3 = 5 - 2\sqrt{6}.$$

∎

B.2 n-th Roots

Definition B.3 The **n-th root** of x, denoted $\sqrt[n]{x}$ is the function defined by the relationship

$$y = \sqrt[n]{x} \quad \text{if} \quad y^n = x$$

for all x and y when n is odd and for all $x \geq 0$ and $y \geq 0$ when n is even.

In the expression $\sqrt[n]{x}$, the value n is called the "index" of the radical.

Notice that when the index of the radical is 2, the radical is the square root function, i.e. $\sqrt[2]{x} = \sqrt{x}$. The square root function is the most well known radical. The second most popular radical is the "cube root" which has an index of 3, i.e. the cube root of x is $\sqrt[3]{x}$.

Example B.7 Compute (a) $\sqrt[3]{8}$, (b) $\sqrt[4]{16}$, and (c) $\sqrt[5]{-32}$.

Solution

(a) Since $2^3 = 8$, we conclude that $\sqrt[3]{8} = 2$.

(b) We have $\sqrt[4]{16} = 2$ because $2^4 = 16$ and $2 \geq 0$.

(c) Due to the fact that $(-2)^5 = -32$, $\sqrt[5]{-32} = -2$. ∎

Proposition B.3 *Suppose $\sqrt[n]{a}$ and $\sqrt[n]{b}$ are real. Then*
$$\sqrt[n]{ab} = \sqrt[n]{a} \cdot \sqrt[n]{b}.$$
Furthermore, if $b \neq 0$,
$$\sqrt[n]{\frac{a}{b}} = \frac{\sqrt[n]{a}}{\sqrt[n]{b}}.$$

Example B.8 Simplify (a) $\sqrt[3]{-54}$ and (b) $\sqrt[4]{\frac{5}{81}}$.

Solution

(a) Notice that $-54 = -27 \cdot 2$. Thus,
$$\sqrt[3]{-54} = \sqrt[3]{-27} \cdot \sqrt[3]{2} = -3\sqrt[3]{2}.$$

(c) We have
$$\sqrt[4]{\frac{5}{81}} = \frac{\sqrt[4]{5}}{\sqrt[4]{81}} = \frac{\sqrt[4]{5}}{3}.$$
∎

Proposition B.4 *For m and n any two positive integers*
$$\sqrt[m]{x} = \sqrt[mn]{x^n}$$
whenever the expressions are real.

Proposition B.4 allows us to substantially simplify the product of radicals. Simply rewrite the radicals so that they have the same index and then use Proposition B.3.

Example B.9 Rewrite
$$\sqrt{3} \cdot \sqrt[3]{4}$$
as a single radical.

Solution We will rewrite the square and cube roots as sixth roots:
$$\sqrt{3} = \sqrt[2\cdot 3]{3^3} = \sqrt[6]{27}$$
and
$$\sqrt[3]{4} = \sqrt[3\cdot 2]{4^2} = \sqrt[6]{16}.$$
Thus,
$$\sqrt{3}\cdot\sqrt[3]{4} = \sqrt[6]{27}\cdot\sqrt[6]{16}$$
$$= \sqrt[6]{27\cdot 16}$$
$$= \sqrt[6]{432}.$$
∎

Appendix C

Transformations

Let f be an arbitrary function, and consider the graph of $y = f(x)$.

Translations

- $y = f(x - h)$ Shifts the graph *right* h units.

- $y = f(x) + k$ Shifts the graph *up* k units.

Vertical Compressions and Stretches

- $y = cf(x)$, $0 < c < 1$ Compresses the graph vertically by a factor of c.

- $y = cf(x)$, $c > 1$ Stretches the graph vertically by a factor of c.

Horizontal Compressions and Stretches

- $y = f(cx)$
  $0 < c < 1$
  Stretches the graph horizontally by a factor of $1/c$.

- $y = f(cx)$
  $c > 1$
  Compresses the graph horizontally by a factor of $1/c$.

Reflections

- $y = f(-x)$ Reflects the graph about the y-axis.

- $y = -f(x)$ Reflects the graph about the x-axis.

Being shifted a negative number of units in a direction is the same as being shifted the absolute value of that number in the opposite direction. For example, our rules tell us that $y = f(x+2)$ is the graph of $y = f(x)$ shifted *right* -2 units, so the graph of $y = f(x+2)$ is the graph of $y = f(x)$ shifted *left* 2 units.

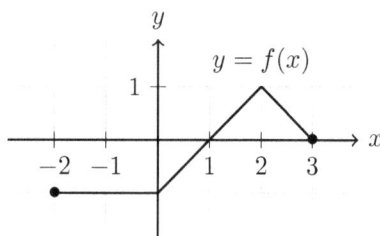

Example C.1 Graph

$$y = f(x+1) + 2.$$

Solution Let us consider a few critical numbers of f and their corresponding y-values.

x	$y = f(x)$
-2	-1
0	-1
2	1
3	0

Our rules tell us that $y = f(x+1) + 2$ is the graph of f shifted 1 unit left and 2 units up. To shift the graph left, we subtract 1 from the x-values in the table. To shift the graph up, we add 2 to the y-values.

x	$y = f(x+1) + 2$
-3	1
-1	1
1	3
2	2

Hence, $y = f(x+1) + 2$ has the following graph.

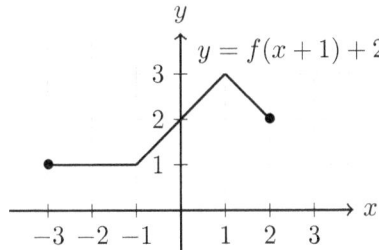

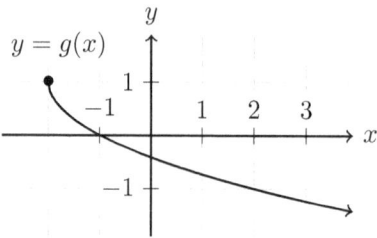

Example C.2 Graph
$$y = 2g(-x).$$

Solution The first step is to obtain some points on the graph of g.

x	$y = g(x)$
-2	1
2	-1

The graph of $y = 2g(-x)$ is the graph of g reflected about the y-axis and then stretched vertically by a factor of 2. To reflect the graph about the y-axis, we negate the x-values. To stretch the graph vertically by a factor of 2, we multiply the y-values by 2.

x	$y = 2g(-x)$
2	2
-2	-2

We conclude the graph of $y = 2g(-x)$ is as follows.

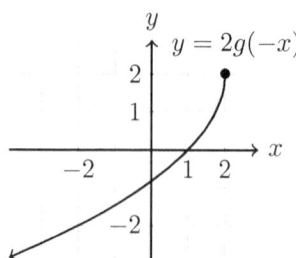

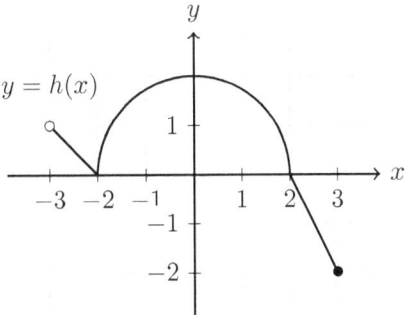

Example C.3 Graph
$$y = -h(2x - 4).$$

Solution Our first step is to construct a table.

x	$y = h(x)$
-3	1
-2	0
0	2
2	0
3	-2

The point $(-3, 1)$ is not included on the graph of h. However, $y = h(x)$ approaches 1 as x approaches -3. In our final graph, we will put an open circle at the corresponding point.

To clarify what transformations are being applied to the graph of h, we will factor a 2 out of $2x - 4$:

$$-h(2x - 4) = -h\Big(2(x - 2)\Big).$$

So, the graph of h is compressed horizontally by a factor of $1/2$, shifted right by a factor of 2, and then reflected about the x-axis.

$$y = h(x) \quad \longrightarrow \quad \overbrace{y = h(2x)}^{\text{horizontal compression}} \quad \longrightarrow \quad \overbrace{y = h\Big(2(x - 2)\Big)}^{\text{right shift}}$$

$$\longrightarrow \quad \overbrace{y = -h\Big(2(x - 2)\Big)}^{\text{reflection about } x\text{-axis}}$$

We are ready to modify our table of points, but we are careful to apply the transformations in the correct order. First, multiply the x-values by $1/2$ to compress the graph horizontally by a factor of $1/2$. Then add 2 to the x-values to shift the graph right 2 units. Then multiply the y-values by -1 to reflect the graph about the x-axis.

x	$y = -h(2x - 4)$
0.5	-1
1	0
2	-2
3	0
3.5	2

Therefore, the graph of $y = -h(2x - 4)$ is as follows.

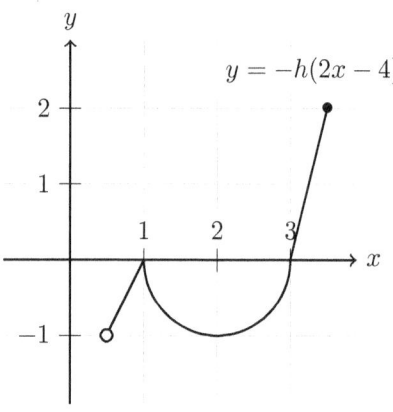

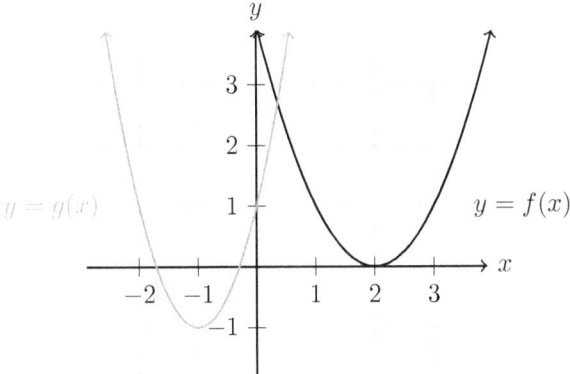

Example C.4 Consider the graphs of f and g above.

(a) Write g in terms of f.
(b) Write f in terms of g.

Solution

(a) The graph of g is the graph of f shifted left 3 units, stretched vertically by a factor of 2, and then shifted down 1 unit. Thus,
$$g(x) = 2f(x+3) - 1.$$

(b) The graph of f is the graph of g shifted right 3 units, shifted up 1 unit, and compressed vertically by a factor of $1/2$. Ergo,
$$f(x) = \frac{1}{2}\Big(g(x-3) + 1\Big).$$

■

Appendix D

Unit Circle

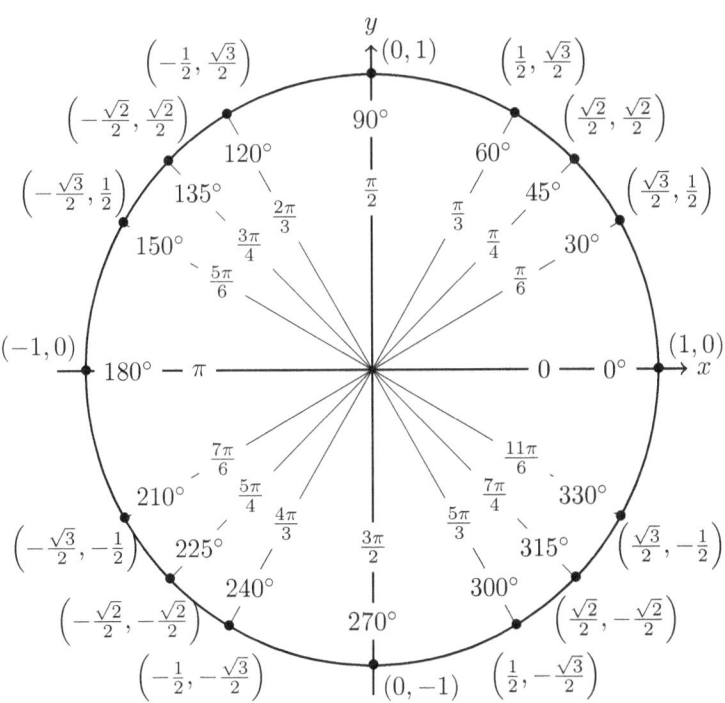

Appendix E

List of Identities

Reciprocal Identities

$$\cos\theta = \frac{1}{\sec\theta}$$

$$\sin\theta = \frac{1}{\csc\theta}$$

$$\tan\theta = \frac{1}{\cot\theta}$$

$$\sec\theta = \frac{1}{\cos\theta}$$

$$\csc\theta = \frac{1}{\sin\theta}$$

$$\cot\theta = \frac{1}{\tan\theta}$$

Quotient Identities

$$\tan\theta = \frac{\sin\theta}{\cos\theta}$$

$$\cot\theta = \frac{\cos\theta}{\sin\theta}$$

Even and Odd Identities

$$\sin(-\theta) = -\sin\theta$$

$$\cos(-\theta) = \cos\theta$$

$$\tan(-\theta) = -\tan\theta$$

$$\sec(-\theta) = \sec\theta$$

$$\csc(-\theta) = -\csc\theta$$

$$\cot(-\theta) = -\cot\theta$$

Pythagorean Identities

$\cos^2 \theta + \sin^2 \theta = 1$

$1 + \tan^2 \theta = \sec^2 \theta$

$1 + \cot^2 \theta = \csc^2 \theta$

Sum and Difference Identities

$\sin(\alpha \pm \beta) = \sin \alpha \cos \beta \pm \cos \alpha \sin \beta$

$\cos(\alpha \pm \beta) = \cos \alpha \cos \beta \mp \sin \alpha \sin \beta$

$\tan(\alpha \pm \beta) = \dfrac{\tan \alpha \pm \tan \beta}{1 \mp \tan \alpha \tan \beta}$

Cofunction Identities

$\sin(90° - \theta) = \cos \theta$

$\cos(90° - \theta) = \sin \theta$

$\tan(90° - \theta) = \cot \theta$

$\sec(90° - \theta) = \csc \theta$

$\csc(90° - \theta) = \sec \theta$

$\cot(90° - \theta) = \tan \theta$

Double Angle Identities

$\sin 2\theta = 2 \sin \theta \cos \theta$

$\cos 2\theta = \begin{cases} \cos^2 \theta - \sin^2 \theta \\ 2\cos^2 \theta - 1 \\ 1 - 2\sin^2 \theta \end{cases}$

$\tan 2\theta = \dfrac{2 \tan \theta}{1 - \tan^2 \theta}$

Half Angle Identities

$\sin \dfrac{\theta}{2} = \pm \sqrt{\dfrac{1 - \cos \theta}{2}}$

$\cos \dfrac{\theta}{2} = \pm \sqrt{\dfrac{1 + \cos \theta}{2}}$

$\tan \dfrac{\theta}{2} = \begin{cases} \pm \sqrt{\dfrac{1 - \cos \theta}{1 + \cos \theta}} \\ \dfrac{1 - \cos \theta}{\sin \theta} \\ \dfrac{\sin \theta}{1 + \cos \theta} \end{cases}$

Power Reducing Identities

$\sin^2 \theta = \dfrac{1 - \cos 2\theta}{2}$

$\cos^2 \theta = \dfrac{1 + \cos 2\theta}{2}$

$\tan^2 \theta = \dfrac{1 - \cos 2\theta}{1 + \cos 2\theta}$

Product to Sum and Difference Identities

$\cos\alpha\cos\beta = \frac{1}{2}\Big(\cos(\alpha+\beta)+\cos(\alpha-\beta)\Big)$

$\sin\alpha\sin\beta = \frac{1}{2}\Big(\cos(\alpha-\beta)-\cos(\alpha+\beta)\Big)$

$\sin\alpha\cos\beta = \frac{1}{2}\Big(\sin(\alpha+\beta)+\sin(\alpha-\beta)\Big)$

$\cos\alpha\sin\beta = \frac{1}{2}\Big(\sin(\alpha+\beta)-\sin(\alpha-\beta)\Big)$

Inverse Trigonometric Identities

$\arcsin x + \arccos x = \dfrac{\pi}{2}$

$\arcsin(-x) = -\arcsin x$

$\arccos(-x) = \pi - \arccos x$

$\arctan(-x) = -\arctan x$

$\operatorname{arcsec} x = \arccos \dfrac{1}{x}$

$\operatorname{arccsc} x = \arcsin \dfrac{1}{x}$

$\operatorname{arccot} x = \arctan \dfrac{1}{x}$

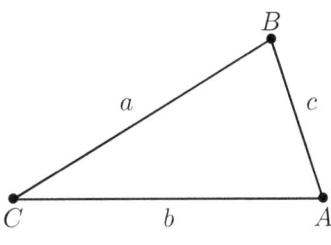

Law of Cosines

$a^2 = b^2 + c^2 - 2bc\cos A$

$b^2 = a^2 + c^2 - 2ac\cos B$

$c^2 = a^2 + b^2 - 2ab\cos C$

Law of Sines

$\dfrac{\sin A}{a} = \dfrac{\sin B}{b} = \dfrac{\sin C}{c}$

Euler's Formula

$e^{i\theta} = \cos\theta + i\sin\theta$

Appendix F

Answers

F.1 Angles

1. (a) right (f) right
   (b) acute (g) acute
   (c) acute (h) obtuse
   (d) right (i) straight
   (e) straight (j) obtuse

2. (a) 90° (e) 60°
   (b) 180° (f) 15°
   (c) 30° (g) 37.5°
   (d) 60° (h) 27.5°

3.

	Comp	Sup
(a)	70°	160°
(b)	15°	105°
(c)	nonsense	88°
(d)	10°	100°
(e)	nonsense	nonsense
(f)	67.5°	157.5°

4. (a) $x = 8°$, (b) $x = 10$, and (c) $x = 20°$.

5. (a) $y = 100°$, (b) $y = 8$, and (c) $y = 80$.

6. (a) 30°, (b) $t = 48°$, (c) $m\angle GHK = 1220°/9$, and (d) $v = 3$.

7. (a) $w = 45°$, (b) $w = 16°$, (c) $w = 3$, and (d) $w = 10$.

8.

9. (a) 160°, (b) 20°, and (c)

449

70°.

10. (a) $y° = 130°$ and (b) $x° = 50°$.

11. Answers vary.

12. (a) 60° (d) 30°
    (b) 30° (e) 150°
    (c) 60° (f) 60°

13. Answers vary.

14. (a) 113°, (b) 68°, (c) 60°, (d) 10°, (e) 1, and (f) 3 or 4.

15. (a) $z = 84$, (b) $y = 65$, and (c) $x = 96$.

16. Answers vary.

17. 124°

18. (a) 50°, (b) 110°, and 6.

19.
(a)	$m\angle D < m\angle A < m\angle B$
(b)	$m\angle B < m\angle D < m\angle C$
(c)	$CD<BC<BD<AB<AD$

F.2 Triangles

1. (a) obtuse, (b) right, and (c) right.

2. (a) right, (b) acute, and (c) obtuse.

3. (a) $x = 60°$, (b) $x = 9$, (c) $x = -1/2$ or $x = 5$, and (d) $x = 11/10$.

4. (a) scalene, (b) isosceles, and (c) equilateral.

5. The Isosceles Triangle Theorem (Theorem 1.2) tells us sides have equal length if and only if the angles opposite are congruent.

6. The Triangle Sum Theorem (Theorem 1.1) says that the sum of the interior angle measures is 180°. This is impossible if more than one angle is obtuse.

7. (a) $\angle Q \cong \angle T$, $\angle R \cong \angle U$, and $\angle S \cong \angle V$

   (b) $QR = TU$, $RS = UV$, and $QS = TV$.

8. (a) $\triangle VWX \cong \triangle VYX$ due to ASA Congruence Postulate.

   (c) $\triangle VWX \cong \triangle VYX$ due to SSS Congruence Postulate.

   (c) $\triangle VWX \cong \triangle VYX$ due to SAS Congruence Postulate.

9. (a) $\triangle ABC \cong \triangle EDC$ due to AAS Congruence Theorem or ASA Congruence Postulate.

   (b) $\triangle ABC \cong \triangle EDC$ due tp SAS Congruence Postulate.

   (c) $\triangle ABC \cong \triangle EDC$ due to AAS Congruence Theorem.

10. (a) $\triangle FGI \cong \triangle HGI$ due to HL Theorem.

    (b) $\triangle FGI \cong \triangle HGI$ due to SAS Postulate.

    (c) $\triangle FGI \cong \triangle HGI$ due to SSS Postulate.

    (d) $\triangle FGI \cong \triangle HGI$ due to AAS Theorem.

11. See section 9.3.3.

12. (a) $\triangle ABC \sim \triangle EBF$ due to AA Similarity Postulate.

    (b) $\triangle ADE \sim \triangle BFE$ due to AA Similarity Postulate.

    (c) $\triangle ADE \sim \triangle BFE$ due to SAS Similarity Theorem.

    (d) $\triangle ABC \sim \triangle EBF$ due to SSS similarity Theorem.

13. (a) 20/3, (b) 2, and (c) 114.

14. (a) $m\angle DEF = 59°$, $m\angle EDF = 31°$, $m\angle F = 90°$, $DF = 15\sqrt{34}/34$, and $EF = 9\sqrt{34}/34$.

    (b) $m\angle GDH = 31°$, $m\angle G = 59°$, $m\angle H = 90°$, $DG = 0.64\sqrt{34}$, and $GH = 1.92$.

15. (a) $h = 10 - 2w$

    (b) $25 - w^2$

    (c) $(5 - w)\sqrt{5}$

16. $16x/9$.

17. (a) $c = 10$, $d = 18/5$, $e = 32/5$, and $f = 24/5$.

    (b) $a = 8$, $b = 15$, $e = 225/17$, and $f = 120/17$.

    (c) $a = 65/12$, $b = 13$, $c = 169/12$, and $d = 25/12$.

    (d) $a = 8$, $b = 15$, $d = 64/17$, and $e = 225/17$, or $a = 15$, $b = 8$, $d = 225/17$, and $e = 64/17$.

    (e) $b = 12$, $c = 13$, $e = 144/13$, and $f = 60/13$.

18. (a) $c = 10$, (b) $a = 8$, (c) $b = 60$, (d) $c = 3\sqrt{13}$, (e) $a = 2\sqrt{30}$, and (f) $b = 4\sqrt{11}$.

19. (a) $t = 15$, (b) $t = 3$, and (c) $t = 3$.
20. (a) yes, (b) yes, and (c) no.
21. Answers vary.
22. (a) $m = 5$ and $n = 5\sqrt{2}$.
    (b) $m = \sqrt{6}$ and $n = 2\sqrt{3}$.
    (c) $\ell = 7$ and $m = 7$.
    (d) $\ell = \sqrt{6}/2$ and $m = \sqrt{6}/2$.
23. (a) $q = 8$ and $r = 4\sqrt{3}$.
    (b) $p = 2$ and $q = 4$.
    (c) $p = 8$ and $r = 8\sqrt{3}$
    (d) $q = 2\sqrt{6}$ and $r = 3\sqrt{2}$.
    (e) $p = 7\sqrt{15}/3$ and $q = 14\sqrt{15}/3$.
    (f) $p = \sqrt{3}/12$ and $r = 1/4$.
24. $2\sqrt{3}$
25. (a) 1:3 and (b) $12\sqrt{3}$
26. $400\pi \left(3 - 2\sqrt{2}\right)$
27. 2:1
28. $3\sqrt{3}$
29. (a) $UV = 7\sqrt{3} - 7$
    (b) $UV = 5\sqrt{3} - 5$
    (c) $TV = 10\sqrt{3}$
    (d) $ST = 12 + 4\sqrt{3}$
30. (a) $UV = 10(\sqrt{3} - 1)$
    (b) $VT = \sqrt{2} + \sqrt{6}$
    (c) $TU = 10(\sqrt{3} + 1)$
    (d) $UV = 20$
31. (a) $VY = 5\sqrt{3} + 5$
    (b) $XZ = 10(\sqrt{6} - \sqrt{2})$
    (c) $WZ = 2$
    (d) 18
32. (a) $VY = 6\sqrt{3} + 18$
    (b) $XY = 36(3 - \sqrt{3})$
    (c) $XZ = 50(\sqrt{6} - \sqrt{2})$
    (d) $8\sqrt{3} + 24$

F.3 Radians, Arc Length, and Rotational Motion

1. (a) $-1080°$ (b) $90°$ (c) $-270°$ (d) $337.5°$ (e) $-360\pi°$ (f) $1020°$
2. (a) $270°$ (b) $90°$ (c) $180°$ (d) $195°$
3. (a) $-30°$ (b) $120°$ (c) $180°$ (d) $-45°$

4. (a) 60° (d) 30°
   (b) 45° (e) 360°
   (c) 180° (f) 90°

5. (a) $\dfrac{\pi}{6}$ (d) $\dfrac{\pi}{4}$
   (b) π (e) 2π
   (c) $\dfrac{\pi}{3}$ (f) $\dfrac{\pi}{2}$

6. (a) $-216°$ (d) $216°$
   (b) $306°$ (e) $-324°$
   (c) $-120°$ (f) $126°$

7. (a) $\dfrac{15\pi}{8}$ (d) $-\dfrac{19\pi}{12}$
   (b) $-\dfrac{19\pi}{18}$ (e) $\dfrac{5\pi}{6}$
   (c) $\dfrac{13\pi}{12}$ (f) $-\dfrac{\pi}{8}$

8. (a) $s = 5\pi/2$, (b) $s = 11\pi/6$, (c) $\theta = 63.75°$, and (d) $r = 270$.

9. (a) 8 km (c) 8 km
   (b) 5 km (d) 3 km

10. (a) $\dfrac{\pi}{5}$ cm (d) $\dfrac{5\pi}{24}$ cm
    (b) $\dfrac{5\pi}{6}$ cm (e) 24π cm
    (c) 3π cm (f) $\dfrac{5\pi}{8}$ cm

11. $2\sqrt{3} + 5\pi/6$

12. (a) $A = \dfrac{25\pi}{4}$

    (b) $A = \dfrac{1575\pi}{8}$

    (c) $\theta = \dfrac{360°}{7}$

    (d) $r = 5$

    (e) $r = \dfrac{11\sqrt{6}}{2}$

    (f) $\theta = 36°$

13. Answers vary.

14. (a) $A = 8\pi/3$, (b) $\theta = 6\pi$, and (c) $r = 7$.

15. Four slices.

16. $x = 4$

17. (a) $60 + 20\pi$ and (b) $600 + 100\sqrt{3} - 200\pi$.

18. (a) $\omega = \dfrac{\pi}{12}$ rad/sec
    (b) $\omega = 12.5°/\text{min}$
    (c) $\omega = 4.3$ rev/day
    (d) $t = \dfrac{2}{17}$ sec
    (e) $t = \dfrac{5}{63}$ min
    (f) $\theta = 4\pi$

19. Answers vary.

20. (a) $v = \dfrac{5\pi}{2}$ ft/min
    (b) $r = \dfrac{75000}{\pi}$ km/sec
    (c) $\omega = 10$ rad/hour
    (d) $v = 20\pi$ in/year

21. Answers vary.

453

22. (a) $\omega = \pi/30$ and $v = \pi/5$

(b) $\omega = \pi/360$ and $v = \pi/72$

23. (a) $\dfrac{2\pi}{15}$ (c) $\dfrac{4\pi}{15}$

(b) $\dfrac{2\pi}{15}$ (d) $\dfrac{4\pi}{5}$

24.

Planet	Rev/day	10^6 km/day
Mercury	0.0114	4.15
Venus	0.00444	3.02
Earth	0.00274	2.58
Mars	0.00146	2.09
Jupiter	2.31×10^{-4}	1.13
Saturn	9.30×10^{-5}	0.859
Uranus	4.83×10^{-5}	0.872
Neptune	1.66×10^{-5}	0.470

25. $\dfrac{5}{8}$ meters

26. $\dfrac{2}{3}$

27. (a) 15.915 and (b) 6.631 revolutions.

F.4 Right Triangle Trigonometry

1. (a) 1.072 (d) 3.716
   (b) 0.292 (e) 0.105
   (c) 0.993 (f) 3.864

2. (a) 0.342 (d) 14.101
   (b) 0.482 (e) 1.286
   (c) 2.351 (f) 4.900

3. (a) 0.6 (d) 0.6
   (b) 0.75 (e) 1.333
   (c) 0.8 (f) 0.8

4. (a) 0.555 (d) 0.555
   (b) 0.667 (e) 1.5
   (c) 0.832 (f) 0.832

5. (a) $b \approx 10.378$
       $c \approx 12.518$
   (b) $a \approx 7.420$
       $c \approx 13.268$
   (c) $a \approx 11.184$
       $b \approx 16.581$
   (d) $b \approx 3.922$
       $c \approx 4.731$

6. (a) $b \approx 21.452$

$c \approx 28.003$

(b) $a \approx 4.195$
$c \approx 6.527$

(c) $a \approx 65.564$
$b \approx 78.137$

(d) $a \approx 1.175$
$c \approx 1.828$

7. (a) (i) 69.136
(ii) 44.250

(b) (i) 262.219
(ii) 78.613

(c) (i) 71.407
(ii) 41.716

8. (a) $DF \approx 1.532$

(b) $EF \approx 1.286$

(c) $GH \approx 1.286$

(d) $DG = 2$

9. (a) $DG \approx 14.619$

(b) $DH \approx 13.737$

(c) $EF = 3.75$

(d) $DE \approx 10.964$

10. 309.982

11. 1.407

12. (a) $\dfrac{\sqrt{3}}{2}$

(b) $\dfrac{\sqrt{2}}{2}$

(c) $\dfrac{\sqrt{3}}{3}$

(d) $\dfrac{\sqrt{3}}{2}$

(e) $\dfrac{1}{2}$

(f) $\sqrt{3}$

13. (a) $\dfrac{1}{2}$

(b) 1

(c) $\dfrac{\sqrt{3}}{2}$

(d) $\sqrt{3}$

(e) $\dfrac{\sqrt{2}}{2}$

(f) $\dfrac{\sqrt{2}}{2}$

14. Answers vary.

15. (a) $UV \approx 6.695$

(b) $VT \approx 8.187$

(c) $TV \approx 15.150$

(d) $ST \approx 11.726$

16. (a) $UV \approx 18.033$

(b) $UV \approx 51.959$

(c) $TV \approx 1.785$

(d) $SV \approx 6.093$

17. (a) $VW \approx 12.483$

(b) $VY \approx 13.486$

(c) The area of $\triangle XYZ$ is about 20.636.

(d) $WZ \approx 12.000$

18. (a) $XZ \approx 21.183$

(b) $VY \approx 26.185$

(c) $XY \approx 57.102$

(d) The area of $\triangle WXZ$ is about 50.150.

19. (a) $\dfrac{\sqrt{6} - \sqrt{2}}{4}$

(b) $\dfrac{\sqrt{6} + \sqrt{2}}{4}$

(c) $2-\sqrt{3}$

20. (a) 11.537° (d) undef
    (b) 67.792° (e) 75.522°
    (c) 78.690° (f) 18.435°

21. (a) 0.841 (d) undef
    (b) undef (e) 0.795
    (c) 0.464 (f) 1.504

22. (a) 25.377° (d) 30°
    (b) 78.463° (e) 45°
    (c) 52.595° (f) 60°

23. (a) 0.287 (d) 0.785
    (b) 0.889 (e) 0.524
    (c) 0.662 (f) 0.524

24. (a) 24.620° (c) 8.213°
    (b) 30.964° (d) 56.251°

25. (a) 1.159 (c) 0.524
    (b) 1.134 (d) 0.983

26.

(a)

x	$\dfrac{1}{2}$	$\dfrac{\sqrt{2}}{2}$	$\dfrac{\sqrt{3}}{2}$
arcsin x	30°	45°	60°
arccos x	60°	45°	30°

(b)

x	$\dfrac{1}{2}$	$\dfrac{\sqrt{2}}{2}$	$\dfrac{\sqrt{3}}{2}$
arcsin x	$\dfrac{\pi}{6}$	$\dfrac{\pi}{4}$	$\dfrac{\pi}{3}$
arccos x	$\dfrac{\pi}{3}$	$\dfrac{\pi}{4}$	$\dfrac{\pi}{6}$

27. (a) $\alpha = 45°$, $\beta = 60°$, and $\gamma = 30°$.
    (b) $\alpha = \pi/4$, $\beta = \pi/3$ and $\gamma = \pi/6$.

28. (a) $m\angle A \approx 65.709°$
    (b) $m\angle ACD \approx 15.412°$
    (c) $m\angle B \approx 10.963°$
    (d) $m\angle A \approx 24.835°$
    (e) $m\angle A \approx 53.130°$

29. 15.722

30. (a) 0.176 mi or 931 ft.
    (b) 1.015 mi or 5361 ft.

31. (a) 42 cm
    (b) 11.254 cm

32. (a) 220.676 ft

(b) 93.262 ft
33. (a) 5.831 ft
(b) 30.968°
34. (a) 6.974 ft
(b) 5.713 ft
(c) 9.567 ft/sec

35. 2.866°
36. 44.427°
37. (a) 347.296 m
(b) 419.550 m
38. 72.471 ft
39. 467.128 m

F.5 Trigonometry of General Angles

1. (a) y-axis (e) QIV
   (b) QIV   (f) QII
   (c) QI   (g) QII
   (d) x-axis (h) QIII

2. (a) QI   (e) QII
   (b) QII (f) y-axis
   (c) QIV (g) QIV
   (d) QIII (h) x-axis

3. (a) $\left(\dfrac{\sqrt{2}}{2}, \dfrac{\sqrt{2}}{2}\right)$

   (b) $\left(\dfrac{1}{2}, \dfrac{\sqrt{3}}{2}\right)$

   (c) $\left(-\dfrac{\sqrt{3}}{2}, -\dfrac{1}{2}\right)$

   (d) $(1, 0)$

   (e) $\left(\dfrac{\sqrt{2}}{2}, -\dfrac{\sqrt{2}}{2}\right)$

   (f) $\left(-\dfrac{1}{2}, \dfrac{\sqrt{3}}{2}\right)$

   (g) $(-1, 0)$

   (h) $\left(\dfrac{\sqrt{3}}{2}, \dfrac{1}{2}\right)$

   (i) $\left(-\dfrac{\sqrt{2}}{2}, \dfrac{\sqrt{2}}{2}\right)$

   (j) $(0, -1)$

4. (a) 30°   (d) 225°
   (b) 60°   (e) 270°
   (c) 90°   (f) 330°

5. (a) $\dfrac{\pi}{4}$ (d) $\dfrac{4\pi}{3}$
   (b) 0   (e) $\dfrac{5\pi}{6}$
   (c) π (f) $\dfrac{7\pi}{4}$

6. (a) $\dfrac{\sqrt{2}}{2}$ (e) -1
   (b) $\dfrac{\sqrt{3}}{3}$ (f) $\dfrac{2\sqrt{3}}{3}$
   (c) undef. (g) 0
   (d) -1 (h) 0

7. (a) $\dfrac{\sqrt{3}}{2}$ (e) 0
   (b) undef. (f) $\dfrac{2\sqrt{3}}{2}$
   (c) $\dfrac{1}{2}$ (g) $\dfrac{2\sqrt{3}}{3}$
   (d) $\sqrt{2}$ (h) undef.

8. (a) $\theta = 90°$
   (b) $\theta = 180°$
   (c) $\theta = 45°$ or $\theta = 225°$
   (d) $\theta = 30°$ or $\theta = 330°$
   (e) $\theta = 60°$ or $\theta = 120°$
   (f) $\theta = 120°$ or $\theta = 300°$

9. (a) $\varphi = \pi/2$ or $\varphi = 3\pi/2$
   (b) $\varphi = \pi/6$ or $\varphi = 5\pi/6$
   (c) $\varphi = 5\pi/6$ or $\varphi = 11\pi/6$
   (d) $\varphi = \pi/3$ or $\varphi = 5\pi/3$
   (e) $\varphi = \pi/4$ or $\varphi = 5\pi/4$
   (f) $\varphi = 7\pi/6$ or $\varphi = 11\pi/6$

10. (a) QI (d) QII
    (b) QIV (e) QIV
    (c) QIII (f) QIII

11. (a) Positive x-axis
    (b) Positive y-axis
    (c) Negative x-axis
    (d) Negative y-axis

12. (a) $33°$ (e) $37°$
    (b) $46°$ (f) $21°$
    (c) $29°$ (g) $44°$
    (d) $53°$ (h) $62°$

13. (a) $\dfrac{\pi}{6}$ (f) $\dfrac{\pi}{6}$
    (b) $\dfrac{\pi}{12}$ (g) $\dfrac{\pi}{3}$
    (c) $\dfrac{\pi}{10}$ (h) $\dfrac{4\pi}{9}$
    (d) $\dfrac{\pi}{4}$ (i) $2\pi - 5$
    (e) $\dfrac{\pi}{8}$ (j) 1

14. (a) $\dfrac{1}{2}$ (c) $-\dfrac{1}{2}$
    (b) $\dfrac{1}{2}$ (d) $-\dfrac{1}{2}$

15. (a) $\dfrac{\sqrt{2}}{2}$ (c) $-\dfrac{\sqrt{2}}{2}$
    (b) $-\dfrac{\sqrt{2}}{2}$ (d) $\dfrac{\sqrt{2}}{2}$

16. (a) $\sqrt{3}$ (c) $\sqrt{3}$
    (b) $-\sqrt{3}$ (d) $-\sqrt{3}$

17. (a) $-\dfrac{\sqrt{3}}{3}$ (f) $-\dfrac{\sqrt{3}}{3}$
    (b) 0 (g) -2
    (c) $-\dfrac{1}{2}$ (h) -1
    (d) undef. (i) $-\dfrac{\sqrt{2}}{2}$
    (e) $-\sqrt{2}$ (j) $-\dfrac{1}{2}$

18. (a) $\dfrac{\sqrt{3}}{2}$ (g) $-\dfrac{\sqrt{3}}{2}$
    (b) -1
    (c) $-\sqrt{2}$ (h) $-\dfrac{\sqrt{3}}{3}$
    (d) 0
    (e) $-\dfrac{2\sqrt{3}}{3}$ (i) $-\dfrac{\sqrt{2}}{2}$
    (f) $-\dfrac{2\sqrt{3}}{3}$ (j) $-\dfrac{1}{2}$

19. (a) $\sin\alpha = 3/5$, $\cos\alpha = 4/5$, $\tan\alpha = 3/4$, $\csc\alpha = 5/3$, $\sec\alpha = 5/4$, and $\cot\alpha = 4/3$.

    (b) $\sin(360°-\alpha)=-3/5$, $\cos(360°-\alpha)=4/5$, $\tan(360°-\alpha)=-3/4$, $\csc(360°-\alpha)=-5/3$, $\sec(360°-\alpha)=5/4$, and $\cot(360°-\alpha)=-4/3$.

    (c) $\sin(\alpha+180°)=-3/5$, $\cos(\alpha+180°)=-4/5$, $\tan(\alpha+180°)=3/4$, $\csc(\alpha+180°)=-5/3$, $\sec(\alpha+180°)=-5/4$, and $\cot(\alpha+180°)=4/3$.

    (d) $\sin(180°-\alpha)=3/5$, $\cos(180°-\alpha)=-4/5$, $\tan(180°-\alpha)=-3/4$, $\csc(180°-\alpha)=5/3$, $\sec(180°-\alpha)=-5/4$, and $\cot(180°-\alpha)=-4/3$.

20. (a) $\sin\beta = 12/13$, $\cos\beta = 5/13$, $\tan\beta = 12/5$, $\csc\beta = 13/12$, $\sec\beta = 13/5$, and $\cot\beta = 5/12$.

    (b) $\sin(\pi - \beta) = 12/13$, $\cos(\pi - \beta) = -5/13$, $\tan(\pi - \beta) = -12/5$, $\csc(\pi - \beta) = 13/12$, $\sec(\pi - \beta) = -13/5$, and $\cot(\pi - \beta) = -5/12$.

    (c) $\sin(2\pi-\beta) = -12/13$, $\cos(2\pi - \beta) = 5/13$, $\tan(2\pi - \beta) = -12/5$, $\csc(2\pi-\beta) = -13/12$, $\sec(2\pi - \beta) = 13/5$, $\cot(2\pi-\beta) = -5/12$.

    (d) $\sin(\beta+\pi) = -12/13$, $\cos(\beta+\pi) = -5/13$, $\tan(\beta+\pi) = 12/5$, $\csc(\beta+\pi) = -13/12$, $\sec(\beta+\pi) = -13/5$, $\cot(\beta+\pi) = 5/12$.

21. (a) $-\dfrac{15}{17}$ (d) $-\dfrac{17}{8}$
    (b) $-\dfrac{8}{15}$ (e) $\dfrac{17}{15}$
    (c) $\dfrac{8}{17}$ (f) $-\dfrac{8}{17}$

22. (a) undef.
    (b) $-\dfrac{\sqrt{3}}{3}$
    (c) $-\sqrt{2}$
    (d) -1
    (e) 1
    (f) -1
    (g) $\sqrt{3}$
    (h) $\dfrac{2\sqrt{3}}{3}$

23. (a) 1
    (b) 0
    (c) $\dfrac{2\sqrt{3}}{3}$
    (d) -1
    (e) $-\dfrac{1}{2}$
    (f) $\sqrt{2}$
    (g) $-\dfrac{2\sqrt{3}}{3}$
    (h) -1

24. Answers vary.

25. (a) D: $\mathbb{R}$ and R: $[-1,1]$
    (b) D: $\mathbb{R}$ and R: $[-1,1]$
    (c) D: $\{x : x \neq \frac{\pi}{2}, -\frac{\pi}{2}, \frac{3\pi}{2}, -\frac{3\pi}{2}, \ldots\}$ and R: $\mathbb{R}$
    (d) D: $\{x : x \neq 0, \pi, -\pi, 2\pi, -2\pi, \ldots\}$ and R: $(-\infty, -1] \cup [1, \infty)$
    (e) D: $\{x : x \neq \frac{\pi}{2}, -\frac{\pi}{2}, \frac{3\pi}{2}, -\frac{3\pi}{2}, \ldots\}$ and R: $(-\infty, -1] \cup [1, \infty)$.
    (f) D: $\{x : x \neq 0, \pi, -\pi, 2\pi, -2\pi, \ldots\}$ and R: $\mathbb{R}$.

26. (a) πn
    (b) $2\pi/3 + 2\pi n$ or $4\pi/3 + 2\pi n$
    (c) $\pi/3 + \pi n$ or $2\pi/3 + \pi n$
    (d) $\pi/2 + 2\pi n$

27. (a) $\pi/18 + 2\pi n/3$ or $5\pi/18 + 2\pi n/3$
    (b) $3\pi/8 + \pi n/2$
    (c) $1/3 + 2n$, $2/3 + 2n$, $4/3 + 2n$, or $5/3 + 2n$
    (d) $2\pi n/5$, $\pi/15 + 2\pi n/5$, or $\pi/3 + 2\pi n/5$

28. (a) $120°$ or $200°$
    (b) $15°$, $45°$, $135°$, $165°$, $255°$, or $285°$
    (c) $90°$ or $270°$
    (d) $105°$, $165°$, $285°$, or $345°$

29. (a) $\pi/6$ or $11\pi/6$.
    (b) $3\pi/4$
    (c) 1, 2, 4, or 5
    (d) 1/2, 5/2, or 9/2

30. (a) $-135°$ or $-45°$.
    (b) $-90°$ or $90°$
    (c) $-168°$, $-24°$ or $120°$
    (d) $-150°$, $-90°$, $-30°$, $30°$, $90°$, or $150°$

31. (a) $\cos\alpha = -3/5$, $\tan\alpha = 4/3$, $\sec\alpha = -5/3$, $\csc\alpha = -5/4$, and $\cot\alpha = 3/4$.
    (b) $\sin\beta = 8/17$, $\cos\beta = 15/17$, $\tan\beta = 8/15$,

$\sec\beta = 17/15$, and $\csc\beta = 17/8$.

(c) $\sin\gamma = -12/13$, $\tan\gamma = -12/5$, $\sec\gamma = 13/5$, $\csc\gamma = -13/12$, and $\cot\gamma = -5/12$.

(d) $\sin\theta = 24/25$, $\cos\theta = -7/25$, $\sec\theta = -25/7$, $\csc\theta = 25/24$, and $\cot\theta = -7/24$.

(e) $\sin\varphi = 0$, $\cos\varphi = -1$, $\tan\varphi = 0$, $\sec\varphi = -1$, and $\cot\varphi$ is undef..

32. (a) $\pi/6 + 2\pi n$, $5\pi/6 + 2\pi n$, or $3\pi/2 + 2\pi n$

(b) $\pi/3 + 2\pi n$ or $5\pi/3 + 2\pi n$

(c) $\pi/4 + 2\pi n$, $3\pi/4 + 2\pi n$, $5\pi/4 + 2\pi n$, or $7\pi/4 + 2\pi n$

33. (a) $2\pi n/3$ or $\pi/3 + 2\pi n/3$

(b) $\pi/6 + \pi n/2$, $\pi/3 + \pi n/2$ or $3\pi/8 + \pi n/2$

(c) $3\pi/20 + \pi n/5$

34. (a) ii, (b) ii, (c) iii, (d) i, and (f) iv.

Answers vary for Exercises 35-41.

F.6 Graphing Trigonometric Functions

		Amplitude	Period	Vertical Shift	Phase Shift
1.	(a)	3	2	0	0
	(b)	1/2	360°	1	30°
	(c)	2	1	$-\pi$	1
	(d)	1	2π	1	π
	(e)	2	4π	0	-2π
	(f)	3/4	6π	$-7/4$	$-\pi$

2.

(a) $f(x) = -3\sin\left(\frac{\pi}{3}x - \frac{2\pi}{3}\right)$ (b) $g(x) = \pi\cos\left(4x + \frac{2\pi}{3}\right) - 2$

Answers 3-4 omitted to save space.

5. Possible answers:

(a) $A = 1$, $B = 1$, and $C = \pi/2$.   (b) $A = 1$, $B = 1$, and $C = -\pi/2$.

6.

	Period	Vertical Shift	Phase Shift	Asymptotes
(a)	360°	0	−90°	$x = 90°(4n+1)$
(b)	180/17	$3\pi/2$	0	$x = 180n/17$
(c)	3π	$2/\pi$	$-\pi$	$x = \pi(6n+1)/2$
(d)	20	1	$-5/6$	$x = 5(12n-1)/6$
(e)	π	$3/4$	π	$x = \pi(2n+3)/2$
(f)	2	1	$3/\pi$	$x = 2n + 3/\pi$

7. Possible answers:

(a) $f(x) = 3\tan\left(\frac{\pi x}{10} - \frac{2\pi}{5}\right) + 1$ (b) $g(x) = -4\cot\left(\frac{x}{5}\right) - 3$

Answers 8-9 omitted to save space.

10. Possible answers:

(a) $A = -1$, $B = 1$, and $C = \pi/2$.   (b) $A = -1$, $B = 1$, and $C = \pi/2$.

	Period	Vertical Shift	Phase Shift	Asymptotes
(a)	3	-3	0	$x = 3(2n+1)/4$
(b)	2π	$\pi/6$	$-\pi/3$	$x = \pi(3n-1)/3$
(c)	$\pi/2$	1	π	$x = \pi(n+4)/4$
(d)	4π	0	-2π	$x = \pi(2n-1)$

11.

12. Possible answers:

(a) $f(x) = 3\sec\left(\frac{\pi x}{2}\right) + 2$
(b) $g(x) = -\frac{23\pi}{2}\csc\left(\frac{x}{2} - \frac{\pi}{2}\right) - 11\pi$

Answers 13-16 omitted to save space.

17. Possible answers:

(a) $y = 2\cos\left(\frac{\pi}{2}x + \pi\right) - 1$
(b) $y = -5\sin(\frac{1}{2}x + \pi)$
(c) $y = \cos\left(3x - \frac{\pi}{2}\right) + 2$
(d) $y = \frac{1}{2}\sin\left(\frac{\pi}{4}x + 2\pi\right) - 1$
(e) $y = 2\tan\left(\frac{\pi}{2}x - \pi\right) + 1$
(f) $y = \frac{1}{2}\cot(x + 135°)$
(g) $y = 3\tan\left(\frac{3}{2}x + \frac{\pi}{4}\right) - 2$
(h) $y = \pi\cot\left(\frac{\pi}{8}x - \frac{\pi}{4}\right) + \pi$
(i) $y = 3\sec\left(\frac{\pi}{4}x\right) + 3$
(j) $y = -2\pi\csc\left(2x + \frac{\pi}{2}\right) - \pi$
(k) $y = \frac{\pi}{2}\sec\left(\frac{2\pi}{3}x + \frac{\pi}{3}\right) + \pi$
(l) $y = \frac{1}{2}\csc\left(\frac{\pi}{2}x - \pi\right) - \frac{3}{2}$

Answers 18-19 omitted to save space.

20. (a) 1, (b) 5, (c) 13, (d) 27, and (e) ∞.

F.7 Using Identities

1. (a) $\dfrac{\sqrt{6}-\sqrt{2}}{4}$

   (b) $2-\sqrt{3}$

   (c) $\dfrac{\sqrt{6}-\sqrt{2}}{4}$

   (d) $\sqrt{3}+2$

   (e) $-\dfrac{\sqrt{6}+\sqrt{2}}{4}$

   (f) $-\sqrt{6}-\sqrt{2}$

   (g) $-\sqrt{6}-\sqrt{2}$

   (h) $\sqrt{3}+2$

2. (a) $\sqrt{3}-2$

   (b) $-\dfrac{\sqrt{6}+\sqrt{2}}{4}$

   (c) $-\sqrt{3}-2$

   (d) $\dfrac{\sqrt{2}-\sqrt{6}}{4}$

   (e) $\sqrt{6}-\sqrt{2}$

   (f) $\sqrt{2}-\sqrt{6}$

   (g) $\dfrac{\sqrt{6}+\sqrt{2}}{4}$

   (h) $-\sqrt{3}-2$

3. (a) $-\dfrac{\sqrt{3}}{2}$

   (b) -1

   (c) $-\dfrac{\sqrt{3}}{2}$

   (d) $-\sqrt{3}$

   (e) $\dfrac{1}{2}$

   (f) $-\dfrac{\sqrt{3}}{3}$

4. (a) -1

   (b) $-\dfrac{\sqrt{3}}{2}$

   (c) $-\dfrac{\sqrt{3}}{3}$

   (d) $\dfrac{\sqrt{2}}{2}$

   (e) 1

   (f) $-\dfrac{\sqrt{3}}{2}$

5. (a) $\pi/6 + \pi n$

   (b) $3\pi/4 + 2\pi n$ or $5\pi/4 + 2\pi n$

   (c) $\pi/2 + \pi n$

6. (a) $\dfrac{96+5\sqrt{17}}{117}$

   (b) $-\dfrac{40+12\sqrt{17}}{117}$

   (c) $\dfrac{96+5\sqrt{17}}{12\sqrt{17}-40}$

   (d) $\dfrac{117}{96-5\sqrt{17}}$

7. (a) $-\dfrac{84}{85}$

   (b) $\dfrac{13}{84}$

   (c) $-\dfrac{85}{36}$

   (d) $\dfrac{36}{77}$

8. (a) $y = 5\sqrt{2}\sin\left(\pi x + \dfrac{\pi}{4}\right)$

   (b) $y = -6\sin\left(x + \dfrac{\pi}{3}\right)$

   (c) $y = -\sqrt{2}\sin\left(x - \dfrac{\pi}{4}\right)$

(d) $y = 2\sin\left(2x - \dfrac{\pi}{6}\right)$

9. Answer omitted to save space.

10. (a) 4.331   (d) 2.145
    (b) 0.292   (e) 0.927
    (c) 1.743   (f) 1.466

11. (a) -0.668 (d) 0.309
    (b) 0.831   (e) 2.305
    (c) 1.556   (f) -2.190

12. Answers vary.

13. Answers vary.

14. $\sin 2\theta = 240/289$
    $\cos 2\theta = -161/289$
    $\tan 2\theta = -240/161$
    $\sec 2\theta = -289/161$
    $\csc 2\theta = 289/240$
    $\cot 2\theta = -161/240$

15. $\sin 2\varphi = 24/25$
    $\cos 2\varphi = 7/25$
    $\tan 2\varphi = 24/7$
    $\sec 2\varphi = 25/7$
    $\csc 2\varphi = 25/24$
    $\cot 2\varphi = 7/24$

16. (a) $\pi/2 + 2\pi n$, $7\pi/6 + 2\pi n$, $3\pi/2 + 2\pi n$, or $11\pi/6 + 2\pi n$

    (b) $2\pi n$, $\pi/6 + 2\pi n$, $5\pi/6 + 2\pi n$, or $\pi + 2\pi n$

    (c) $\pi + 2\pi n$

    (d) $2\pi n$, $\pi/3 + 2\pi n$, $2\pi/3 + 2\pi n$, πn, $4\pi/3 + 2\pi n$, or $5\pi/3 + 2\pi n$

17. (a) $-\dfrac{\sqrt{2+\sqrt{3}}}{2}$

    (b) $-2 - \sqrt{3}$

    (c) $\dfrac{\sqrt{2-\sqrt{2}}}{2}$

    (d) $\dfrac{2}{\sqrt{2-\sqrt{3}}}$

    (e) $2 - \sqrt{3}$

    (f) $-\dfrac{2}{\sqrt{2+\sqrt{2}}}$

    (g) $\dfrac{\sqrt{2-\sqrt{2+\sqrt{3}}}}{2}$

    (h) $-\dfrac{\sqrt{2+\sqrt{2+\sqrt{2}}}}{2}$

18. (a) $\dfrac{\sqrt{2+\sqrt{2}}}{2}$

    (b) $\sqrt{3} - 2$

    (c) $\dfrac{\sqrt{2+\sqrt{2}}}{2}$

    (d) $-\dfrac{2}{\sqrt{2-\sqrt{3}}}$

    (e) $1 + \sqrt{2}$

    (f) $\dfrac{2}{\sqrt{2-\sqrt{3}}}$

    (g) $\dfrac{2 - \sqrt{2+\sqrt{3}}}{\sqrt{2-\sqrt{3}}}$

(h) $-\dfrac{\sqrt{2-\sqrt{2+\sqrt{2}}}}{2}$

19. $\sin(\theta/2) = \sqrt{2}/10$
$\cos(\theta/2) = -7\sqrt{2}/10$
$\tan(\theta/2) = -1/7$
$\sec(\theta/2) = -5\sqrt{2}/7$
$\csc(\theta/2) = 5\sqrt{2}$
$\cot(\theta/2) = -7$

20. $\sin(\varphi/2) = 5\sqrt{34}/34$
$\cos(\varphi/2) = 3\sqrt{34}/34$
$\tan(\varphi/2) = 5/3$
$\sec(\varphi/2) = \sqrt{34}/3$
$\csc(\varphi/2) = \sqrt{34}/5$
$\cot(\varphi/2) = 3/5$

21. (a) $\dfrac{1-\sqrt{3}}{4}$

(b) $\dfrac{\sqrt{6}-\sqrt{2}}{8}$

(c) $-\dfrac{1}{4}$

(d) $\dfrac{\sqrt{2}-1}{4}$

22. (a) $\dfrac{1-\sqrt{3}}{4}$

(b) $\dfrac{\sqrt{6}+\sqrt{2}}{8}$

(c) $\dfrac{\sqrt{3}-\sqrt{2}}{4}$

(d) $-\dfrac{1}{4}$

23. (a) iv   (e) i
(b) i   (f) ii
(c) iii (g) ii
(d) iii (h) i

Answers vary for Exercises 24-33.

F.8 Inverse Trigonometric Functions

1. (a) −1   (e) 2
   (b) −5   (f) 5
   (c) 1   (g) −3
   (d) −5   (h) 0

2. (a) not invertible
   (b) not invertible
   (c) invertible
   (d) not invertible
   (e) invertible
   (f) invertible
   (g) invertible
   (h) invertible

   (f) not invertible

3. (a) not invertible
   (b) not invertible
   (c) not invertible
   (d) not invertible
   (e) invertible

(i) not invertible

(j) not invertible

(k) not invertible

(l) not invertible

4. Possible Answers:

   (a) $(-2, 0]$ or $[0, 2]$

   (b) $[0.5, 2]$ or $[2, 3.5]$

   (d) $[0, 3]$ or $[3, 4]$

   (f) $(-2, -1]$ or $[1, 2]$

5. Answers vary.

6. (a) invertible

   (b) not invertible

   (c) not invertible

   (d) not invertible

7. (a) 45°   (d) 90°
   (b) 60°   (e) 30°
   (c) 0   (f) 30°

8. (a) 0   (d) $\frac{\pi}{3}$
   (b) $\frac{\pi}{4}$ (e) 0
   (c) $\frac{\pi}{2}$ (f) $\frac{\pi}{3}$

9. (a) 90°   (d) −90°
   (b) −60°   (e) 120°
   (c) −60°   (f) 30°

10. (a) $-\frac{\pi}{4}$ (d) undef.
    (b) $\frac{\pi}{4}$ (e) $-\frac{\pi}{6}$
    (c) $-\frac{\pi}{3}$ (f) undef.

11. (a) $\frac{2}{5}$ (d) $\frac{5}{2}$
    (b) $\frac{2}{5}$ (e) $\frac{2}{5}$
    (c) $\frac{2}{5}$ (f) $\frac{5}{2}$

12. (a) 25°   (c) 65°
    (b) 25°   (d) 65°

13. (a) 155°   (c) −65°
    (b) 25°   (d) 65°

14. (a) $\frac{4\pi}{5}$ (d) $-\frac{3\pi}{10}$
    (b) $\frac{\pi}{5}$ (e) $\frac{3\pi}{10}$
    (c) $\frac{4\pi}{5}$ (f) $\frac{7\pi}{10}$

15. (a) $\frac{\pi}{5}$ (d) $\frac{3\pi}{10}$
    (b) $-\frac{\pi}{5}$ (e) $-\frac{3\pi}{10}$
    (c) $\frac{\pi}{5}$ (f) $\frac{7\pi}{10}$

16. (a) 191.537° or 348.463°.
    (b) 44.415° or 315.585°.
    (c) 99.462° or 279.462°.
    (d) no solution

(e) 19.471°, 160.529°, 199.471°, or 340.529°.

17. (a) −0.340

    (b) 0.464 or 0.785

    (c) −0.142 or −0.785

    (d) −0.201

18. (a) 3.481 or 5.943.

    (b) 0.464, 0.785, 3.605, or 3.927.

    (c) 2.356, 3.000, 5.498, or 6.141.

    (d) 3.343 or 6.082.

19. (a) 70.529°

    (b) 98.213°

    (c) no solution

    (d) 83.621° or 99.594°

20. (a) −70.529° or 70.529°

    (b) −98.213° or 98.213°

    (c) no solution

    (d) −99.594°, −83.621°, 83.621°, or 99.594°

21. (a) $y \approx 5\sin(2x + 0.927)$

    (b) $y \approx -13\sin(x - 1.176)$

    (c) $y \approx 15\sin(x/2 - 0.644)$

    (d) $y \approx -25\sin(x + 1.287)$

22. $\sin\alpha = x$
    $\cos\alpha = \sqrt{1 - x^2}$
    $\tan\alpha = \dfrac{x\sqrt{1 - x^2}}{1 - x^2}$
    $\sec\alpha = \dfrac{\sqrt{1 - x^2}}{1 - x^2}$
    $\csc\alpha = \dfrac{1}{x}$
    $\cot\alpha = \dfrac{\sqrt{1 - x^2}}{x}$

23. $\sin\beta = \dfrac{\sqrt{1 + 4x^2}}{1 + 4x^2}$
    $\cos\beta = \dfrac{2x\sqrt{1 + 4x^2}}{1 + 4x^2}$
    $\tan\beta = \dfrac{1}{2x}$
    $\sec\beta = \dfrac{\sqrt{1 + 4x^2}}{2x}$
    $\csc\beta = \sqrt{1 + 4x^2}$
    $\cot\beta = 2x$

24. (a) $\dfrac{2x\sqrt{9 - x^2}}{9}$

    (b) $\dfrac{2x^2 - 9}{9}$

    (c) $\dfrac{2x\sqrt{9 - x^2}}{2x^2 - 9}$

    (d) $\dfrac{9\sqrt{9 - x^2}}{2x(9 - x^2)}$

25. (a) $\dfrac{\sqrt{2x^2 - 2x\sqrt{x^2 - 1}}}{2x}$

    (b) $\dfrac{\sqrt{2x^2 + 2x\sqrt{x^2 - 1}}}{2x}$

    (c) $x - \sqrt{x^2 - 1}$

    (d) $x + \sqrt{x^2 - 1}$

26. (a) $-\dfrac{\sqrt{2x^2 - x}}{2x}$

(b) $-\dfrac{\sqrt{2x^2+x}}{2x}$

(c) $-\dfrac{\sqrt{4x^2-1}}{2x+1}$

(d) $-\dfrac{\sqrt{4x^2-1}}{2x-1}$

27. (a) $20x^2+\sqrt{(1-16x^2)(1-25x^2)}$

(b) $5x\sqrt{1-16x^2}+4x\sqrt{1-25x^2}$

(d) $\dfrac{20x^2+\sqrt{(1-16x^2)(1-25x^2)}}{5x\sqrt{1-16x^2}-4x\sqrt{1-25x^2}}$

28. (a) $\pi/6$ and (b) $-\pi/4$.

29. Answers vary.

30. Answers vary.

31. Answers omitted to save space.

32. Answers omitted to save space.

33. (a) $f(x) = \arcsin\dfrac{x}{2} + \dfrac{\pi}{2}$

(b) $g(x) = 7\arccos(-x)$

(c) $h(x) = -\arctan(x+3)$

34. Possible answers:

(a) $y = -2\arcsin x$

(b) $y = -\arctan(x+1)$

(c) $y = \arccos(-x)$

(d) $y = \dfrac{1}{3}\arccos\left(x+\dfrac{1}{2}\right)$

(e) $y = \arctan\left(\dfrac{x}{\pi}\right) - \dfrac{\pi}{2}$

(f) $y = \arcsin(x-1) + \dfrac{\pi}{4}$

F.9 Oblique Triangles

1. (a) SAS (e) ASA (g) Law of Sines
   (b) AAS (f) SAS (h) Law of Sines
   (c) SSS (g) SSA
   (d) SSA (h) AAS

2. (a) Law of Cosines
   (b) Law of Sines
   (c) Law of Cosines
   (d) Law of Sines
   (e) Law of Sines
   (f) Law of Cosines

3. (a) $m\angle X \approx 38.213°$, $m\angle Y = 60°$, and $m\angle Z \approx 81.787°$.

   (b) $m\angle X \approx 0.900°$, $m\angle Y \approx 5.101°$, and $z \approx 19.986$.

   (c) $m\angle X \approx 48.275°$, $m\angle Z \approx 61.725°$, and $y \approx 62.953$.

   (d) $m\angle X \approx 34.048°$, $m\angle Y \approx 44.415°$, and

$m\angle Z \approx 101.537°$.

4. (a) $m\angle Y = 45°$, $x \approx 1.726$, and $z \approx 3.961$.

   (b) $m\angle X = 43°$, $y \approx 8.304$, and $z \approx 10.187$.

   (c) $m\angle X = 36°$, $x \approx 1.229$, and $y \approx 1.977$.

   (d) $m\angle Y = 73°$, $x \approx 0.875$, and $y \approx 4.817$.

5. (a) $m\angle U \approx 22.793°$, $m\angle V \approx 2.207°$, and $v \approx 1.094$.

   (b) No triangle.

   (c) No triangle.

   (d) $m\angle T \approx 7.540°$, $m\angle V \approx 64.460°$, and $t \approx 5.381$.

   (e) $m\angle T \approx 38.682°$, $m\angle U \approx 51.318°$, and $u \approx 6.245$.

6. (a) yes, (b) no, and (c) yes.

7. (a) $m\angle E \approx 85.642°$, $m\angle F \approx 59.358°$, and $e \approx 17.384$ or $m\angle E \approx 24.358°$, $m\angle F \approx 120.642°$, and $e \approx 7.190$.

   (b) $m\angle D \approx 80.103°$, $m\angle F \approx 34.897°$, and $f \approx 14.519$, or $m\angle D \approx 99.897°$, $m\angle F \approx 15.103°$, and $f \approx 6.612$.

   (c) $m\angle D \approx 108.137°$, $m\angle E \approx 49.863°$, and $d \approx 124.305°$, or $m\angle D \approx 27.863°$, $m\angle E \approx 130.137°$, and $d \approx 61.132$.

   (d) $m\angle E \approx 65.174°$, $m\angle F \approx 55.826°$, and $f = 16.408$, or $m\angle E \approx 114.826°$, $m\angle F \approx 6.174°$, and $f \approx 2.133$.

8. (a) 1   (e) 2
   (b) 1   (f) 1
   (c) 0   (g) 1
   (d) 0   (h) 1

9. (a) 1   (e) 1
   (b) 0   (f) 1
   (c) 0   (g) 2
   (d) 0   (h) 1

10. (a) $m\angle P \approx 114.026$, $m\angle R \approx 114.026°$, and $p \approx 62.309$.

    (b) $m\angle P \approx 22.686°$, $m\angle Q \approx 62.314°$, and $p \approx 3.484$.

    (c) $m\angle Q \approx 22.620°$, $m\angle R \approx 90°$, and $q = 5$.

    (d) $m\angle P \approx 105.605°$, $m\angle R \approx 51.395°$, $p \approx 17.255$, or $m\angle P \approx$

$28.395°$, $m\angle R \approx 128.605°$, and $p \approx 8.519$.

(e) $m\angle P \approx 31.404°$, $m\angle R \approx 40.596°$, and $p \approx 10.410$.

(f) $m\angle Q \approx 105.332°$, $m\angle R \approx 29.668°$, and $q \approx 27.278$.

(g) No triangle.

(h) $m\angle P \approx 69.572°$, $m\angle R \approx 85.428°$, and $r \approx 54.249$, or $m\angle P \approx 110.428°$, $m\angle R \approx 44.572°$, and $r \approx 38.194$.

(i) No triangle.

11. (a) $m\angle A \approx 36.443°$, $m\angle B \approx 33.557°$, and $a \approx 10.747$.

(b) $m\angle B = 55°$, $a \approx 23.303°$, and $b \approx 19.383$.

(c) $m\angle B \approx 66.674°$, $m\angle C \approx 73.326°$, and $c \approx 31.296$, or $m\angle B \approx 113.326°$, $m\angle C \approx 26.674°$, and $c \approx 14.666$.

(d) $m\angle B \approx 36.870°$, $m\angle C \approx 53.130°$, and $c = 20$.

(e) $m\angle A = 110°$, $b \approx 17.990$, and $c \approx 30.099$.

(f) $m\angle A \approx 33.184°$, $m\angle C \approx 124.816°$, and $b \approx 6.844$.

(g) $m\angle A \approx 120.549°$, $m\angle B \approx 28.451°$, and $a \approx 66.884$.

(h) $m\angle A \approx 35.296°$, $m\angle B \approx 43.897°$, and $m\angle C \approx 100.807°$.

(i) $m\angle B = 30°$, $m\angle C = 90°$, and $b = 5$.

F.10 Area and Perimeter

1. (a) $A \approx 55.902$
$P \approx 36.180$

(b) $A \approx 197.818$
$P \approx 80.055$

(c) $A \approx 101.180$
$P \approx 49.657$

(d) $A \approx 29.790$
$P \approx 26.497$

(e) $A \approx 494.433$
$P \approx 107.387$

(f) $A \approx 26.806$
$P \approx 27.554$

2. (a) $A = 17.5$
$P \approx 22.268$

(b) $A \approx 14.697$
$P = 18$

(c) $A \approx 72.444$
$P \approx 20.066$

(d) $A \approx 425.440$
$P = 182$

3. (a) $A \approx 6.495$
$P = 15$

(b) $A \approx 98.740$
$P \approx 50.938$

(c) $A \approx 2185.333$
$P \approx 236.986$

(d) $A \approx 5.539$
$P \approx 20.243$

(e) $A \approx 129.352$
$P \approx 57.757$

4. Multiply each expression by $2/(abc)$.

5. 0.464

6. (a) $A \approx 279.808$
$P = 60$

(b) $A \approx 709.454$
$P \approx 94.594$

(c) $A \approx 636.396$
$P \approx 91.844$

(d) $A \approx 51.987$
$P \approx 25.994$

(e) $A = 75$
$P \approx 31.058$.

(f) $A \approx 28227.781$
$P = 600$

7. (a) $10\sqrt{3}$, (b) 20, and (c) 20.

8. (a) 13.858, (b) 15, and (c) 11.481

9. (a) $A \approx 129.904$
$P \approx 51.962$

(b) $A = 200$
$P \approx 56.569$

(c) $A \approx 237.764$
$P \approx 58.779$

(d) $A \approx 259.807$
$P = 60$

(e) $A \approx 282.843$
$P \approx 61.229$

(f) $A \approx 300$
$P \approx 62.117$

10. $A \longrightarrow 100\pi \approx 314.159$
$P \longrightarrow 20\pi \approx 62.832$

11. (a) $A \approx 519.615$
$P \approx 103.923$

(b) $A = 400$
$P = 80$

(c) $A \approx 363.271$
$P \approx 72.654$

(d) $A \approx 346.410$
$P \approx 69.282$

(e) $A \approx 331.371$
$P \approx 66.274$

(f) $A \approx 321.539$
$P \approx 64.308$

12. $A \to 100\pi \approx 314.159$
$P \to 20\pi \approx 62.832$

13. (a) $x = 5\sqrt{3}$
(b) $\theta = 90°$

(c) $r \approx 15.419$

(d) $x = 16\sqrt{2}$

(e) $\theta \approx 14.362°$

(f) $r \approx 6.527$

14. (a) $r = 4$

(b) $A = \dfrac{27(5\pi - 3)}{4}$

(c) $r \approx 2.999$

(d) $A = \dfrac{100\left(2\pi - 3\sqrt{3}\right)}{3}$

(e) $r = 20$

(f) $A = \dfrac{361\left(\pi - 2\sqrt{2}\right)}{8}$

15. (a) $A \approx 10.295$, (b) $x \approx 18.456$, and (c) $A \approx 5.118$.

16. 12.320

17. 0.424

F.11 Vectors

1. (a) $(6, 0)^T$ (c) $(0, -8)^T$
   (b) $(-4, 2)^T$ (d) $(-10, -10)^T$
   (f) $\boldsymbol{u} = \left(-5\sqrt{3\pi}, -5\sqrt{\pi}\right)^T$

2. (a) 30° (e) 225°
   (b) 341.565° (f) 0
   (c) 120° (g) 201.801°
   (d) 270° (h) 315°

3. (a) 5 (d) 1
   (b) $3\sqrt{5}$ (e) $17\sqrt{10}$
   (c) 13 (f) 1

4. (a) $\boldsymbol{u} = (22, 0)^T$
   (b) $\boldsymbol{u} = \left(\dfrac{\sqrt{2}}{2}, -\dfrac{\sqrt{2}}{2}\right)^T$
   (c) $\boldsymbol{u} = (0, -2.1)^T$
   (d) $\boldsymbol{u} = \left(-\dfrac{11}{10}, \dfrac{11\sqrt{3}}{10}\right)^T$
   (e) $\boldsymbol{u} = (\sqrt{15}, \sqrt{5})$

5. Answer omitted to save space.

6. (a) $(-2, 1)^T$, (b) $-6\boldsymbol{j}$, and (c) $\boldsymbol{0}$.

7. Answer omitted to save space.

8. Answer omitted to save space.

9. (a) $-2\boldsymbol{v} = (2, -6)^T$
   $3\boldsymbol{v} = (-3, 9)^T$.
   (b) $\boldsymbol{v} = -2\boldsymbol{i} - 2\boldsymbol{j}$
   $3\boldsymbol{v} = 3\boldsymbol{i} + 3\boldsymbol{j}$.
   (c) $-2\boldsymbol{v} = (6, 4)^T$
   $3\boldsymbol{v} = (-9, -6)^T$.

10. (a) (i) $-\dfrac{78}{5}\boldsymbol{i} + \boldsymbol{j}$,
    (ii) $3\boldsymbol{i} - 15\boldsymbol{j}$

(iii) $-\dfrac{336}{5}i + 63j$.

(b) (i) $5i + 7j$

(ii) $-15i - 15j$

(iii) $3j$

(c) (i) $15i - 7j$

(ii) $10j$

(iii) $\dfrac{228}{5}i - 93j$

11. (a) $\overrightarrow{BA}$ (c) $\overrightarrow{DB}$

(b) $\overrightarrow{AC}$ (d) $\overrightarrow{BD}$

12. Answers vary.

13. (a) $\left(-\dfrac{5}{13}, -\dfrac{12}{13}\right)^T$

(b) $\dfrac{7\sqrt{2}}{10}i + \dfrac{\sqrt{2}}{10}j$

(c) $\left(\dfrac{40}{41}, -\dfrac{9}{41}\right)^T$

(d) $\dfrac{3}{5}i - \dfrac{4}{5}j$

(e) $\left(\dfrac{\sqrt{69}}{9}, \dfrac{2\sqrt{3}}{9}\right)^T$

(f) $-\dfrac{\sqrt{2}}{2}i + \dfrac{\sqrt{2}}{2}j$

14. (a) $u = \left(-\dfrac{5\sqrt{3}}{2}, \dfrac{5}{2}\right)^T$

(b) $u = \dfrac{3}{10}i - \dfrac{2}{5}j$

(c) $u = (-12, -5)^T$

(d) $u = -i - j$

(e) $u = (9, 12)^T$

15. (a) $\left(10, 10\sqrt{3}\right)^T$

(b) $(-6, 6)^T$

(c) $\left(-50, -50\sqrt{3}\right)^T$

(d) $\left(\dfrac{15}{2}, -\dfrac{5\sqrt{3}}{2}\right)^T$

16. (a) $S45°E$, (b) $N60°W$, (c) $S45°W$, (d) $N30°E$, (e) about $S32.471°E$, and (f) about $N39.232°E$.

17. About (a) $N33.690°E$, (b) $N11.310°E$, (c) $N25.017°W$, (d) $S33.690°W$, (e) $S11.310°W$, and (f) $S25.017°E$.

18. The distance between ships A and B is 42.321 kilometers and the bearing from Ship A to Ship B is $N52.316°E$.

19. $N48.153°E$

20. (a) 49.378 kilometers per hour and (b) $S4.277°W$.

21. (a) 841.828 kilometers per hour and (b) $N54.610°E$.

22. (a) 100 N, (b) 196.2 N, (c) 1.02 kg, (d) 49.1 N, and (e) 8.14 kg.

23. (a) 100 N, (b) 98.1 N, (c) 2.55 kg, (d) 19.6 N, and (e) 2.55 kg.

24. (a) $a \approx 8.50 \ m/s^2$, (b) $\theta \approx 45.0°$, (c) $a \approx 3.36 \ m/s^2$, and (d) $\theta \approx 24.0°$.

25. Answers vary.

26. (a) About 508 N in the left cable and 718 N in the right cable.

    (b) About 498 N in the left cable and about 763 N in the right.

    (c) $\alpha \approx 27.1°$.

    (d) About 868 N.

27. The mass of the box is about 108 kilograms.

28. (a) $\alpha = 60°$ and $\beta = 30°$.

    (b) $\alpha \approx 30.0°$ and $\beta \approx 14.5°$.

29. (a) 50 kilograms, (b) 48.6°, (c) about 35.6 kilograms, and (d) 29.6°.

30. (a) -9 (c) -18

    (b) 13 (d) 18

31. (a) 0 (c) $-\dfrac{15}{2}$

    (b) 90 (d) 45

32. Answers vary.

33. (a) -1 (c) $-\dfrac{15\sqrt{2}}{2}$

    (b) $\dfrac{35\sqrt{3}}{2}$ (d) 0

34. (a) 0 (d) 60°
    (b) 180° (e) 63.435°
    (c) 125.538° (f) 104.250°

35. (a) parallel, (b) neither, (c) parallel, (d) orthogonal, (e) neither, and (f) orthogonal.

36. Possible answers:

    (a) $5i + 3j$
    $-5i - 3j$

    (b) $(-7, 2)^T$
    $(7, -2)^T$

    (c) $i - j$
    $-i + j$

    (d) $\left(\cos 112.5°, \sin 112.5°\right)^T$
    $\left(\cos 67.5°, -\sin 67.5°\right)^T$

37. (a) $\sqrt{5}/5$, (b) $3\sqrt{13}/13$, and (c) 0.

38. (a) $\left(\dfrac{1}{5}, -\dfrac{2}{5}\right)^T$

    (b) $\dfrac{9}{13}i + \dfrac{6}{13}j$

    (c) $\mathbf{0}$

39. (a) $(-12, 9)^T$

    (b) $\dfrac{14}{5}i - \dfrac{7}{5}j$

    (c) $(-27, -36)^T$

(d) $i + \sqrt{3}j$

40. (a) 5 J
    (b) −17 J
    (c) 14 J
    (d) 0
    (e) 4 J
    (f) $\dfrac{13\sqrt{\pi}}{6}$ J

41. (a) 36 J
    (b) $50\sqrt{2}$ J
    (c) 0
    (d) $-\dfrac{13\sqrt{2}}{2}$ J

42. About (a) 2730 J, (b) 1470 J, and (c) 2770 J.

F.12 Complex Numbers

1. (a) $8 + 7i$ (c) $75 + 11i$
   (b) $-9 - 46i$ (d) $5 + 3i$

2. (a) $9/2$ (c) $9 + 5i$
   (b) $26 - 19i$ (d) $-1 + i$

3. (a) $-23 + 145i$
   (b) $42 + 236i$
   (c) $42 + 236i$
   (d) $-213 - 233i$
   (e) $-213 - 233i$

4. (a) $-\dfrac{3}{2} - i$ (d) $4i$
   (b) $-\dfrac{3}{2} - i$ (e) $11 - 8i$
   (c) $\dfrac{1}{2}$ (f) $11 - 8i$

5. (a) $\dfrac{8}{25} + \dfrac{6}{25}i$ (c) $\dfrac{22}{25} + \dfrac{54}{25}i$
   (b) $\dfrac{5}{34} - \dfrac{3}{34}i$ (d) $\dfrac{87}{68} + \dfrac{9}{68}i$

6. (a) 4 (d) 3
   (b) 2 (e) 29
   (c) $\dfrac{5}{2}$ (f) 4

7. (a) 3 (e) 2
   (b) 11 (f) 1
   (c) $\sqrt{2}$ (g) $3\sqrt{5}$
   (d) 25 (h) $\dfrac{3\sqrt{5}}{2}$

8. (a) 2.6 (d) $\dfrac{5\sqrt{5}}{2}$
   (b) $\sqrt{5}$ (e) $\dfrac{13\sqrt{5}}{25}$
   (c) 6.5 (f) $\dfrac{13}{29}$

9. (a) $\dfrac{1}{10}+\dfrac{3}{10}i$ (c) $-8-6i$

(b) 1 (d) $-26+18i$

10. (a) $\dfrac{3}{25}-\dfrac{4}{25}i$ (c) 1

(b) $\dfrac{2}{5}-\dfrac{1}{5}i$ (d) $3+4i$

11. (a) $-i$ (c) -1

(b) $2i$ (d) 17

12. (a) $-5-6i$

(b) $-2-56i$

(c) $-\dfrac{83}{145}+\dfrac{19}{145}i$

13. (a) $1+i$

(b) -2

(c) $-5+5i$

14.

	Real Part	Imaginary Part
(a)	2	-3
(b)	0	4
(c)	-7	0
(d)	1/3	2/3

15. Answer omitted to save space.

16. (a) -2

(b) $2+2i$

(c) $2-3i$

(d) $-5+i$

(e) 0

17. (a) 5 (d) $\sqrt{2}$

(b) 13 (e) 7

(c) 1 (f) $\dfrac{\sqrt{5}}{2}$

Answers very for Exercises 18-20.

21. (a) $6e^{5\pi i/3}$

(b) $\dfrac{2\sqrt{5}}{7}e^{5\pi i/6}$

(c) $e^{3\pi i/2}$

(d) $5\sqrt{2}e^{\pi i/4}$

(e) $\sqrt{6}e^{5\pi i/4}$

(f) $2e^{\pi i}$

(g) $42e^{\pi i/2}$

22. (a) $\sqrt{15}e^{-\pi i/2}$

(b) $2e^{3\pi i/4}$

(c) $e^{-5\pi i/6}$

(d) $\sqrt{3}e^{-\pi i/6}$

(e) $\dfrac{\pi}{2}e^{\pi i/3}$

(f) $7e^{\pi i}$

(g) $\sqrt{2}e^{\pi i/4}$

23. (a) $\dfrac{\pi}{2}+i\dfrac{\pi\sqrt{3}}{2}$

(b) $2\sqrt{3}-2i$

(c) $\dfrac{\sqrt{6}}{2}-i\dfrac{3\sqrt{2}}{2}$

(d) $-\dfrac{3\sqrt{2}}{2}+i\dfrac{3\sqrt{2}}{2}$

(e) $-\dfrac{\sqrt{2}}{2} - i\dfrac{\sqrt{6}}{2}$

(f) 1

(g) $\dfrac{\pi}{3} + i\dfrac{\pi}{3}$

(h) $-6\sqrt{3} + 6i$

(i) $3i$

(j) $-\dfrac{\sqrt{15}}{2} + i\dfrac{\sqrt{5}}{2}$

24. (a) $-\dfrac{3}{2}i$

(b) $-3\sqrt{3} + 3i$

(c) $-27i$

(d) $16 + 16i\sqrt{3}$

25. (a) $\dfrac{3\sqrt{6}}{4} + i\dfrac{3\sqrt{2}}{4}$

(b) $\dfrac{3i\sqrt{2}}{4}$

(c) $-\dfrac{27}{8}$

(d) $-\dfrac{\sqrt{6}}{16} + i\dfrac{\sqrt{2}}{16}$

26. (a) $-4 - 4i$ (c) $-\dfrac{1}{64}$

(b) $\dfrac{i}{8}$ (d) $8 - 8i$

27. (a) $-8 + 8i\sqrt{3}$

(b) $\dfrac{1}{64} - i\dfrac{\sqrt{3}}{64}$

(c) $\dfrac{1}{256} + i\dfrac{\sqrt{3}}{256}$

(d) 64

Answers vary for Exercises 28-30.

31.

(a) $\dfrac{\sqrt{2}}{2} + i\dfrac{\sqrt{2}}{2}$ and $-\dfrac{\sqrt{2}}{2} - i\dfrac{\sqrt{2}}{2}$.

(b) $5, -\dfrac{5}{2} + i\dfrac{5\sqrt{3}}{2}$, and $-\dfrac{5}{2} - i\dfrac{5\sqrt{3}}{2}$.

(c) $\dfrac{\sqrt{2}}{2} + i\dfrac{\sqrt{2}}{2}, -\dfrac{\sqrt{2}}{2} + i\dfrac{\sqrt{2}}{2}, -\dfrac{\sqrt{2}}{2} - i\dfrac{\sqrt{2}}{2}$, and $\dfrac{\sqrt{2}}{2} - i\dfrac{\sqrt{2}}{2}$.

(d) $2, 1 + i\sqrt{3}, -1 + i\sqrt{3}, -2, -1 - i\sqrt{3}$, and $1 - i\sqrt{3}$.

32.

(a) $\dfrac{\sqrt{3}}{2} + \dfrac{1}{2}i, i, -\dfrac{\sqrt{3}}{2} + \dfrac{1}{2}i, -\dfrac{\sqrt{3}}{2} - \dfrac{1}{2}i, -i$, and $\dfrac{\sqrt{3}}{2} - \dfrac{1}{2}i$.

(b) $\dfrac{\sqrt{2}}{2} + i\dfrac{\sqrt{2}}{2}, \dfrac{-\sqrt{2+\sqrt{3}} + i\sqrt{2-\sqrt{3}}}{2}$, and $\dfrac{\sqrt{2-\sqrt{3}} - i\sqrt{2+\sqrt{3}}}{2}$.

(c) $\frac{\sqrt[4]{2}}{2}\left(\sqrt{2-\sqrt{3}}+i\sqrt{2+\sqrt{3}}\right)$, $\frac{\sqrt[4]{2}}{2}\left(-\sqrt{2+\sqrt{3}}+i\sqrt{2-\sqrt{3}}\right)$, $\frac{\sqrt[4]{2}}{2}\left(-\sqrt{2-\sqrt{3}}-i\sqrt{2+\sqrt{3}}\right)$, and $\frac{\sqrt[4]{2}}{2}\left(\sqrt{2+\sqrt{3}}-i\sqrt{2-\sqrt{3}}\right)$.

(d) $\sqrt{2}+i\sqrt{2}$, $-\sqrt{2-\sqrt{3}}+i\sqrt{2+\sqrt{3}}$, $-\sqrt{2+\sqrt{3}}+i\sqrt{2-\sqrt{3}}$, $-\sqrt{2}-i\sqrt{2}$, $\sqrt{2-\sqrt{3}}-i\sqrt{2+\sqrt{3}}$, and $\sqrt{2+\sqrt{3}}-i\sqrt{2-\sqrt{3}}$.

33.

(a) 1 and -1.

(b) 1, $-\frac{1}{2}+i\frac{\sqrt{3}}{2}$, and $-\frac{1}{2}-i\frac{\sqrt{3}}{2}$.

(c) 1, i, -1, and $-i$.

34. 1

35. 0

F.13 Polar Coordinates and Equations

1. Answer omitted to save space.

2. Answer omitted to save space.

3. Possible answers:

(a) $\left(\frac{3}{2}, 0\right)$

(b) $\left(2, \frac{\pi}{4}\right)$

(c) $\left(\frac{1}{2}, \frac{\pi}{2}\right)$

(d) $\left(\frac{5}{2}, \frac{9\pi}{8}\right)$

(e) $\left(\frac{3}{2}, \frac{7\pi}{4}\right)$

(f) $\left(2, \frac{5\pi}{8}\right)$

(g) $\left(3, \frac{3\pi}{2}\right)$

(h) $(3, \pi)$

4. Possible answers:

(a) (i) $(1, 690°)$

(ii) $(1, -30°)$

(iii) $(-1, 150°)$

(iv) $(-1, -210°)$

(b) (i) $\left(\frac{1}{2}, 2\pi\right)$

(ii) $\left(\frac{1}{2}, -2\pi\right)$

(iii) $\left(-\frac{1}{2}, \pi\right)$

(iv) $\left(-\dfrac{1}{2}, -\pi\right)$

(c) (i) $(5, 495°)$
(ii) $(5, -225°)$
(iii) $(-5, 315°)$
(iv) $(-5, -45°)$

(d) (i) $\left(2.5, \dfrac{10\pi}{3}\right)$
(ii) $\left(2.5, -\dfrac{2\pi}{3}\right)$
(iii) $\left(-2.5, \dfrac{\pi}{3}\right)$
(iv) $\left(-2.5, -\dfrac{5\pi}{3}\right)$

(e) (i) $(2, 195°)$
(ii) $(2, -165°)$
(iii) $(-2, 15°)$
(iv) $(-2, -705°)$

(f) (i) $\left(1, \dfrac{3\pi}{2}\right)$
(ii) $\left(1, -\dfrac{5\pi}{2}\right)$
(iii) $\left(-1, \dfrac{\pi}{2}\right)$
(iv) $\left(-1, -\dfrac{3\pi}{2}\right)$

5. (a) $(0, -3)$
(b) $\left(-\dfrac{3\sqrt{2}}{2}, -\dfrac{\sqrt{6}}{2}\right)$
(c) $(0, -2)$
(d) $(-7, 7)$
(e) $(-2, -2\sqrt{3})$
(f) $\left(-\dfrac{3\sqrt{2}}{4}, \dfrac{3\sqrt{2}}{4}\right)$

6. (a) $\left(2\sqrt{2}, \dfrac{3\pi}{4}\right)$
(b) $\left(\sqrt{3}, \dfrac{4\pi}{3}\right)$
(c) $\left(4, \dfrac{\pi}{2}\right)$
(d) $\left(2\sqrt{3}, \dfrac{11\pi}{6}\right)$
(e) $(1, \pi)$
(f) $\left(6, \dfrac{\pi}{4}\right)$

7. (a) $r = 1$
(b) $r = -4\sin\theta$
(c) $r = 5\csc\theta$
(d) $\theta = \dfrac{\pi}{4}$
(e) $r = -\dfrac{\sec\theta\tan\theta}{2}$
(f) $r = \dfrac{1}{\sin\theta - \cos\theta}$
(g) $r = 6\cos\theta - 2\sin\theta$
(h) $r^2 = 2\csc 2\theta$
(i) $r^2 = -\sec 2\theta$

8. (a) $x^2 + y^2 = 4$
(b) $(x-1)^2 + y^2 = 1$
(c) $y = -\dfrac{x\sqrt{3}}{3}$

(d) $(x^2+y^2)^{3/2} = y$

(e) $\tan\sqrt{x^2+y^2} = \dfrac{y}{x}$

(f) $x = -3$

(g) $(x^2+y^2)^2 = 2xy$

(h) $y = \dfrac{x^2-1}{2}$

(i) $x^2 - y^2 = 1$

Answers vary for Exercises 9-11.

12. (a) 10, (b) 15, and (c) $2\sqrt{3}$.

13. Answer omitted to save space.

14. (a) $\theta = 15°$, (b) $r = -\sec\theta$, and (c) $r = 2\csc\theta$.

15. Answer omitted to save space.

16. Answer omitted to save space.

17. (a) Symmetric about the polar axis.

(b) Symmetric about the pole.

(c) Symmetric about $\theta = \pi/2$ and the polar axis.

(d) Symmetric about $\theta = \pi/2$.

(e) Symmetric about the polar axis.

(f) Symmetric about $\theta = \pi/2$.

(g) Symmetric about $\theta = \pi/2$ and the pole.

(h) Contains all symmetric (note $r = 3 \Leftrightarrow r = -3$).

18. (a) 20 (c) 3
    (b) 7   (d) 44

19. Answer omitted to save space.

20. (a) (i) circle and (ii) $r = 1.5\cos\theta$.

(b) (i) dimpled limaçon and (ii) $r = 1.5 + \sin\theta$.

(c) (i) inner loop limaçon and (ii) $r = 1 - 2\cos\theta$.

(d) (i) rose curve and (ii) $-2\cos 3\theta$.

(e) (i) circle and (ii) $r = -2\sin\theta$.

(f) (i) cardioid and (ii) $r = 1 + \sin\theta$.

(g) (i) circle and (ii) $r = 1.5$.

(h) (i) rose curve and (ii) $r = 1.5\cos 4\theta$.

(i) (i) rose curve and (ii) $r = 2.5\sin 5\theta$.

(j) (i) lemniscate and (ii) $r^2 = 2.25\sin 2\theta$.

(k) (i) convex lemiçon and (ii) $r = 2 + \cos\theta$.

(l) (i) rose curve and (ii) $r = \sin 2\theta$.

21. Answer omitted to save space.

22. Answer omitted to save space.

Glossary

Acute angle An angle of measure strictly between 0 and 90°. In other words, $\angle W$ is acute if and only if $0 < m\angle W < 90°$.

Acute triangle All three interior angles are acute.

Adjacent angles Two angles are adjacent if and only if

- they share a common vertex,
- they share a common side, and
- they are non-overlapping.

Altitude of a triangle The shortest segment which extends from a vertex of a triangle to the line containing its opposite side. Because the altitude is the shortest segment, it must be perpendicular to the line which contains the side opposite the vertex.

Amplitude The amplitude of a perodic function f is

$$\frac{\max\{f(x)\} - \min\{f(x)\}}{2}.$$

Angle The figure formed by two rays which meet at a common endpoint.

Angle bisector A line, line segment, or ray, which divides an angle into two adjacent angles of the same measure.

Angle measure A number which quantifies how much an angle opens up. Angle measures are obtained from protractors, and their units are usually either degrees, radians, or revolutions.

Angle of depression An angle formed by a horizontal ray and another ray below the horizontal.

Angle of elevation An angle formed by a horizontal ray and another ray above the horizontal.

Angular velocity When an object rotates about a point, the angular velocity is the rate of change of θ with respect to time, where θ is the central angle subtended by the initial and final position of the object.

Apothem A line segment from the center of a regular n-gon to the midpoint of one of its sides. An apothem is always perpendicular to the side it intersects.

Arc A portion of a circle.

Arc cosecant The arc cosecant of x, denoted $\operatorname{arccsc} x$, is the function defined by the relationship

$$y = \operatorname{arccsc} x \quad \text{if} \quad \csc y = x$$

for $x \leq -1$ or $1 \leq x$ and $-\pi/2 \leq y < 0$ or $0 < y \leq \pi/2$.

Arc cosine The arc cosine of x, denoted $\arccos x$, is the function defined by the relationship

$$y = \arccos x \quad \text{if} \quad \cos y = x$$

for $-1 \leq x \leq 1$ and $0 \leq y \leq \pi$.

Arc cotangent The arc cotangent of x, denoted $\operatorname{arccot} x$, is the function defined by the relationship

$$y = \operatorname{arccot} x \quad \text{if} \quad \cot y = x$$

for x any real number and $-\pi/2 < y < 0$ or $0 < y \leq \pi/2$.

Arc secant The arc secant of x, denoted $\operatorname{arcsec} x$, is the function defined by the relationship

$$y = \operatorname{arcsec} x \quad \text{if} \quad \sec y = x$$

for $x \leq -1$ or $1 \leq x$ and $0 \leq y < \pi/2$ or $\pi/2 < y \leq \pi$.

Arc sine The arc sine of x, denoted $\arcsin x$, is the function defined by the relationship

$$y = \arcsin x \quad \text{if} \quad \sin y = x$$

for $-1 \leq x \leq 1$ and $-\pi/2 \leq y \leq \pi/2$.

Arc tangent The arc tangent of x, denoted $\arctan x$, is the function defined by the relationship

$$y = \arctan x \quad \text{if} \quad \tan y = x$$

where x is any real number and $-\pi/2 < y < \pi/2$.

Argument Consider the complex number $z \neq 0$. An argument of z is a value of θ which makes the equation $z = re^{i\theta}$ true for some real number $r > 0$.

Chord A line segment connecting two points on a circle.

Circumradius A line segment between the center of a regular n-gon and one of its vertices.

Complementary angles Two acute angles whose measures sum to $90°$.

Complex conjugare The complex conjugate of $z = a + bi$ is

$$\bar{z} = a - bi.$$

Complex numbers The set

$$= \{a + bi : a, b \in \mathbb{R}\}.$$

Component form Suppose

$$\boldsymbol{v} = \begin{pmatrix} a \\ b \end{pmatrix}.$$

Then component form of $\boldsymbol{v}$ is

$$a\boldsymbol{i} + b\boldsymbol{j}.$$

Congruent angles Two angles of the same measure. If $\angle X$ and $\angle Y$ are congruent, we write $\angle X \cong \angle Y$. So,

$$m\angle X = m\angle Y \quad \text{if and only if} \quad \angle X \cong \angle Y.$$

Congruent triangles There is a correspondence between the vertices of two triangles such that corresponding angles are congruent and corresponding sides have the same length. If $\triangle ABC$ is congruent to $\triangle XYZ$, we write $\triangle ABC \cong \triangle XYZ$.

Coordinate vector Suppose v has a position vector whose tip is at (a, b). Then the coordinate vector of v is

$$\begin{pmatrix} a \\ b \end{pmatrix}.$$

Coplanar Two or more geometric objects are coplanar if they are both contained within some particular plane.

Cosecant Suppose θ is an angle in standard position whose terminal side intersects the unit circle at the point (x, y). Then define cosecant of θ to be the function such that $\csc \theta = 1/y$.

Cosine Suppose θ is an angle in standard position whose terminal side intersects the unit circle at the point (x, y). Then define cosine of θ to be the function such that $\cos \theta = x$.

Cotangent Suppose θ is an angle in standard position whose terminal side intersects the unit circle at the point (x, y). Then define cotangent of θ to be the function such that $\cot \theta = x/y$.

Denominator The value or function b in the expression a/b.

Directed angle Consider rays $\overrightarrow{OX}$ and $\overrightarrow{OY}$. A rotation of $\overrightarrow{OX}$ about O which terminates at $\overrightarrow{OY}$ is the directed angle $\angle XOY$.

Displacement vector The vector which represents the path traveled by an object when calculating work.

Domain The domain of a function f is the set of x such that $f(x)$ is defined. In other words, the domain of a function is the set of inputs of the function.

Dot product Suppose

$$u = \begin{pmatrix} u_1 \\ u_2 \end{pmatrix} \quad \text{and} \quad v = \begin{pmatrix} v_1 \\ v_2 \end{pmatrix}.$$

The dot product of u and v is

$$u \bullet v = u_1 v_1 + u_2 v_2.$$

Equiangular triangle All the interior angles are congruent.

Equilateral triangle All sides have the same length.

Even function The function f is even if

$$f(-x) = f(x).$$

Gravitational force The force exerted by gravity, denoted by G. Sometimes subscripts are used to indicate different gravitational forces. We assume that objects are "close" to earth. In which case, the gravitational force exerted on an object of mass m is mg, where $g \approx 9.81$ meters per square second.

Hypotenuse The side of a right triangle opposite the right angle. The hypotenuse is always the longest side of the triangle.

i Define i to be a solution of $x^2 + 1 = 0$.

Identity A statement of equality between mathematical expressions, which holds for all values of the variables contained within the domains of each expression.

Imaginary part of a complex number The imaginary part of the complex number $z = a + bi$ is the real number b.

Initial side of an angle Within the context of a directed angle, the initial side is the side where the angle begins its rotation.

Interior angle An angle formed within a polygon by a vertex and its adjacent sides.

Inverse function The function g is the inverse of f if
$$f\Big(g(x)\Big) = x \quad \text{and} \quad g\Big(f(x)\Big) = x.$$

Irrational number When a real number cannot be written as a ratio of integers, we way that it is an irrational number.

Isosceles triangle A triangle with two or more sides of the same length.

Leg Either of the shorter two sides of a right triangle. That is, a side of a right triangle which is not the hypotenuse.

Line The geometric figure contained within a plane which extends infinity in both directions and does not bend.

Line segment The portion of a line contained between two points.

Linear pair Adjacent angles such that the rays not shared between the two angles form a straight angle.

Magnitude The magnitude of a vector $\boldsymbol{v}$ is defined to be the length of $\boldsymbol{v}$. It is denoted $|\boldsymbol{v}|$.

Modulus The modulus of a complex number $z = a + bi$ is
$$|z| = \sqrt{a^2 + b^2}.$$

n-th root The n-th root of x, denoted $\sqrt[n]{x}$ is the function defined by the relationship
$$y = \sqrt[n]{x} \quad \text{if} \quad y^n = x$$
for all x and y when n is odd and for all $x \geq 0$ and $y \geq 0$ when n is even.

Normal force The force perpendicular to a surface, denoted by $\boldsymbol{N}$. A subscript is sometimes needed when there are multiple normal forces.

Numerator The value or function a in the expression a/b.

Oblique triangle No interior angle is right. In other words, a triangle that is either acute or obtuse.

Obtuse angle An angle of measure strictly between 90° and 180°. That is, $\angle Y$ is obtuse if and only if $90° < m\angle Y < 180°$.

Obtuse triangle Contains one obtuse interior angle.

Odd function The function f is odd if

$$f(-x) = -f(x).$$

One-to-one A function f is one-to-one if

$$f(u) = f(v) \quad \text{implies} \quad u = v.$$

Orthogonal The vectors $\boldsymbol{u}$ and $\boldsymbol{v}$ are orthogonal if

$$\boldsymbol{u} \cdot \boldsymbol{v} = 0.$$

Parallel lines Coplanar lines which never intersect.

Parallel vectors The vectors $\boldsymbol{u}$ and $\boldsymbol{v}$ are parallel if there is a nonzero scalar c such that

$$\boldsymbol{u} = c\boldsymbol{v}.$$

Parent function The function considered to be the most basic within a family of functions. For example, $f(x) = x^2$ is the parent function of functions of the form $f(x) = a(x-h)^2 + k$.

Periodic The function f is periodic with period $p > 0$ if p is the smallest number such that

$$f(x+p) = f(x)$$

for all values of x within the domain.

Perpendicular Two geometric objects are perpendicular if they intersect at a right angle.

Phase shift How much the principal period of a periodic function is shifted left or right relative to its parent function.

Polar form The polar form of the complex number $z \neq 0$ is
$$z = re^{i\theta},$$
where $r = |z|$ and θ is an argument of z.

Pole The central point of the polar coordinate system. We usually denote the point by the letter O.

Polar axis On the polar coordinate system, the polar axis is the horizontal ray with endpoint the pole. By convention, the polar axis points right.

Polygon A closed geometric figure that is bounded by line segments.

Position vector The representation of a vector which has its tail at the origin.

Principal period The period of a periodic function considered to be most fundamental, e.g. the principal period of $f(x) = \sin x$ is usually considered to be the interval $[0, 2\pi)$.

Pythagorean Triple An ordered triplet of positive integers (a, b, c) such that
$$a^2 + b^2 = c^2.$$

Radian measure Place a circle of radius 1 around $\angle A$ such that the center of the circle is the point A. The length of the arc which subtends $\angle A$ is the radian measure of $\angle A$.

Range The range of a function f is the set of y such that $y = f(x)$ for some x. In other words, the range is the set of outputs of a function.

Ratio The ratio of a and b is a/b.

Rational equation A rational equation is an equation which contains one or more rational expressions.

Rational expression A ratio of polynomials. In other words, a rational expression is any expression of the form
$$\frac{a_m x^m + a_{m-1} x^{m-1} + \ldots + a_1 x + a_0}{b_n x^n + b_{n-1} x^{n-1} + \ldots + b_1 x + b_0},$$

where $a_m, a_{m-1}, \ldots, a_0, b_n, b_{n-1}, \ldots,$ and b_0 are constants.

Rational number A number is rational if it can be written in the form
$$\frac{a}{b},$$
where $a = 0, 1, -1, 2, -2, \ldots$ and $b = 1, -1, 2, -2, \ldots$.

Ray The portion of a line which has an endpoint and extends infinitely in a direction.

Real part of a complex number The real part of the complex number $z = a + bi$ is the real number a.

Reference angle Consider the directed angle θ in standard position. Suppose the terminal side of θ lies within a quadrant. The reference angle θ_R is the acute angle that shares the terminal side of θ and whose other side is either the positive or negative x-axis.

Regular All interior angles are congruent and all sides have equal length.

Right angle An angle of measure exactly $90°$ is a right angle. That is, $\angle X$ is right if and only if $m\angle X = 90°$.

Right triangle One interior angle of the triangle is right.

Root of unity An n-th root of unit is a solution of
$$z^n = 1.$$

Scalars Real numbers, most notably within the context of vectors.

Scalene triangle All of the sides of the triangle have different lengths.

Secant Suppose θ is an angle in standard position whose terminal side intersects the unit circle at (x, y). Then define secant to be the function such that $\sec\theta = 1/x$.

Sector of a circle A region bounded between two radii and their intercepted arc.

Segment bisector A line, line segment, or ray, which divides a line segment into two parts of equal length.

Segment of a circle The region bounded between a chord and the arc with the same endpoints.

Similar triangles There is a correspondence between the vertices such that corresponding angles are congruent and the ratio of corresponding side lengths is fixed. To indicate that $\triangle ABC$ is similar to $\triangle XYZ$ we write $\triangle ABC \sim \triangle XYZ$.

Simple A polygon whose boundary does not intersect itself. We assume all polygons are simple within this text.

Sine Suppose θ is an angle in standard position whose terminal side intersects the unit circle at (x, y). Then define sine to be the function such that $\sin \theta = y$.

Solve a triangle To find all of the triangle's side lengths and interior angle measures.

Square root The square root of x, denoted $\sqrt{x}$, is the function defined by the relationship

$$y = \sqrt{x} \quad \text{if} \quad y^2 = x$$

for $x \geq 0$ and $y \geq 0$.

Standard form The standard form of a complex number is

$$a + bi,$$

where a and b are real numbers.

Standard position angle The initial side of the angle lies on the positive x-axis.

Straight angle An angle of measure exactly $180°$. That is, $\angle Z$ is straight if and only if $m\angle Z = 180°$.

Supplementary angles Two angles whose measures sum to $180°$.

Tangent Suppose θ is an angle in standard position whose terminal side intersects the unit circle at (x, y). Then define tangent to be the function such that $\tan \theta = y/x$. It is useful to note

$$\tan \theta = \frac{\sin \theta}{\cos \theta},$$

wherever the expressions are defined.

Tension The force exerted by a cable. Tension travels along the cable, and is obtain via the tightness of the cable. We denote tension by T. When there are multiple tension vectors, subscripts are used.

Terminal side of an angle For a directed angle, the terminal side of the angle is the side where the angle ends its rotation about the angle's vertex.

Transversal A line which intersects two coplanar lines at distinct points.

Unit circle The set of points (x, y) such that $x^2 + y^2 = 1$.

Unit vector A vector of magnitude 1.

Vector A nonzero vector is a mathematical expression that shows magnitude and direction.

Vertex of a polygon A common endpoint of two sides of the polygon's boundary.

Vertex of an angle The common endpoint shared between the two rays that form an angle.

Vertical angles The angles on the opposite sides of intersecting lines.

Vertical asymptote The function f has a vertical asymptote of $x = a$ if $f(x)$ goes to $\pm\infty$ as x goes to a from the left or the right. Within sketches of graphs, asymptotes are usually denoted by dashed lines.

Work The work done by a force F which moves an object from point P to point Q is

$$W = \mathbf{F} \cdot \overrightarrow{PQ}.$$

Zero vector The vector which has no magnitude or direction. Denote it by $\mathbf{0}$. Its coordinate vector is

$$\mathbf{0} = \begin{pmatrix} 0 \\ 0 \end{pmatrix}.$$

www.ingramcontent.com/pod-product-compliance
Lightning Source LLC
Chambersburg PA
CBHW050045230526
45470CB00004B/1408